CELLULAR IMAGING TECHNIQUES FOR NEUROSCIENCE AND BEYOND

CELLULAR IMAGING TECHNIQUES FOR NEUROSCIENCE AND BEYOND

FLORIS G. WOUTERLOOD

VU University Medical Center
Department of Anatomy
Amsterdam
The Netherlands

AMSTERDAM • BOSTON • HEIDELBERG • LONDON • NEW YORK • OXFORD
PARIS • SAN DIEGO • SAN FRANCISCO • SINGAPORE • SYDNEY • TOKYO
Academic Press is an imprint of Elsevier

Academic Press is an imprint of Elsevier
32 Jamestown Road, London NW1 7BY, UK
225 Wyman Street, Waltham, MA 02451, USA
525 B Street, Suite 1800, San Diego, CA 92101-4495, USA

First edition 2012

Notice
No responsibility is assumed by the publisher for any injury and/
or damage to persons or property as a matter of products liability,
negligence or otherwise, or from any use or operation of any methods,
products, instructions or ideas contained in the material herein.
Because of rapid advances in the medical sciences, in particular,
independent verification of diagnoses and drug dosages should be
made

British Library Cataloguing-in-Publication Data
A catalogue record for this book is available from the British Library

Library of Congress Cataloging-in-Publication Data
A catalog record for this book is available from the Library of Congress

ISBN: 978-0-12-385872-6

For information on all Academic Press publications
visit our website at elsevierdirect.com

Typeset by MPS Limited, Chennai, India
www.adi-mps.com

CONTENTS

List of Contributors...xi

**Chapter 1 Confocal Laser Scanning: of Instrument,
            Computer Processing, and Men 1**

Introduction...2

Pinhole, Depth of Focus, and Laser Illumination 2

When/Why Does One Need a CLSM? 4

Abbe, Shannon, and Nyquist..8

Imaging of a 2D Line and Deblurring 9

Axial Resolution...11

Resolution and Sampling..12

Signal Separation, Orders of Magnitude, and Resolution Limits.........13

Confocal Microscopy Further Considered 14

Cross Talk Awareness..16

Elimination of Cross Talk 18

Biological Objects Translated to Pixels.....................................19

High-probability Determination of Diameter.................................20

Why Does a 3D Reconstructed Cell Resemble a Pancake? 24

Touch..25

Actual Experiment 26

Synaptic Contacts: Extra Marker 29

Colocalization..30

Conclusion..32

Acknowledgments 32

References...33

Chapter 2 Beyond Abbe's Resolution Barrier: STED Microscopy...... 35

Introduction...36

A New Wave of Imaging...37

STED Microscopy: The Basic Concept.......................................39

Implementation of STED Microscopy 42

Sine Qua Non: Speed, Color, Depth, Live Imaging.................... 47

Summary and Outlook.. 52

References.. 52

**Chapter 3 Enhancement of Optical Resolution by 4pi Single
and Multiphoton Confocal Fluorescence Microscopy...... 55**

Introduction.. 56

The 4pi Principle and Setup... 56

Microscope Alignment 58

4pi Imaging 60

4pi Deconvolution... 61

Sample Preparation... 62

Microtubule and Microtubule Plus End Imaging 65

Visualization of DNA.. 66

Single-photon Excitation (Measurement of the Redox State in
Dopamine Neurons)... 68

SYCP3 Axis as a Marker for Chromatin Organization in Mouse
Spermatocytes... 70

Microbubbles with Medicine 75

Future of 4pi Imaging 77

Acknowledgment.. 77

References.. 77

**Chapter 4 Nano Resolution Optical Imaging Through
Localization Microscopy 81**

Introduction.. 82

Superresolution Microscopy Techniques 82

The Main Approaches to Single-molecule Localization-based
Superresolution Microscopy 87

Fluorescent Probes... 89

Fluorescent Proteins... 89

Multicolor Localization Microscopy 90

Outlook 93

Conclusion.. 96

Acknowledgments 96

References... 97

Chapter 5 Optical Investigation of Brain Networks Using
Structured Illumination **101**

Introduction.. 101

Structuring Light by Phase Modulation Using SLMs 103

Wavefront Engineering Using SLMs: The Optical Setup 104

Light-sensitive Molecular Tools for the Investigation
of the Central Nervous System 111

SLM-based Approaches for the Optical Dissection
of Brain Microcircuits 112

Conclusions..116

Acknowledgments 116

References...116

Chapter 6 Multiphoton Microscopy Advances Toward
Super Resolution **121**

Introduction.. 121

Point Spread Function for Single- and Multiphoton Imaging............. 123

Super Resolution Techniques for Multiphoton
Fluorescence Microscopy 126

Conclusions.. 134

Acknowledgments 134

References... 134

Chapter 7 The Cell at Molecular Resolution: Principles
and Applications of Cryo-Electron Tomography **141**

Introduction: Cellular Landscapes at Molecular Resolution 141

The Cryo-ET Method 143

Detection, Identification, and Hybrid Methods 151

Conclusions 175

Acknowledgments 178

References 178

Chapter 8 Cellular-Level Optical Biopsy Using Full-Field Optical Coherence Microscopy **185**

Introduction 185

The FF-OCM Technique 187

Detection Sensitivity 189

Spatial Resolution 191

Sample Motion Artifacts 192

FF-OCM for High-Resolution "Optical Biopsy" 193

Conclusion 195

Acknowledgment 196

References 196

Chapter 9 Retroviral Labeling and Imaging of Newborn Neurons in the Adult Brain **201**

Techniques to Label and Detect Newborn Neurons
    in the Adult Brain 202

Retrovirus-mediated Labeling of Adult-born Neurons 203

Single-cell Genetic Manipulation in Adult-born Neurons 205

Retrovirus Production and Delivery 207

Viral-labeled Cell Toxicity and Physiological Changes 209

Imaging Newborn Neurons in the Adult Brain 210

In Vivo Live Animal Imaging of Adult Neurogenesis 210

In Vivo Window Preparation 211

In Vivo Imaging Setup and Acquisition 213

Postacquisition Image Processing and Analysis 213

Future Directions in Live Animal Imaging of Adult Neurogenesis 214

Acknowledgments 216

References 216

Chapter 10 Study of Myelin Sheaths by CARS Microscopy 221

Traditional Myelin Imaging Methods 221

Principle and History of CARS Microscopy 224

Technical Characteristics of CARS Microscopy 226

CARS Microscopy for *Ex Vivo* and *In Vivo* Myelin Imaging 227

Mechanistic Understanding of Demyelination and Remyelination
    Enabled by CARS Imaging 231

Other Methods for *In Vivo* Imaging of Myelin 236

Outlook for Myelin Imaging by CARS Microscopy 238

Acknowledgments 240

References 240

**Chapter 11 High-Resolution Approaches to Studying Presynaptic
        Vesicle Dynamics Using Variants of FRAP and Electron
        Microscopy 247**

Introduction 248

Quantifying Dynamic Events at the Macromolecular Scale 249

FRAP for Studying Mobility 251

Variations on FRAP Using Photoswitchable Fluorophores 255

Linking Fluorescence and Ultrastructure: Correlative
    Approaches for Assaying Presynaptic Function 259

Structure–Function Relationships of Vesicle Pools in
    Hippocampal Synapses 263

Combining FRAP with Correlative Electron Microscope 265

Concluding Remarks 269

Acknowledgments 269

References 270

Index 275

Color Plates

LIST OF CONTRIBUTORS

Jeroen A.M. Beliën
Department of Pathology, VU University Medical Center, Amsterdam, The Netherlands

Riccardo Beltramo
Department of Neuroscience and Brain Technologies, Istituto Italiano di Tecnologia, Genova, Italy

Paulo Bianchini
Department of Nanophysics, Istituto Italiano di Tecnologia, Genova, Italy

Axel Blau
Department of Neuroscience and Brain Technologies, Istituto Italiano di Tecnologia, Genova, Italy

Tiago Branco
Wolfson Institute for Biomedical Research, and Department of Neuroscience, Physiology, and Pharmacology, University College London, London, United Kingdom

Francesca Cella Zanacchi
Department of Nanophysics, Istituto Italiano di Tecnologia, Genova, Italy

Ji-Xin Cheng
Weldon School of Biomedical Engineering, Purdue University, West Lafayette, Indiana, USA

Marco Dal Maschio
Department of Neuroscience and Brain Technologies, Istituto Italiano di Tecnologia, Genova, Italy

Angela Michela De Stasi
Department of Neuroscience and Brain Technologies, Istituto Italiano di Tecnologia, Genova, Italy

Alberto Diaspro
Department of Nanophysics, Istituto Italiano di Tecnologia, Genova, Italy

Francesco Difato
Department of Neuroscience and Brain Technologies, Istituto Italiano di Tecnologia, Genova, Italy

Shilpa Dilipkumar
Department of Instrumentation and Applied Physics, Indian Institute of Science, Bangalore, India

Arnaud Dubois
Laboratoire Charles Fabry, UMR 8501, Institut d'Optique, Centre National de la Recherche Scientifique, Palaiseau Cedex France

Helge Ewers
Institute of Biochemistry, ETH Zurich, Zürich, Switzerland

Tommaso Fellin
Department of Neuroscience and Brain Technologies, Istituto Italiano di Tecnologia, Genova, Italy

Rubén Fernández-Busnadiego
Yale University School of Medicine, New Haven, Connecticut, USA

A.B. Houtsmuller
Department of Pathology, and Erasmus Optical Imaging Centre, Erasmus Medical Centre, Rotterdam, The Netherlands

Bing Hu
CAS Key Laboratory of Brain Function and Disease, and School of Life Sciences, University of Science and Technology of China, Hefei, China

Chun-Rui Hu
CAS Key Laboratory of Brain Function and Disease, and School of Life Sciences, University of Science and Technology of China, Hefei, China

Vladan Lucic
Max-Planck-Institute of Biochemisty, Martinsried, Germany

Guo-li Ming
Institute for Cell Engineering, The Solomon H. Snyder Department of Neuroscience, and Department of Neurology, Johns Hopkins University School of Medicine, Baltimore, Maryland, USA; Diana Helis Henry Medical Research Foundation, New Orleans, Louisiana, USA

Partha P. Mondal
Department of Instrumentation and Applied Physics, Indian Institute of Science, Bangalore, India

U. Valentin Nägerl
Interdisciplinary Institute for Neuroscience, Université Bordeaux Segalen, and UMR 5297, Centre National de la Recherche Scientifique, Bordeaux, France

A. Nigg
Department of Pathology, and Erasmus Optical Imaging Centre, Erasmus Medical Centre, Rotterdam, The Netherlands

Emiliano Ronzitti
Department of Nanophysics, Istituto Italiano di Tecnologia, Genova, Italy

Kurt A. Sailor
Institute for Cell Engineering, and The Solomon H. Snyder Department of Neuroscience, Johns Hopkins University School of Medicine, Baltimore, Maryland, USA

Hongjun Song
Institute for Cell Engineering, The Solomon H. Snyder Department of Neuroscience, and Department of Neurology, Johns Hopkins University School of Medicine, Baltimore, Maryland, USA; Diana Helis Henry Medical Research Foundation, New Orleans, Louisiana, USA

Kevin Staras
School of Life Sciences, University of Sussex, Brighton, United Kingdom

W.A. van Cappellen
Department of Reproduction and Fertility, Department of Pathology, and Erasmus Optical Imaging Centre, Erasmus Medical Centre, Rotterdam, The Netherlands

Floris G. Wouterlood
Department of Anatomy and Neurosciences, VU University Medical Center, Amsterdam, The Netherlands

CONFOCAL LASER SCANNING: OF INSTRUMENT, COMPUTER PROCESSING, AND MEN

Jeroen A.M. Beliën[1], and Floris G. Wouterlood[2]

[1]Department of Pathology, VU University Medical Center, Amsterdam, The Netherlands, [2]Department of Anatomy and Neurosciences, VU University Medical Center, Amsterdam, The Netherlands

CHAPTER OUTLINE

Introduction 1
Pinhole, Depth of Focus, and Laser Illumination 2
When/Why Does One Need a CLSM? 4
Abbe, Shannon, and Nyquist 8
Imaging of a 2D Line and Deblurring 9
Axial Resolution 11
Resolution and Sampling 12
Signal Separation, Orders of Magnitude, and Resolution Limits 13
Confocal Microscopy Further Considered 14
Cross Talk Awareness 16
   Excitation Cross Talk 18
Elimination of Cross Talk 18
Biological Objects Translated to Pixels 19
High-probability Determination of Diameter 20
   Best-Fit Object 3D Recognition 20
   Automated Objective Threshold Analysis 22
Why Does a 3D Reconstructed Cell Resemble a Pancake? 24
Touch 25
Actual Experiment 26
   Computer Software to Define a Contact 29
Synaptic Contacts: Extra Marker 29
Colocalization 30
Conclusion 32
Acknowledgments 32
References 33

Cellular Imaging Techniques for Neuroscience and Beyond.
DOI: http://dx.doi.org/10.1016/B978-0-12-385872-6.00001-5

Introduction

The confocal laser scanning microscope (CLSM) has evolved in the past 25 years from a fancy contraption full of optical wizardry into an indispensable, image acquisition tool for the biomedical sciences. By its physical design the instrument generates images that, compared with those obtained with a conventional fluorescence microscope, are razor sharp and offer higher radial and, in particular, higher axial resolution. Before the first commercial instruments became available in the early 1980s, experimental instruments had been built by Egger and Petran (1967) and by Brakenhoff et al. (1979). Since the introduction of commercial CLSMs from the 1980s onward, a growing number of applications have emerged in cell biology and medicine that rely on imaging of both fixed and living cells and tissues.

In the following sections we will outline the technical principles and present applications of CLSM in combination with 3D visualization and quantification.

The interested reader is further pointed to an excellent Web site on microscopy in general and CLSM in particular, at http://micro.magnet.fsu.edu/primer/. A Java-based CLSM simulator can be found at www.olympusconfocal.com/java/confocalsimulator/index.html.

Pinhole, Depth of Focus, and Laser Illumination

The core feature of the instrument is the confocal principle, proposed by Minsky (1957), which reduces the depth of focus and sharply diminishes out-of-focus haze. Confocality is produced by a pinhole positioned in the light path between the objective lens and the detector that rejects out-of-focus light (Figure 1.1).

The smaller the pinhole the thinner is the depth of focus. Reduction of the pinhole's size is bound to a limit since the ultimate small pinhole is a completely closed pinhole (zero pinhole diameter) that will not allow any light to pass; hence, it is useless. Consequently, the focal plane always has a finite, although reduced, thickness. The optimal diameter of the pinhole is governed by Abbe's diffraction equation (see in the next paragraph), which features the wavelength of the light that is supposed to be detected. The consequence of a tiny pinhole is a low photon efficiency of the instrument (the pinhole is made specifically to *reject* light, and by virtue of this less than a fraction of 1% of all the light emitted from or reflected by an object passes the pinhole and reaches the detector; Pawley, 1995). Low photon efficiency can be compensated for through several workarounds. One of these is to collect more light in the detector by opening the pinhole,

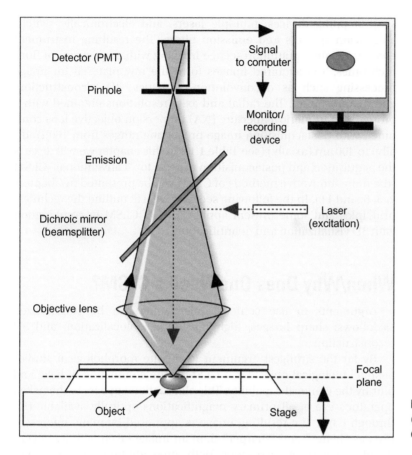

Figure 1.1 Principal components and the light path of a CLSM.

at the cost of depth of focus and the reappearance of haze. Another approach is to apply multiple pinholes arranged on a Nipkow disk (as proposed by Egger and Petran, 1967), while an alternative approach anticipating the loss of large amounts of light is to use high-intensity laser light at the excitation side to provide overwhelming illumination. At the detection side a highly sensitive photomultiplier collects the incidental photons that manage to pass the pinhole. Laser light has the additional advantage of being monochromatic as well as two more "golden advantages": first that the physical diameter of laser beams can easily be made very narrow, and second that lasers can be pulsed. Combined with scanning mirrors in the illumination path and, in the image acquisition device, an analog-to-digital converter attached to a computer, a synergistic relationship exists between scanning instrument and computer bitmapped imaging. Through this marriage of various physical principles and technology an opto-electronic instrument was born that we recognize today as a CLSM. Further development included combination with accurate stage-stepping devices, multiple lasers, and electronically switchable beam splitters and emission filters. The resulting instrument enables reliable multifluorescence imaging with a spectrum of fluorochromes. The acquired images form the raw material for image processing such as deconvolution, statistics, 3D reconstruction, and quantification. The radial and axial resolutions obtained with a proper high numerical aperture (NA) immersion objective lens combined with postacquisition image processing ranges from 160 (radially) to 400 nm (axially) (see Table 1.1). In this chapter we will discuss the acquisition and postacquisition aspects for "conventional" CLSM (the more innovative methods of CLSM will be presented in Chapters 2, 3, 6, and 11). In the following sections we will outline the technical principles and some present applications of CLSM in combination with 3D visualization and quantification.

When/Why Does One Need a CLSM?

Arguments to use confocal microscopy can be summarized as follows: sharp images, high resolution, colocalization, and 3D reconstruction.

By far the strongest argument to initiate morphological studies of small biological objects in a CLSM is the elimination of haze and blur by the confocal capability. This elimination appears to the naked operator's eye at all primary magnifications optically available (5× through 63×). The resulting image is appraised by many microscopists as "better" and "sharper" than an image taken with a conventional fluorescence microscope. With some additional measures and postacquisition processing, a razor-sharp image can be obtained at very high magnification, such as illustrated in Figure 1.2.

It is often overlooked by inexperienced CLSM operators that for low-power imaging (less than 40× objective lens magnification), it is the dramatic increase of the microscope's depth of field (a thicker part of the specimen is seen as "sharp"; an increase of haze), which "makes the image" at low magnifications while reduction of blur is not an issue at these magnifications. "Haze" is considered as the diminished visibility in a classical fluorescence microscope of individual structures in sections due to fluorescence emitted by nearby structures (radially and especially axially, that is above and below the structure of interest). "Blur" in this chapter equals diffraction. Abbe's well-known diffraction equation $r = 0.61 \ \lambda/NA_{obj}$ that governs resolution in an optical system plays a negligible role at low magnifications. Feeding Abbe's equation with real-life numbers — a 40× NA 0.85 dry objective lens with no spherical or chromatic aberration; and a 488 nm excitable fluorochrome, which has a fluorescence

Table 1.1 Abbe-type Diffraction: Theoretical Exercise with Various Popular Fluorochromes/Emission Wavelengths to Calculate the Parameter r in Abbe's Equation in an Ideal Instrument*

40x Dry Lens NA 0.85
1 Pixel = 91 × 91 × 244 nm
Refractive Index Medium = 1

Dye	Excitation	Emission	Before Deconvolution Radial r (nm)	Radial r in 91 × 91 nm Pixels	After Deconvolution Radial r (nm)	Radial r in 91 × 91 nm Pixels	Before Deconvolution Axial r (nm)	Axial r in 224 nm Pixels	After Deconvolution Axial r (nm)	Axial r in 224 nm Pixels
Alexa 350	346	442	317	3	227	2	1224	5	874	
Alexa 405	402	421	302	3	216	2	1165	5	832	7
Alexa 430	434	539	387	4	276	3	1492	6	1066	9
Alexa 488	488	519	372	4	266	3	1437	6	1026	8
Cy2	492	510	366	4	261	3	1412	6	1008	8
Lucifer yellow	488	544	390	4	279	3	1506	6	1076	9
Alexa 514	518	540	388	4	277	3	1495	6	1068	9
Alexa 532	531	554	398	4	284	3	1534	7	1095	9
Alexa 546	546	573	411	5	294	3	1586	7	1133	9
TRITC	550	570	409	4	292	3	1578	7	1127	9
Cy3	550	570	409	4	292	3	1578	7	1127	9
Alexa 555	555	565	405	4	290	3	1564	7	1117	9
Alexa 568	568	603	433	5	309	3	1669	7	1192	10
Alexa 594	594	617	443	5	316	3	1708	7	1220	10
Texas Red	596	620	445	5	318	3	1716	7	1226	10
Alexa 610	612	628	451	5	322	4	1738	7	1242	10
Alexa 633	632	647	464	5	332	4	1791	8	1279	10
Alexa 647	650	680	488	5	349	4	1882	8	1345	11
Cy5	650	670	481	5	343	4	1855	8	1325	11
Alexa 660	663	690	495	5	354	4	1910	8	1364	11

(Continued)

Table 1.1 (Continued)

Dye	Excitation	Emission	Before Deconvolution		After Deconvolution		Before Deconvolution		After Deconvolution	
63× Immersion Lens NA 1.4 **1 Pixel = 58 × 58 × 122 nm** **Refractive Index Medium = 1.5**			Radial r (nm)	Radial r 58 × 58 nm Pixels	Radial r (nm)	Radial r 58 × 58 nm pixels	Axial r (nm)	Axial r in 122 nm Pixels	Axial r (nm)	Axial r in 122 nm Pixels
Alexa 350	346	442	193	4	138	2	451	4	322	3
Alexa 405	402	421	183	3	131	2	430	4	307	3
Alexa 430	434	539	235	4	168	3	550	5	393	3
Alexa 488	488	519	226	4	162	3	530	4	378	3
Cy2	492	510	222	4	159	3	520	4	372	3
Lucifer yellow	488	544	237	4	169	3	555	5	397	3
Alexa 514	518	540	235	4	168	3	551	5	394	3
Alexa 532	531	554	241	4	172	3	565	5	404	3
Alexa 546	546	573	250	5	178	3	585	5	418	3
TRITC	550	570	248	5	177	3	582	5	415	3
Cy3	550	570	248	5	177	3	582	5	415	3
Alexa 555	555	565	246	4	176	3	577	5	412	3
Alexa 568	568	603	263	5	188	3	615	5	440	4
Alexa 594	594	617	269	5	192	3	630	5	450	4
Texas Red	596	620	270	5	193	3	633	5	452	4
Alexa 610	612	628	274	5	195	3	641	5	458	4
Alexa 633	632	647	282	5	201	3	660	5	472	4
Alexa 647	650	680	296	5	212	4	694	6	496	4
Cy5	650	670	292	5	209	4	684	6	488	4
Alexa 660	663	690	301	5	215	4	704	6	503	4

*Theoretical resolutions for various wavelengths and objectives—calculated for emissions with Abbe's radial and axial diffraction equations.

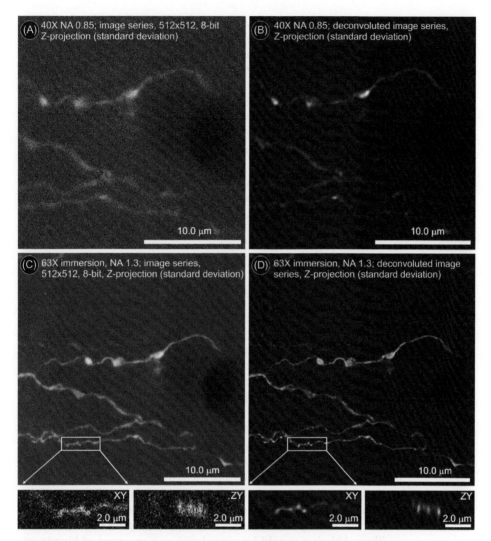

Figure 1.2 Images of the same structures obtained with a 40× dry objective (Leica HCX plan apo, NA 0.85) and a 63× glycerin immersion objective (Leica HCX plan apo CS, NA 1.30). Thin axons in rat hippocampus, labeled with a neuroanatomical tracer (BDA), and visualized via streptavidin-Alexa Fluor 488. Forty micron sections cut on a freezing sliding microtome of formaldehyde/glutaraldehyde fixed brain. (A) Projection image acquired through the 40× dry objective: *Z*-scan of 20 frames, each frame averaged from two scans (a total of 40 exposures), *Z*-increment 122 nm. (B) Same projection image after deconvolution of the dataset with Huygens Professional software. (C) Afterwards we rescanned the same structures with the 63× glycerin objective: projection image of 53 frames, each frame averaged from two scans (a total of 106 exposures), *Z*-increment 122 nm. (D) Same projection image after deconvolution of the dataset with Huygens Professional software. Details of a small labeled fiber in *XY* and *ZY* are shown below (C and D) to illustrate differences in radial and axial resolution.

emission peak at 519 nm (Alexa Fluor 488; Panchuk-Voloshina et al., 1999) (Table 1.1), that is, a commonly used "green" fluorochrome; and provided the microscope is ideal — $r = 372$ nm. Thus, two points, 372 nm or 0.372 μm, separated from each other is the Abbe limit of resolution in this ideal microscope with this 40× dry lens and this excitation source/fluorochrome combination. Figure 1.2A shows a Z-projection of a series of images taken in the CLSM at 488 nm laser excitation, without any postprocessing. Figure 1.2B shows the same dataset, after postacquisition processing. An average mitochondrion (diameter 0.5 μm) will appear approximately 5 pixels in diameter in an image at this magnification with 4 pixels of blur at the edges and be perceived as one "dot," as in a regular microscope. Purchasing a CLSM to study mitochondria becomes interesting because of the resolution offered by high-power immersion objective lenses, especially those with a high NA.

Abbe, Shannon, and Nyquist

Abbe's equation becomes really important when the dimensions of the structure under study are of the same magnitude or smaller than the dimension of Abbe's r; that is, structures *inside* mitochondria, and organelles such as chromosomes, ribosomes, Golgi apparatus, filaments, endoplasmic reticulum, and transport vesicles. The argument supporting this statement is that the parameter r in Abbe's equation is a radius describing a statistical photon distribution (the point spread function; PSF); that is, a normal distribution with a primary maximum intensity in the center and a primary minimum intensity at a distance r from the center, such as in our 40× dry lens example at 372 nm (488 nm excitation/519 nm emission fluorochrome) or 0.5 Airy unit. One Airy unit equals the *diameter* of the photon distribution disk whose edge is the first or primary minimum in the concentric photon distribution "ripple" (Figure 1.3).

Thus, Abbe's r should be considered as a measure of blurriness. Note also that in order to observe Airy distribution of photons one needs a microscope configuration capable of sampling pixels small enough to cover the distance r. At this point, Shannon's sampling theorem becomes important; it states that the digitizing device must utilize a sampling interval that is no greater than one-half the size of the smallest resolvable feature of the optical image. Thus, in order to record an Airy distribution the very minimum sampling is theoretically 6 pixels (2 times min-max-min), and in the dry lens example, one pixel then should not be bigger than $372 \times 2/6 = 124$ nm, and should preferably be smaller. In our microscope with a 40× dry objective NA 0.85, zoom = 8, pixels measure standard 91×91 nm in *XY* and

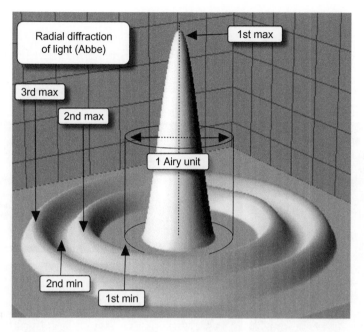

Figure 1.3 Three-dimensional representation of radial diffraction according to Abbe. Photons collected by the detector originating from a point source distribute according to a diffraction pattern. The distance between the primary maximum and the first diffraction minimum is called 1 Airy disk radius. Note that Airy units are wavelength and lens dependent.

234 nm in Z, which is in compliance with Shannon's sampling theorem. An alternative way of defining the sampling frequency is given by the Nyquist criterion: sampling should be done at twice the highest spatial frequency contained in the image (f/2).

Imaging of a 2D Line and Deblurring

Assume a 2D line (Figure 1.4A, left). This line is imaged in the CLSM at high magnification as a distribution of photons according to a PSF depending on the wavelength of the emitted light and the NA of the objective. The CLSM image would consist of a blurry stripe built up from photons arriving in the CLSM's detector as a linear sequence of PSFs with a width of $2 \times r$, with one r of photon distribution on one side of the axis of the center maximum band and the other r photon distribution on the other (Figure 1.4A, middle). The original line, although in this example with a width of 0 pixels, will occur in the CLSM image as a *gradient* of pixels with diminishing intensities, one on each side of an intensity axis (which coincides with the position of the line it is representing; Figure 1.4A, middle) similar to the decrease of the slope in the statistical photon distribution portrayed by Abbe. The number of pixels in each gradient is dictated by the photon's energy (the wavelength of the light), the size of the pinhole, and the NA value of the objective lens. This is Abbe diffraction or true blur.

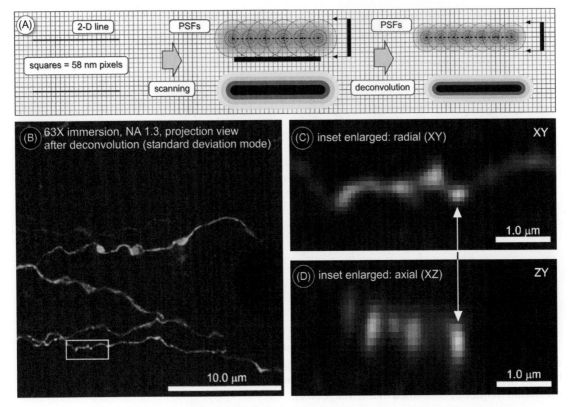

Figure 1.4 Simulation of the radial representation. (A) Simulation of the radial representation of a 2D line (left, , no thickness) in a high-magnification CLSM image, before and after deconvolution. Photons emitted by the line source distribute according to a linear series of PSFs (middle). In the computer bitmap the line is therefore represented as gradients of pixels on each side of the line's original position (right). (B–D). Maximal radial (**XY**) and axial (**ZY**) resolution in our CLSM. Image acquisition conditions: 63× NA 1.3 glycerin immersion objective, Z-increment 122 nm, zoom = 8, pixels dimensions 58×58 nm in **XY** and 122 nm in **Z**. (C) Enlargement of the inset in (A); (D) is an orthogonal reconstruction in **YZ** made from the deconvoluted **Z**-image stack with Amira software with identical magnification as in (C). This illustrates the poorer resolution in the axial direction (D) compared with the radial direction (C).

A solution to improve the image is to reduce the width of these intensity pixel gradients by making their slope steeper. This can be done by a calculation called "deconvolution," a statistical exercise reversing the course of the photons from the blurred image back to the focal plane where they came from (van der Voort and Strasters, 1995; Wallace et al., 2001; Boccacci and Bertero, 2002). Typically, deconvolution uses CLSM instrument parameters (emission wavelength,

NA of the objective lens, refractive index of the immersion medium) to calculate from the acquired image a new image that, with high statistical reliability, resembles the original object imaged as much as possible. Deconvolution narrows the photon distribution by a factor of 1.4 (Sheppard and Choudhury, 1977; Inoué, 1995). This means that r becomes 1.4 times smaller and the pixel intensity gradient slope at the edges of structures steeper. Through the use of a high-quality, high-NA immersion objective lens (NA 1.4), a radial resolution at 488 nm excitation and 519 nm emission can be obtained of $r = 0.61 \times 519/1.5 = 226$ nm (Table 1.1) which, after deconvolution (factor 1.4), becomes a resolution of 162 nm, that is, a >50% better image than with the NA 0.85 dry lens. The effect in practice of using a high-NA immersion lens and (optionally) the combination with post-acquisition deconvolution is shown in Figure 1.2 C and D, and especially in Figure 1.4A (right), B, and C.

The insets in Figure 1.2 show a thin fiber in XY and XZ to illustrate the differences between radial (XY) and axial (ZY) resolution in a CLSM. In Figure 1.4C–D the same fiber is shown at higher magnification and after deconvolution in the XY and XZ planes. Note that these fibers are extremely thin structures, with shafts approximately 0.1 μm in diameter; they are the thinnest structures available in our histological slide collection for this purpose.

Axial Resolution

The above example should be part of the basic knowledge of any CLSM operator. Less well known is that Abbe's equation, $r = 0.61$ λ/NA_{obj}, applies only to *radial* resolution that is in the XY plane. In the *axial*, or Z direction, the resolution is less because in this direction the mathematical expression for r equals $2\lambda\acute{\eta}/(NA_{obj})^2$ where $\acute{\eta}$ is the refractive index of the mounting medium/immersion medium. Again feeding real-life numbers (NA 0.85 lens, 519 nm emission, $\acute{\eta}$ for air is 1), the axial r becomes 1437 nm for the 40× NA 0.85 dry lens and 530 nm for the 63× NA 1.4 immersion lens. Deconvolution reduces the above "dry" resolution down to $1437/1.4 = 1026$ nm and the "immersion" axial resolution to 378 nm. Thus, a 63× immersion objective lens can be expected to perform in the axial direction three times better than a 40× dry lens. This effect is clearly visible in the insets in Figure 1.2. Note that even for the high-NA immersion lens axial resolution remains 2.3 times as poor as its radial resolution (Figure 1.4C and D), while for the dry lens the deficit in axial resolution is even more dramatic (3.8 times). Thus, as a rule of thumb, for the imaging of details of very small biological objects a high-NA immersion lens is compulsory plus a proper immersion medium, because under these circumstances the CLSM produces superior images compared with a conventional microscope in terms of sharpness, richness of detail, and especially axial resolution. Postacquisition deconvolution considerably improves the image and enhances resolution (Figure 1.2). Don't expect to see with visible light details smaller than 160 nm radially and 372 nm axially. A detailed mathematical description of radial and axial resolution is provided by Jonkman and Stelzer (2002).

Resolution and Sampling

Images in CLSMs are obtained as samples that are rows and columns of pixels that are presented to the operator as bitmapped images built up of square pixels. Typical sample frequencies are 512×512 and 1024×1024 pixels. For each pixel the detector produces an analog signal in proportion to the number of the detected photons and the energy they contain (Pawley, 2006). This analog signal is converted by an AD converter for each pixel into a gray intensity level: 8-bit conversion produces 256 gray levels and 16-bit conversion produces 4096 gray levels. Sampling referred to in most manuals does not take analog signal sampling by the AD converter into consideration, yet focuses on the size of an object and the number of samples (pixels) necessary to produce a confident image of the biological object. Let's use the 2D line (above) with the PSF-induced blur down to the first intensity minimum (r). Sampling of this line is conducted with an NA 1.4 immersion lens and 488 nm excitation/519 nm emission coverage plus postacquisition deconvolution. In the XY plane the Airy distribution will take 2×162 nm and in the XZ plane it will take 2×378 nm (Table 1.1; Figure 1.4A). According to the Shannon/Nyquist criterion one would need radially 162/2 = 81 nm pixel size and radially 378/2 = 189 nm pixel size. In our practice, with an instrument that is an engineer's compromise of theoretical design and the practically attainable, and equipped with an NA 1.3 immersion lens, we oversample approximately two times, that is with a pixel size of 58 nm X and Y, and a 122 nm Z-increment axially. Smaller pixels and shorter Z-increment are pertinent "overkill." At this magnification a 1024×1024 bitmap consisting of 29×29 nm pixels contains exactly the same information as a 512×512 bitmap made up of 58×58 nm pixels, yet takes 4 times as much file size on hard disk, and in the event of 10-bit scanning instead of 8-bit, files that are 64 times bigger *per frame and per channel.* "Smaller" pixels may be tempting even to the experienced CLSM operator; however, bigger bitmap sizes, for example, 1024×1024 or 2048×2048 have, based on the above, their best applications at lower magnifications where they can support the details in mapping studies. At high magnification pixel size has to match resolution and the effectiveness of smaller pixel dimensions is reflected only in successfully draining computer resources.

Signal Separation, Orders of Magnitude, and Resolution Limits

Almost equally important as the improved imaging compared with conventional microscopy is the ability of a CLSM to provide reliable and complete separation of multiple fluorescence emission signals with a minimum or absence of cross-channel cross talk. This property has little to do with the confocality of the instrument. Conditions for enhanced separation are provided by using monochromatic excitation light combined with proper fluorochromes with narrow excitation and emission spectra and completed with proper emission-filtering practices. Resulting from this high confidence and accuracy is the ability to study colocalization of biological agents in the same (small) compartments of cells.[1] What makes a CLSM unique is the combination of the instrument's confocality, monochromatic laser light excitation, and imaging followed by postacquisition deconvolution. This combination allows, under the proper excitation and filtering conditions, the study of colocalization of fluorescence-tagged proteins in small cell compartments such as axon terminals (radial diameter 0.5–0.8 μm; Wouterlood et al., 2007). With conventional fluorescence microscopy, colocalization can only be studied at one order of magnitude higher; that is, in structures the size of complete cells or in gross cellular compartments (typically nucleus vs. cytoplasm) with diameters of, say, 5 μm and more. A particular challenge is at the level of one order of magnitude below the small cellular compartment level. In this microrealm that occupies the transition zone between optical and electron microscopy, wherein the biggest structure has the size of a single ribosome, one might question whether receptor molecules are attached to the

---

[1] Physical, absolute colocalization of molecules is a contradiction. Colocalization means that two molecules occur very close to each other in the same cell compartment, for example, a vesicle or in/on a membrane. In terms of CLSM, colocalization should be considered as a function of voxel dimensions: in our example: if two molecules are spaced less than 58 nm apart they will be imaged in the same voxel otherwise they will be imaged in neighboring voxels. At this size scale Abbe diffraction plays an overwhelming role. Statistical testing of the exact overlap of the PSFs belonging to each marker is mandatory in such circumstances.

outside or to the inside of a cell membrane. This requires a resolution of the thickness of a biological membrane (~20 nm). Such a resolution is far below the best resolution of the routine CLSM. Solutions at this point to overcome the barrier thrown up by Abbe's equation ("Abbe's diffraction barrier") are provided in Chapters 2, 3, 4, 6, and 11.

Confocal Microscopy Further Considered

The "essential" component of the CLSM is its pinhole in front of the detector, and Figure 1.1 shows why. All light emanating from planes below or above the focal plane is rejected. Pinholes should therefore be given special attention.

From Abbe's radial equation, $r=0.61\ \lambda/NA_{obj}$, it is easy to follow that the diameter of the pinhole must be proportional to the wavelength of the used light: a pinhole for "red" light should have a greater diameter than when "green" light is used. Or if a pinhole's diameter is set at a value that accommodates red light it is too wide for green light. Conversely, if a pinhole's diameter is set at a value that accommodates green light it is too narrow for red light. The diameter of the pinhole also has consequences for the focal plane from which emanated photons may pass the pinhole and reach the detector. This focal "plane" is not a 2D plane, but it always has a certain thickness governed by Abbe's "axial" resolution equation, $r=2\lambda\acute{\eta}/(NA_{obj})^2$. Thus the observed thickness of the recorded focal plane depends on the wavelength of the recorded light. In layman's words, a green focal plane is thinner than a red focal plane. This phenomenon has been mathematically underscored by Wilson (2002). The detector in the CLSM, however, does not detect color; instead it counts photons with a certain amount of energy that hit the detector. The computer "categorizes" the analog signal from the detector into 256 (8-bit) or 4096 (16-bit) gray levels. The detector will record photons from a thicker "focal" plane when red laser light is used for excitation than when the excitation light is green or blue. This effect can be calculated for an ideal instrument equipped with a 63× NA 1.4 immersion lens and using 519 nm (green) emitted light (peak after excitation with 488 nm laser light) compared, for instance, with 617 nm (red) emitted light (peak after excitation with 594 nm laser light). Abbe's r calculated axially at 519 nm emission is 530 nm while at 617 nm emission, the equation produces a thickness of 630 nm, or 100 nm extra on each side of the focal plane compared with the 519 nm emitted light (after deconvolution, the axial r becomes 378 nm for 519 nm light and 450 nm for 610 nm light). It should be mentioned that in a CLSM *only* the excitation light is monochromatic, whereas the emitted light has an emission spectrum of its own. Even with an ideal pinhole size and a circular pinhole this already and inevitably leads with single monochromatic laser illumination to the addition of "rainbow" photons at the upper and lower faces of the focal plane. This is represented in confocal Z-images as a "flaring" instrument-measured PSF (Hiesinger et al., 2001). A square or rectangular pinhole adds additional small distortions of the instrument's PSF.

Actually, the focal plane's geometry is disturbed by the pinhole *by design*. A [con]focal plane is not exactly flat. The disturbance is easy to understand if one assumes an elongated fluorescing structure such as a rod, which is positioned in the specimen perpendicular to the coverslip and with an alignment exactly along the optical axis of the instrument. This rod is special because its diameter is exactly that of the Airy distribution of the emitted light, for example, $2 \times r$ for the particular emission wavelength. *All* photons emitted from this rod will be detected by the detector of the CLSM whether or not they come from the focal "plane." The cause for this phenomenon is that photons originating from any position along the optical axis are not subjected to refraction by the objective lens. As far as the "eccentric" photons are concerned, these are emitted from a position within a distance of Abbe's r from the optical axis and thus refracted such that they project on the detector a spot exactly the diameter of the pinhole. Thus, all photons emitted from a cylindrical area in the specimen aligned with the instrument's optical axis and with a diameter of $2 \times r$ will pass the pinhole. Consequently, the confocal principle is valid for the entire focal plane except for a "punch" spot with a diameter of one Airy unit at the very center of the imaging field where the optical axis passes the specimen. All photons emitted from that spot in the specimen are not subjected to the pinhole principle. The defect of the confocal plane is proportional to the size of the Airy disk. Unfortunately it is situated at a place exactly where confocality is most urgently needed and where human operators intuitively position their biological object of interest (the center of their field of view). The defect, though, is very small. In the center of the image an axial resolution dip can be expected (in our instrument with a 63× immersion lens NA 1.3 with 488 nm excitation light, zoom = 8 and after deconvolution in a spot with a radius of 3 pixels on a 512 × 512 bitmap, or 0.6%).

From the above it follows that (1) the size of an ideal pinhole should match the value of Abbe's r (×2) for the particular wavelength of light emitted by the specimen and that (2) the diameter of the pinhole should be changed when a different laser illumination wavelength is being used, such as in multiple fluorochrome colocalization experiments. A pinhole preferably should be a real hole, which is a circular opening. Manufacturers of CLSMs are sparse in providing information with respect to size and geometry of the pin-holes applied in their equipment and whether the diameter of these pinholes changes when a different excitation wavelength is selected or when sequential Z-scanning is performed. (Actually, a manufacturer should concentrate on the fluorochrome's emission peak wavelength and emission spectrum width and less on the fluorochrome's peak excitation wavelength; we suspect that practical considerations tempt manufacturers' engineers to favor the first option.) The innate defect of a confocal microscope is that it has reduced resolution in the very center of the imaging field where the optical axis traverses the object. This spot has a radius equal to r. This defect may be masked by software measures or it may be resolved by constructing a "darklight" confocal microscope. As a rule of thumb, the best resolution in a CLSM is available a fraction eccentric from the bitmapped image's center.

Cross Talk Awareness

Channel cross talk (sometimes called bleed-through) is an undesired phenomenon that may interfere in multilabel fluorescence studies with the results. Cross talk is the occurrence of fluorescence signal in a channel[2] configured for imaging a second different fluorochrome. Cross talk therefore is a source of false-positive observations. The caveat here is that cross talk is considered by many to occur only in a "higher" channel, such as an instrument configuration around a laser-fluorochrome combination with longer wavelengths and lower energy of the emitted light. This is because, according to Rayleigh, emitted light always has a lower energy than the excitation light. We call this type of cross talk "emission cross talk." Yet another phenomenon may occur, which might seem at first paradoxical, that we call "excitation cross talk" (Figure 1.5).

Emission cross talk is the well understood phenomenon that occurs during a multiple fluorescence preparation when the excitation of a fluorochrome by its associated laser produces the inadvertent detection of some emission of this fluorochrome in an adjacent, longer wavelength channel. Under emission cross talk conditions the appropriate channel will produce a nice image while a faint copy of the same image occurs in the inappropriate channel. For instance, excitation by 594 nm laser light of a 594 nm fluorochrome may

---

[2] A "channel" in a multifluorescence imaging session is a subselection/configuration for each laser, (acousto-optic) beam splitter setting and emission band filtering with the single purpose of obtaining an exclusive emission signal for each individual fluorochrome.

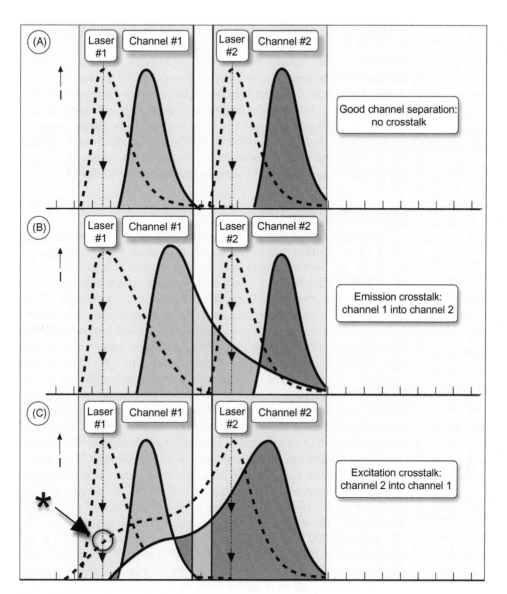

Figure 1.5 Channel separation in fluorescence. (A) Correct channel separation.
The lasers excite their associated fluorochromes and emitted light is only detected
in the "own" channel. (B) Emission cross talk: emitted light belonging to channel
1 is detected in channel 2. This is the type of cross talk most often encountered.
(C) Excitation cross talk. In this type of cross talk, emitted light that apparently
belongs to channel 2 is detected in channel 1. This is at first paradoxical because
according to Raleigh emitted light always has a longer wavelength than excited
light. In this case fluorochrome 2 has an excitation "tail" in the wavelength domain
belonging to channel 1 and thus can both be excited by laser 1 (asterisk) and
produce emission in channel 1. Rayleigh's theory is not compromised.

produce some emission detected in the channel configured around a 633 nm fluorochrome. Emission cross talk is easy to detect (in a 594–633 nm laser double fluorescence experiment this type of cross talk produces an image in the 633 nm channel while the laser for that channel is switched off and the 594 nm laser is on). This type of cross talk can be avoided by reducing the excitation light intensity for the first channel (in this example the 594 nm laser) in combination with modifying the detection bandwidth for the second channel (where in this example the 633 nm fluorochrome is detected). Additional effective measures are listed in the section Elimination of Crosstalk.

Excitation Cross Talk

This type of cross talk is the unexpected and often unanticipated effect in a multifluorochrome experiment that a particular laser excites in its "own" channel and both its own fluorochrome *and* a second fluorochrome (e.g., one that belongs to an adjacent, longer wavelength channel). The effect becomes more pertinent when the second fluorochrome has an excitation spectrum "shoulder" at wavelengths assigned to the channel configured around the first fluorochrome. Consequently, a faint copy of the image in the second channel is produced in the first channel. This type of cross talk resembles emission cross talk yet has a completely different cause (Figure 1.5): excitation of a tail of the second fluorochrome excitation spectrum by the first laser and emission of light of the second fluorochrome objects in the first channel. An example is the excitation of Alexa Fluor 633 and emission of that fluorochrome in the 594 nm channel by a 594 nm laser used in a double-fluorochrome experiment wherein the combination Alexa Fluor 594 and Alexa Fluor 633 is used. The occurrence of this type of cross talk can be checked by configuring channels first, followed by the insertion and systematic inspection of three control slides: one stained with fluorochrome #1, one stained with fluorochrome #2, and the third stained with both fluorochromes. The caveat with the combination of 594 nm/633 nm fluorochromes in double fluorescence experiments is that observations with this pair may suffer from *both* emission and excitation cross talk.

Elimination of Cross Talk

In neuroscience applications, a frequent question is whether a particular protein occurs in particular processes or, inside neuronal processes, in particular compartments. Typically such a question is attacked with a combination of immunofluorescence and additional fluorescence staining to resolve the individual proteins and the surrounding biological structure (cell body of a neuron, process, axon terminal, outer membrane, vesicle, etc.). Cross talk is extremely annoying in such a situation and must be completely eliminated. While emission cross talk can be safely disposed of by sequential scanning procedures, it is much more difficult to eliminate excitation cross talk. If changing the excitation laser intensity and emission filtering bandwidth does not work (to be checked with the three control slides mentioned in the previous section), a workaround is by means of postacquisition image subtraction (so-called linear unmixing). Most manufacturers offer a linear unmixing function in their standard CLSM software package. This measure, however, presents a cure and not prevention. A truly preventive measure consists of the application of fluorochrome combinations whose excitation spectra show no overlap, such as the combination of Alexa Fluor 488 and Alexa Fluor 594 nm. In our instrument with complete channel separation in double-fluorescence experiments we use the combination of Alexa Fluor 488 and Alexa Fluor 594, and in triple fluorescence experiments we use the combination of Alexa Fluor 488, Alexa Fluor 546, and Alexa Fluor 633. The selection of a particular fluorochrome combination depends on the lasers and filters available in the used confocal instrument. It is advisable to be familiar with the excitation/emission spectra of the applied fluorochromes, because false positivity caused by both types of cross talk can be embarrassing. Any serious image acquisition session on a CLSM should start with "testing the channels" via a battery of single- and double-fluorescent test sections prepared for and kept in storage for this purpose.

In addition to channel cross talk caused by the selection of combinations of fluorochromes that do not exactly match the instrument's channel configurations, instrument-related errors may produce cross talk-like effects, such as internal reflections in the confocal instrument. Incomplete cutoff of bandpass filters may produce inadvertent ghost images as well (i.e., signal leakage). A CLSM operator should always be aware of the possibility of cross talk and cross talk-like phenomena.

Biological Objects Translated to Pixels

Assume a small, discrete biological object is imaged in a CLSM at high magnification. The object has been labeled with a fluorochrome that provides a diffuse, yet homogeneous filling. For this purpose we use the axon terminal in the CNS of Figure 1.2 as an example. In the electron microscope, a typical axon terminal is a unit membrane-bound swelling at the distal end of a thin unmyelinated fiber. The thinnest of these fibers have diameters ranging from 0.1 to 0.5 μm (Peters et al., 1991), whereas the diameter of the terminal swellings ranges from 0.5 to 1.0 μm. Big swellings like hippocampal mossy fiber terminals and "giant" corticothalamic terminals (Hoogland et al., 1991) may reach 5 μm or more. These swellings have no uniform shape. Single or repetitive swellings may occur on a fiber on its course (so-called boutons *en passant*), while at the distal end complicated rosettes of lollipop-like structures may exist ("terminal rosettes" with "terminal boutons"). Shape can easily be determined in confocal micrographs, but how do we determine the exact size or, more specifically, how do we determine where the limiting membranes are exactly located?

Assume a geometrical, unit membrane-bound biological structure 580 nm across (Figure 1.6). We limit ourselves to observations in the *XY* plane, and we have labeled the structure with a 488 nm excitation probe that emits light at 519 nm. The readout of our CLSM using a 63× NA 1.3 immersion objective and 8× electronic zoom is that 1 pixel measures 58 nm in *X* and *Y*. Image formation is as follows. According to Abbe, the radius *r* of the PSF, after deconvolution, is 162 nm, or ~3 pixels. The axial *r* is 378 nm, or ~3 pixels at a stepping increment of 122 nm. As the fluorochrome marker fills the inside of the structure homogeneously, the image of the outer membrane of the structure will be built up as a series of PSFs with the membrane as their centers. The matrix of the biological structure will occur as a population of PSFs filling the structure with "spillover" of 1*r* at the edges (Figure 1.6).

High-probability Determination of Diameter

Based on the previous information, we have adopted a strategy to determine with best confidence from the CLSM images the position of the outer membrane of the biological structure and hence the diameter of that biological structure. Needless to say the histology should be of the highest quality available, mounting medium and coverslipping should have been properly applied, and the imaging instrument should be in mint condition. Determination of diameter is based on our strategy in 3D object recognition. For this purpose, sufficient objects should be available in the image. We employ two methods: best-fit and automated "objective" threshold analysis.

Best-Fit Object 3D Recognition

Z-imaging is performed in a standard way. After deconvolution, the image stack is copied to a computer running the 3D reconstruction/surface rendering program Amira (Visage Imaging; www.amira.

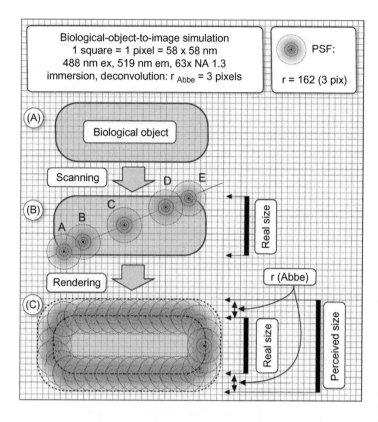

Figure 1.6 Simulation of the translation process of a biological object into pixels in a high-magnification (deconvoluted) CLSM image. (A) Simple biological object being scanned; 10 pixels in diameter (580 nm). (B) The CLSM's photomultiplier collects photons according to the PSFs calculated from their points of origin. Photons originating from positions A and E contribute to 3 pixels (r_{Abbe}) of haze. Fluorochrome molecules at positions B and D actually inside the object distribute photons inside the cytoplasm yet with PSF spillover outside the limiting membrane. Only photons emitted from position C distribute according to a PSF entirely inside the boundaries of the biological object. (C) The result of the image acquisition: a perceived size of the object one Airy distribution larger ($2 \times r_{Abbe}$) for that wavelength.

com). Amira is available in Linux and Windows versions. The software considers an image Z-series as a 3D pixel matrix. In order to render properly the program requests a Z-dimension which should be set in proportion to the Z-stepping increment during image acquisition. Two particular functions of the program are important: orthoslice and isosurface. Orthoslice is a function that shows any of the frames loaded, and it is capable of slicing, that is, showing a slice of the 3D pixel matrix in any of the orientations X, Y, or Z. Simultaneously the function isosurface is applied. This interactive function creates a wireframe connecting all pixels in X, Y, and Z that possess equal gray intensity levels. The intensity level selected for calculating the wireframe is interactively supplied by the operator as a "threshold" gray intensity value. The function isosurface can be executed in overlay with orthoslice. Thus, the operator selects in isosurface a gray intensity level (threshold) for which the wireframe matches *as well as possible* the outline of the object visible in orthoslice. This "best match" is quite a subjective exercise depending on the interpretation of the object by the operator. The human operator observes in orthoslice structures with fuzzy edges, and with the isosurface function he is faced with the job to produce "hard" edges. Confronted with this decision most human operators have a tendency to contain both structure *and* their fuzzy edges by the iso-surface's wireframes. Humans perceive objects with fuzzy outlines as being bigger than they in reality are (Figures 1.4A and 1.6). With complicated and especially irregularly shaped structures the outcome of the orthoslice–isosurface procedure in Amira by two different human operators performing blindly can be considerable. As a rule of thumb and based on considerations explained in Figure 1.5, a "tight fit" of the wireframe around the orthoslice object is recommended.

How tight is a tight fit? The previous calculation demonstrates that tight fit means the wireframe in orthoslice is 1*r*, or in this case 3 pixels narrower radially than the onset of the pixel gradient from dark background to the object. After the object as it appears in orthoslice is matched minus 3 pixels with the wireframe, another Amira function can be called "measure" that provides the desired numerical values, for example, distances from any point of the wireframe to another point. The advantage of the best-fit procedure is that it is fast. A serious disadvantage is that a human operator is involved whose decisions regarding thresholding are highly subjective and may be biased. A human operator also gets tired and in a blind experiment may select different thresholds in a dataset early in the morning and again at the end of the workday.

Automated Objective Threshold Analysis

The subjectiveness of the best-fit procedure hampers the reliability of the results. Feeling unsatisfied with this approach, we developed a procedure that avoids human interference in the determination of the gray level intensity threshold and which, in images that cover many biological objects (particles), produces consistent results that can be replicated in any laboratory by any computer operator. For this we process the images with the plug-in "3D object counter" (Fabrice Cordelières, Institut Curie, Orsay, France; Bolte and Cordelières, 2006) of the image analysis program ImageJ (version 1.37c; Rasband, 1997–2006). The concept of determining thresholds is in analogy to counting sandbars on a beach by watching the tide come and go. An image series (a 3D pixel matrix) is loaded into

Figure 1.7 Automatic thresholding and counting of VGluT1 and VGluT2 immunofluorescent "particles" (clusters of synaptic vesicles). (A and B) *Z*-projections from a two-fluorochrome, two-channel CLSM experiment: VGluT2 (488 nm fluorochrome) and VGluT1 (633 nm fluorochrome). (C) Both datasets subjected to ImageJ threshold-particle analysis with the plug-in "3D object counter." Threshold 35 (eight step in a 5 gray intensity level stepping series) distinguishes most particles (*n* = 604). (D) This threshold is used to 3D reconstruct ▶

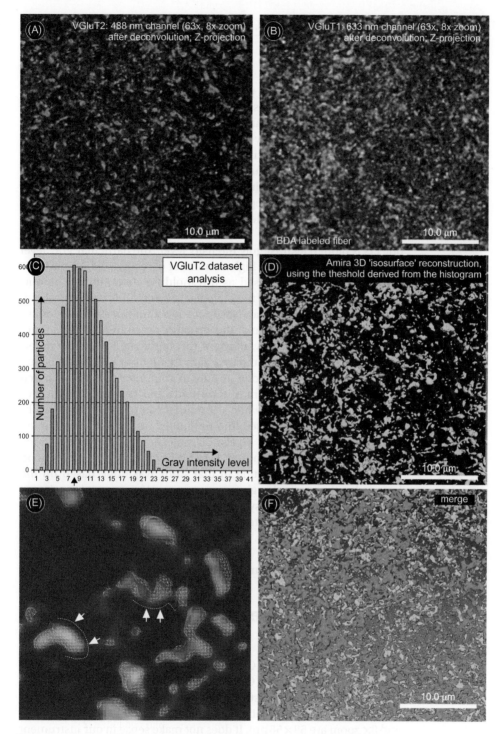

Figure 1.7 (Continued) in Amira. (E) Combination of isosurface and orthoslice functions in Amira to show the fit of the isosurface wireframes with pixel aggregates (enlarged and brightness artificially increased). The dashed lines indicate tight fit as a human operator could have decided. (F) Merge of the VGluT2 and VGluT1 3D reconstructions (the VGluT1 dataset identically processed as the VGluT2 dataset). Please see color plates at the back of the book.

ImageJ (Figure 1.7). The plug-in requests a threshold. Using this threshold it then identifies aggregates of high-intensity pixels in the 3D pixel matrix that occur within an isodensity envelope of pixels whose gray level intensity equals the threshold (similar to Amira's isosurface function). By using a script that runs the procedure n times where n stands for the number of gray levels (8-bit, 0–256; 16-bit, 0–4096; to reduce processing time the script carries a line that introduces increments, e.g., steps of five gray intensity levels) the plug-in calculates for every gray level the number of identified 3D objects. Next, a histogram can be built where the number of identified 3D objects is plotted against gray intensity level. The particular gray level that is associated with the highest number of recognized 3D particles (Figure 1.7C) is further used as threshold for determining the outlines of the membranes of the biological objects. This peak threshold, if introduced in Amira as an isosurface function threshold, produces wireframes that tightly match the structures seen in orthoslice function (Figure 1.7E) The advantage of this automated objective determination of thresholds compared with the tight fit procedure is that the human factor is excluded. A human computer operator is much more variable in determining thresholds and becomes fatigued or loses concentration soon while a computer can run indefinitely without varying criteria. We used this automated procedure to determine densities of aggregates of pixels (3D objects or boutons containing vesicular glutamate transporter 1 and 2 in rat striatum; Wouterlood et al., 2012). Object counting in that study was done based on the knowledge that vesicular glutamate transporters are proteins incorporated in the walls of synaptic vesicles. As one synaptic vesicle measures 50 nm in diameter, that is, far below the resolution offered by our optical instrument, only clusters of vesicles are reported. We assumed a 1:1 proportion between clusters of synaptic vesicles and the axon terminals that contain these clusters.

Why Does a 3D Reconstructed Cell Resemble a Pancake?

Most scientists new to CLSM assume by default that the shape of a voxel resembles a cube. Obviously this is not the case in biological confocal scanning. For instance, the Z-step increment in our instrument is usually 122 nm while the XY dimensions at 63× magnification and 8× zoom are 58×58 nm. It does not make sense in our instrument to reduce the Z-step increment to, say, 61 nm because of the reduced axial resolution of the instrument. First, step sizes smaller than 122 nm contribute to Nyquist oversampling, and second, scanning at, say, Z-step increment 61 nm might even lead to a poorer image recording because of bleaching that occurs while scanning the oversampled focal planes. The calculations for resolution (Table 1.1) show that r radially with 488 nm excitation and 519 nm emission, and after deconvolution, is 163 nm radially and 378 nm axially. If voxels were cubes with identical XYZ dimensions of 58 nm, then this would translate into an inaccuracy of 3 pixels radially and 7 pixels axially, or more than two times oversampling according to Nyquist in the axial direction.

Because most 3D reconstruction programs allow the introduction of a numerical value for the Z-increment next to that for X and Y dimensions, the distorted Z-compressed 3D rendering as experienced by many can be compensated for. As a rule of thumb the value of the Z-dimension in high-resolution CLSM scanning can be taken as twice as big as the values of the X and Y dimensions.

Touch

In functional neuroanatomy many questions concern the configuration of neuronal networks. These networks consist of numerous projection neurons and interneurons that individually express their own particular neurotransmitter and whose axon terminals are engaged in synaptic contacts with other neurons in the network that may utilize any neurotransmitter available. One of the least complicated questions is whether a neuron (neuron type 1) expressing neurotransmitter A, observed in a neuronal network, is synaptically engaged with another neuron (neuron type 2) that expresses neurotransmitter B. Are there synaptic contacts between processes of neuron type 1 with those of neuron type 2? For this purpose we need to identify neurons that belong to type 1, neurons that belong to type 2, and we need to identify synapses between neurons of type 1 and 2. Please note that packing of neurons in the CNS of cellular processes is so dense (comparable to sardines in a can) that processes are often in contact with each other without any evidence of synaptic exchange of information. As this chapter deals with high-resolution scanning we will focus first on how we define contacts and then ask when a contact is a synapse.

To identify contacts at least two markers are necessary, one for each participating neuron. For practical reasons we often use a neuroanatomical tracer to label fibers and axon terminals of type 1 neurons, while for the purpose of labeling type 2 neurons a population-specific marker may be applied, which can be a general category marker or a subcategory marker. Examples of general category markers are gamma amino butyric acid (GABA) or its metabolizing enzyme, glutamic acid decarboxylase, while examples of subcategory markers are polypeptide markers, calcium binding proteins (highlighting subclasses of GABAergic neurons), choline acetyltransferase (cholinergic neurons), and tyrosine hydroxylase (monoaminergic neurons, i.e., dopaminergic neurons). While the advantage of these category markers is that they label discrete populations of neurons, the members of these populations can be so densely packed that individual neurons are difficult to distinguish. In our field of research (substantia nigra) this is the case with tyrosine hydroxylase-positive dopaminergic neurons, especially in the so-called pars compacta (hence its name). To distinguish individual cells in the substantia nigra pars compacta we have successfully applied intracellular, single-cell injection with a fluorescent dye (Lucifer yellow; Rho and Sidman, 1986; Buhl and Lübke, 1989).

Actual Experiment

In the example shown here we have labeled in rats neurons of type 1 with a neuroanatomical tracer, biotinylated dextran amine (BDA). The transported BDA was detected via a reaction with fluorochromated streptavidin (streptavidin-Alexa Fluor 546). The presumed type 2 neurons were intracellularly injected with Lucifer yellow. As intracellular injection works only in intact neurons, this part of the procedure was performed with thick slices cut from lightly fixed brain. After a series of neurons had been intracellularly injected, these slices were resected into thinner sections that were then subjected to incubation with a cocktail of streptavidin-Alexa Fluor 546 to stain the transported BDA, and with a cocktail of antibodies: anti-Lucifer yellow raised in rabbit and anti-tyrosine hydroxylase raised in mouse. Finally, an Alexa Fluor 488 conjugated goat anti-rabbit IgG was applied to detect the Lucifer yellow, and an Alexa Fluor 633 conjugated goat anti-mouse IgG to label dopaminergic neurons and their processes. In the CLSM we configured three independent channels around the laser excitation wavelengths 488, 543, and 633 nm, and checked channel separation with our kit with control slides, and we scanned in sequential mode.

In the confocal microscope we noted in the 488 and 543 nm channels close apposition of some fibers labeled with the neuroanatomical tracer and the cell processes containing the intracellular marker. A contact is shown in Figure 1.8. In the 633 nm channel we did a secondary check to see whether the latter processes expressed tyrosine hydroxylase. This check consists of looking for colocalization of fluorescence in structures visible in both 488 and 633 nm channels.

Two questions arise here: (1) Are these "contacts" true in the sense that the limiting membrane of neuron type 1 apposes the limiting membrane of neuron and (2) do these contacts represent synapses?

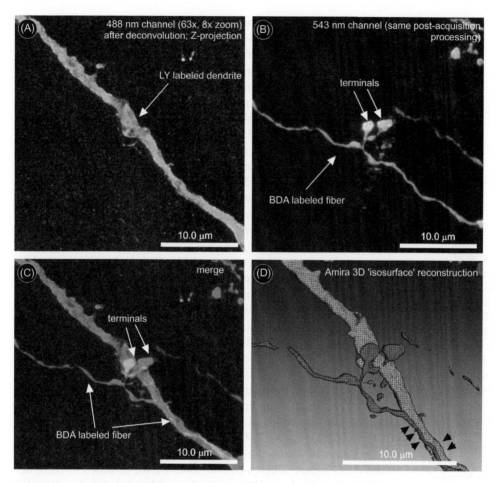

Figure 1.8 Contacts in the substantia nigra of the rat between labeled fibers and a dendrite of an intracellularly injected nigra neuron. Two-channel scanning with the 63× NA 1.3 immersion objective, 8× zoom. (A) *Z*-projection of the 488 nm dataset: a dendrite of a Lucifer yellow-injected cell. (B) *Z*-projection of thin labeled fibers that give rise to two big terminals (arrows). (C) Merge of both datasets. Note the intimate relationship of the BDA fiber and part of the dendrite. (D) 3D reconstruction isosurface in Amira. Please see color plates at the back of the book.

The presence of a contact can be assessed by a human computer operator by the best-fit method mentioned earlier. However, it has also been argued previously that an image obtained via the best-fit method of a 3D reconstructed biological structure has a tendency to be larger than the biological object it represents. In the radial direction the human uncertainty involved in deconvoluted images in pinpointing the exact position of the limiting membrane can be as large as 3 pixels for Lucifer yellow and also 3 pixels for the 546 nm fluoro-chrome (Table 1.1). This results in a reduced confidence to distinguish contacts. A matter of deep concern here is that thin wraps of glial processes may encapsulate processes or separate neuronal processes from each other. These wraps have in electron micrographs dimensions in the order of magnitude of 200 nm thickness (4 pixels), and they lack packets of glial fibrils characteristic for larger astroglial processes. Usually, glial fibrillary acidic protein (astroglial cell marker) is absent in small glial wraps. For this reason, the possible presence of thin glial wraps around neuronal processes, invisible in the CLSM and with dimensions at or just below detection resolution level, may interfere in an annoying way with the observations.

In an attempt to distinguish contacts we took the following approach. Assume that there are two membranes, each labeled with a different fluorochrome with Abbe's r in the order of 3 pixels. When these membranes are in true apposition, that is they lie parallel to each other with only extracellular space intervening, the imaging and following 3D reconstruction in images acquired in a two-channel CLSM at high magnification will show partial overlap of pixels that still exists after deconvolution (Figure 1.9).

The cause for this overlap is that the imaging of both apposing membranes produces Abbe-type diffraction where the r of the Airy distributions of the emitted photons partially overlaps. The solution here is to use this overlap as an extra criterion to ascertain the existence of a contact. A small calculation leads to the following: an active zone of a synaptic contact is on average a spot with a diameter of 0.3 μm (6 pixels across). Let's assume a volume with this diameter where Abbe-diffraction overlap exists. Such a volume takes $4/3 \pi r^3$ voxels = 113 voxels. Albeit a coarse criterion, we included a "compulsory" overlap of at least 100 voxels as an additional criterion into the acceptance of an apposition of two structures as a contact. If there is no voxel overlap between the channels, there must be nonfluorescing

Figure 1.9 Simulation of two-channel imaging at high resolution of "biological" objects that are in physical contact. Lens = 63× immersion NA 1.3; zoom = 8; each square represented on a pixel. In the acquired images, each structure has a diffraction-induced "edge" of 3 pixels (green object, 488 nm fluorochrome) or 5 pixels (red object, 647 nm fluorochrome). The merged channels produce voxel overlap at the site where, in the specimen, the biological objects are in contact. Please see color plates at the back of the book.

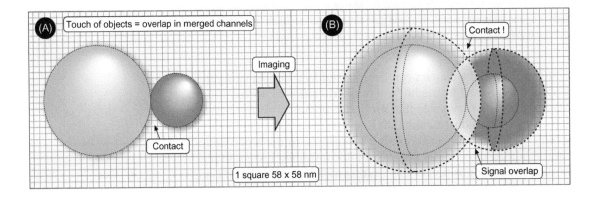

matter present in between both fluorescing structures and, consequently, the structures involved must be located some distance apart.

Computer Software to Define a Contact

Thus we established two criteria based upon which we decide that two apposing structures — each visualized in a different channel in the CLSM — share a contact. The first is that processes from neurons 1 and 2 must appose each other, after application of best-fit 3D reconstruction. The second is that there must be voxel overlap of both 3D reconstructed processes.

A computer program can be instructed to distinguish isodensity wireframes in a 3D pixel matrix. It can do this in two related 3D pixel matrices (a two-channel Z-image series), and next the program can calculate for each aggregate in each channel the number of appositions it has with aggregates in the other channel's dataset and reject those appositions that do not fulfill the minimum of 100 voxel overlap criterion. For this purpose we use dedicated software written in SCIL_Image that can be called using scripts that can be run unattended (Wouterlood et al., 2007, 2008a). The program provides two readouts: the number of recognized contacts of 3D objects present in the first channel versus those in the second (or third) channel, and vice versa. If desired, binarized images are provided as output, reporting the positions of the identified contacts in a graphical way.

Synaptic Contacts: Extra Marker

Even when there is apposition of two neuronal processes and the software does report overlap of signal according to the above 100 voxel criterion, it may not be 100% clear that there is a synapse. In the extremely tightly packed brain tissue, most contacts between neurons can be considered casual. An extra criterion, for example, a synapse-specific signal, would be welcome. We have experimented in the past with such a third marker. The postsynaptic density in glutamatergic excitatory synapses is composed of a large number of structural proteins whose role is to anchor postsynaptic receptors in the postsynaptic density and to connect the postsynaptic receptors to the cytoskeleton of the postsynaptic cell (Gundelfinger and Fejtova, 2011). Thus, the postsynaptic density is a site where highly specific proteins occur that mark the presence of synapses in a way that is suitable for high-resolution CLSM. However, this postsynaptic density carries molecular machinery at such a particularly high concentration (hence its name) that antibodies against component proteins or polypeptides encounter difficulty in the intact terminal bouton to penetrate into the protein lattice and bind to their epitope. Experience shows that only a few antibodies that "work" in homogenates, synaptosome fractions, or thin cryostat sections also produce staining in preparations consisting of brain sections with intact terminal boutons. We tried several of these postsynaptic density-reporting molecules and found one antibody against ProSAP2/Shank3 that binds in sufficient quantity to postsynaptic sites in brain sections to provide sufficient signal in the CLSM (Wouterlood et al., 2003). Indeed, fluorescence signal associated with ProSAP2/Shank3 was detected in several cases sandwiched in between the presynaptic neuron label and the postsynaptic neuron marker, thus enhancing the evidence that a glutamatergic excitatory synaptic contact is present is this site.

As the alternative to optical recognition of synaptic contacts is electron microscopy, which involves complicated, extremely tedious time- and resource-consuming procedures, we feel that through the use of proper criteria plus an extra marker, fast mass optical synapse recognition has progressed toward claiming much of the scientific territory that was once the exclusive domain of electron microscopic identification.

Colocalization

Colocalization means that structures labeled with a particular fluorochrome and visible in a particular CLSM channel coincide in cell compartments with structures labeled with another fluorochrome and made visible in a parallel CLSM channel. Every CLSM on the market today is equipped with colocalization software, that is, a program module capable of calculating Pearson and Spearman correlation coefficients at the pixel level. Also in the public domain, software of this type is available as a plug-in for ImageJ (French et al., 2008). Our interest, however, focuses on biological structures containing a particular marker. As an example we present here synaptic vesicles. Electron micrographs show that these small spherical (20 nm across) structures occur in groups or aggregations inside terminal boutons. Functionally, synaptic vesicles accumulate neurotransmitters, move toward the presynaptic specialization of a synapse, dock and release their neurotransmitter, and undock and move back to the synaptic vesicle pool for recycling. This functionality requires a differentiated molecular machinery, and indeed antibodies such as synaptophysin, synaptotagmin, vesicular transporters, etc., have become available against components of this machinery. Thus we are interested in 3D reconstructing objects labeled with a particular fluorochrome inside (larger) structures labeled with another fluorochrome. The example shown here consists of terminal boutons on nerve fibers containing a synaptic vesicle-associated protein, vesicular glutamate transporter 2 (VGluT2), and a specific calcium binding protein, calretinin. Axonal endings of neocortical projection neurons express VGluT1 (Fremeau et al., 2001; Kaneko and Fujiyama, 2002), whereas endings of thalamostriatal and subcortical fibers express a different vesicular glutamate transporter named VGluT2 (Herzog et al., 2001; Fremeau et al., 2004). As VGluT2 occurs in considerable quantities in temporal cortex of the rat we combined neuroanatomical tracing with VGluT2 immunofluorescence to determine whether the VGluT2 is associated with thalamocortical input fibers. Calretinin is present both in neocortical interneurons and in some thalamic projection neurons.

The neuroanatomical tracer (again BDA) was injected in a thalamic nucleus (nucleus reuniens), which resulted in labeling of fibers and boutons in temporal cortex. Sections of this cortex were subjected to VGluT2 immunohistochemistry (rabbit anti-VGluT2 detected with goat anti-rabbit IgG Alexa Fluor 488) in combination with calretinin immunofluorescence (mouse anti-calretinin/goat anti-mouse IgG Alexa Fluor 594) and finally treated with streptavidin-Alexa Fluor 546 to visualize the thalamic labeled fibers and their endings. In the CLSM we configured 488, 543, and 594 nm channels and double-checked with our kit of single-stained control sections that cross talk was completely absent. Then we scanned the sections in sequential mode with the 63× NA 1.3 immersion lens at 8× zoom and acquired Z-series of images in all three channels. These images were deconvoluted and analyzed for colocalization of aggregates of VGluT2 and calretinin inside axon terminals labeled with BDA. First we subjected the VGluT2 and calretinin datasets to 3D object analysis using automated objective threshold analysis, while the BDA dataset was processed with the best-fit procedure (see earlier). Using the thresholds derived from that analysis we instructed software to search in all 3D pixel matrices for isodensity-bounded aggregates of voxels where the center of gravity in both channels coincided, with 100% overlap of the 3D objects identified in the 488 nm 3D dataset with those in the 543 nm dataset (assuming that clusters of VGluT2-expressing synaptic vesicles fit axon terminals, i.e., the diameters of VGluT2 3D objects are smaller than those of the encapsulating axon terminals). Indeed, we found aggregates of VGluT2 voxels completely internalized in swellings of fibers of thalamic origin (Figure 1.10; Wouterlood et al., 2008b). We even found triple colocalization, that is, VGluT2 and calretinin in endings of labeled thalamocortical fibers (Figure 1.10C). An unexpected finding was that numerous aggregates of VGluT2 voxels were only partially enclosed in swellings on BDA-labeled thalamocortical fibers (e.g., in Figure 1.10C). Whereas we assumed coincidence of centers of gravity of both markers, pools of synaptic vesicles in the real world may not often occur positioned in

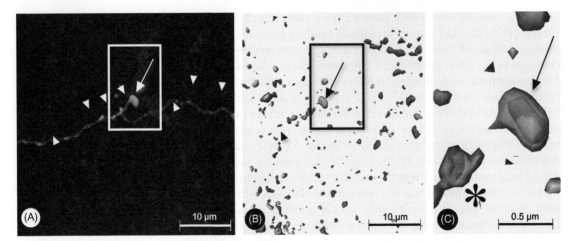

Figure 1.10 Neuroanatomical tracing with BDA combined with double-immunohistochemistry (detecting VGluT2 and calretinin). Three-channel scanning with the 63× NA 1.3 immersion objective, 8× zoom. (A) Z-projection of images acquired in the 543 nm channel combined with isosurface 3D reconstruction. The BDA labeled fiber (arrowheads) has a terminal bouton (arrow). (B) Amira 3D isosurface reconstructions of all three channels merged. VGluT2 shown in green, BDA in red, and calretinin in blue. The arrow indicates the BDA-labeled terminal (red). (C) Enlarged part of the 3D reconstruction of (B) where the BDA-3D reconstruction has been made transparent. Inside the terminal are small clusters of voxels associated with VGluT2 (green) and calretinin (blue). Note that the BDA and calretinin in the terminal on the left (asterisk) only partially overlap. Please see color plates at the back of the book.

the very center of a synaptic bouton. If a pool of synaptic vesicles is in the "live" situation eccentrically located in the axonal bouton, its representation in two-channel CLSM, based on Abbe diffraction, may produce VGluT2 voxel aggregates partially "outside" the BDA-labeled fiber terminal. Thus at this order of magnitude a 100% overlap of structures between both channels seems unrealistic, and we reduced the overlap criterion to a level of 80%, which better coincides with the statistical nature of image formation at the edge of the resolving power of the optical imaging instrument.

Conclusion

For studying processes at subcellular resolution in neuroscience and in wider perspective, within bioscience applications the CLSM is our proposed instrument of choice as long as one performs the proper control experiments and is aware of applying the optimal instrument settings. High-magnification, high-resolution CLSM imaging is influenced by the statistical behavior of the emitted photons through Abbe-type diffraction that imposes limits. Through the application of several measures, these limits can be pushed back to a certain degree.

Acknowledgments

The authors deeply appreciate the assistance in experimental animal and immunohistological procedures by Amber Boekel, Angela Engel, Wolfgang Härtig, and Peter Goede. Nico Blijleven's ongoing technical support with the Leica confocal instrument is especially acknowledged, cheerfully keeping all the different types of hardware and software running and communicating with each other, updating software on all auxiliary computers, and providing sample management advice and deconvolution software scripts. We also thank Drs. Verena Aliane and Riichi Kajiwara for writing additional computer scripts that when combined made automated 3D object recognition proceed with efficiency and speed.

References

Boccacci, P., Bertero, M., 2002. Image restoration methods: basics and algorithms. In: Diaspro, A. (Ed.), Confocal and Two-Photon Microscopy: Foundations, Applications and Advances. Wiley-Liss, New York, pp. 253–269.

Bolte, S., Cordelières, P., 2006. A guided tour into subcellular colocalization analysis in light microscopy. J Microsc 224, 213–232.

Brakenhoff, G.J., Blom, P., Barends, P., 1979. Confocal scanning light microscopy with high aperture immersion lenses. J Microsc 117, 219–232.

Buhl, E.H., Lübke, J., 1989. Intracellular Lucifer yellow injection in fixed brain slices combined with retrograde tracing, light and electron microscopy. Neuroscience 28, 3–16.

Egger, M.D., Petrăn, M., 1967. New reflected-light microscope for viewing unstained brain and ganglion cells. Science 157, 305–307.

Fremeau Jr., R.T., Troyer, M.D., Pahner, I., Nygaard, G.O., Tran, C.H., Reimer, R.J., et al., 2001. The expression of vesicular glutamate transporters defines two classes of excitatory synapse. Neuron 31, 247–260.

Fremeau Jr., R.T., Voglmaier, S., Seal, R.P., Edwards, R.H., 2004. VGLUTs define subsets of excitatory neurons and suggest novel roles for glutamate. Trends Neurosci 27, 98–103.

French, A.P., Mills, S., Swarup, R., Bennett, M.J., Pridmore, T.P., 2008. Colocalization of fluorescent markers in confocal microscope images of plant cells. Nat Protoc 3, 619–628.

Gundelfinger, E.D., Fejtova, A., 2011 October 24. Molecular organization and plasticity of the cytomatrix at the active zone. Curr Opin Neurobiol (Epub ahead of print).

Herzog, E., Bellenchi, G.C., Gras, C., Bernard, V., Ravassard, P., Bedet, C., et al., 2001. The existence of a second vesicular glutamate transporter specifies subpopulations of glutamatergic neurons. J Neurosci 21, U1–U6.

Hiesinger, P.R., Scholz, M., Meinertzhagen, I.A., Fischbach, K.-F., Obermayer, K., 2001. Visualization of synaptic markers in the optic neuropil of *Drosophila* using a new constrained deconvolution method. J Comp Neurol 429, 277–288.

Hoogland, P.V., Wouterlood, F.G., Welker, E., Van der Loos, H., 1991. Ultrastructure of giant and small thalamic terminals of cortical origin; A study of the projections from the barrel cortex in mice using *Phaseolus vulgaris*-leucoagglutinin. Exp Brain Res 87, 159–172.

Inoué, S., 1995. Foundations of confocal scanned imaging in light microscopy. In: Pawley, J.B. (Ed.), Handbook of Biological Confocal Microscopy. Plenum Press, New York, pp. 1–4.

Jonkman, J.E.N., Stelzer, E., 2002. Resolution and contrast in confocal and two-photon microscopy. In: Diaspro, A. (Ed.), Confocal and Two-Photon Microscopy: Foundations, Applications and Advances. Wiley-Liss, New York, pp. 101–125.

Kaneko, T., Fujiyama, F., 2002. Complementary distribution of vesicular glutamate transporters in the central nervous system. Neurosci Res 42, 243–250.

Minsky, M., 1957. U.S. patent #301467, Microscopy Apparatus.

Panchuk-Voloshina, N., Haugland, R.P., Bishop-Stewart, J., Bhalgat, M.K., Millard, P.J., Mao, F., et al., 1999. Alexa dyes, a series of new fluorescent dyes that yield exceptionally bright, photostable conjugates. J Histochem Cytochem 47, 1179–1188.

Pawley, J., 1995. Fundamental limits in confocal microscopy. In: Pawley, J. (Ed.), Handbook of Biological Confocal Microscopy. Plenum, New York, pp. 19–37. (Chapter 2).

Pawley, J., 2006. Points, pixels and gray levels: Digitizing image data. In: Pawley, J. (Ed.), Handbook of Biological Confocal Microscopy. Plenum Press, New York, pp. 59–79. (Chapter 4).

Peters, A., Palay, S.L., Webster, deF.H., 1991. The Fine Structure of the Nervous System: Neurons and their Supporting Cells. Oxford University Press, New York.

Rasband, W.S., 1997–2006. ImageJ. U.S. National Institutes of Health, Bethesda, Maryland, <http://rsb.info.nih.gov/ij/>.

Rho, J.H., Sidman, R.L., 1986. Intracellular injection of Lucifer yellow into lightly fixed cerebellar neurons. Neurosci Lett 72, 21–24.

Sheppard, C.J.R., Choudhury, A., 1977. Image formation in the scanning microscope. Opt Acta 24, 1051–1073.

van der Voort, H.T.M., Strasters, K.C., 1995. Restoration of confocal images for quantitative image analysis. J Microsc 158, 43–45.

Wallace, W., Schaefer, L.H., Swedlow, J.R., 2001. A workingperson's guide to deconvolution in light microscopy. Biotechniques 31, 1076–1078.

Wilson, T., 2002. Confocal microscopy: Basic principles and architectures. In: Diaspro, A. (Ed.), Confocal and Two-Photon Microscopy: Foundations, Applications and Advances. Wiley-Liss, New York, pp. 19–38.

Wouterlood, F.G., Böckers, T., Witter, M.P., 2003. Synaptic contacts between identified neurons visualized in the confocal laserscanning microscope. Neuroanatomical tracing combined with immunofluorescence detection of postsynaptic density proteins and target neuron-markers. J Neurosci Methods 128, 129–142.

Wouterlood, F.G., Boekel, A.J., Kajiwara, R., Beliën, J.A., 2008a. Counting contacts between neurons in 3D in confocal laser scanning images. J Neurosci Methods 171, 296–308.

Wouterlood, F.G., Boekel, A.J., Meijer, G.A., Beliën, J.A.M., 2007. Computer assisted estimation in the CNS of 3D multimarker overlap or touch at the level of individual nerve endings. A confocal laser scanning microscope application. J Neurosci Res 85, 1215–1228.

Wouterlood, F.G., Härtig, W., Groenewegen, H.J., Voorn, P., 2012. Density gradients of vesicular glutamate- and GABA transporter immunoreactive boutons in calbindin- and µ-opiod receptor defined compartments in the rat striatum. J Comp Neurol 520, 2123–2142.

Wouterlood, F.G., Aliane, V., Boekel, A.J., Hur, E.E., Zaborszki, L., Barroso-Chinea, P., et al., 2008b. Origin of calretinin containing, vesicular glutamate transporter 2-coexpressing fiber terminals in the entorhinal cortex of the rat. J Comp Neurol 506, 359–370.

2

BEYOND ABBE'S RESOLUTION BARRIER: STED MICROSCOPY

U. Valentin Nägerl

Interdisciplinary Institute for Neuroscience (IINS), Université Bordeaux Segalen, and UMR 5297, Centre National de la Recherche Scientifique (CNRS), Bordeaux, France

CHAPTER OUTLINE
Introduction   36
A New Wave of Imaging   37
STED Microscopy: The Basic Concept   39
Implementation of STED Microscopy   42
    The Microscope Base   42
    Laser Sources   42
    The Doughnut   42
    Synchronization   44
    Objective Lenses   44
    Emission Detection   44
    Scanning Schemes   45
    Software   46
    Stability Considerations   46
    Checking the Resolution   47
***Sine Qua Non:* Speed, Color, Depth, Live Imaging   47**
    Temporal Resolution and Imaging Speed   47
    Labeling Strategies   48
    Multicolor Imaging   49
    Depth Penetration and Spatial Resolution   50
    Spatial Resolution in z 50
    Live-cell Imaging   51
Summary and Outlook   52
    Reversible Saturable Optical Fluorescence Transitions:
        A More General Principle for Nanoscopy   52
    Other Areas of Development   52
References   52

Cellular Imaging Techniques for Neuroscience and Beyond.
DOI: http://dx.doi.org/10.1016/B978-0-12-385872-6.00002-7

Introduction

Stimulated emission depletion (STED) microscopy is a recent optical microscopy technique that breaks the classic diffraction barrier of light microscopy (Hell, 2007), which for a long time was believed to define a hard and immutable limit for the spatial resolution of a light microscope. It offers the ability to shed light on dynamic processes inside living cells and biological tissue that have been out of reach for conventional light microscopy such as confocal and two-photon microscopy, whose spatial resolution is limited by the diffraction of light.

Because it is based on laser-scanning fluorescence microscopy, STED microscopy naturally stands to inherit many advantages of confocal microscopy regarding labeling and sensitivity, contrast and optical sectioning, scanning, and speed.

Similar to confocal microscopy, a STED microscope uses focused laser light to scan the distribution of fluorescent molecules in 3D samples. In addition, a second laser is used to reduce the size of the focal spot and increase the spatial resolution of the microscope. As STED implies, the process of stimulated emission is used to achieve this effect.

The idea that stimulated emission could be used to break the diffraction resolution limit of optical microscopy was proposed by Stefan Hell in 1994 (Hell and Wichmann, 1994). In this paper a basic scheme for how a STED fluorescence scanning microscope might work is presented and estimates for the gain in resolution over conventional fluorescence microscopes are given. The first successful implementation of this scheme to break the diffraction limit in one dimension was published in 1999 using nanocrystals (Klar and Hell, 1999). In the following year the STED principle was demonstrated in living yeast cells in 3D (Klar et al., 2000). Since then STED microscopy has experienced huge progress on many levels. Nevertheless, these seminal papers still offer a fresh read.

Fluorescence microscopy is one of the most powerful and widely used imaging techniques in the biosciences, because it allows us to visualize dynamic processes inside living cells with exquisite sensitivity and specificity. As literally thousands of publications each year rely on fluorescence microscopy, there is intense pressure to make it even more useful and powerful.

A series of technological developments came together to facilitate the ascendancy of modern fluorescence microscopy, including in laser and detector technology, fluorescent probes, and molecular biology. As a result it is becoming increasingly possible to study large and complex biological systems like whole brains or embryos at the single-cell level under realistic conditions and diverse physiological contexts.

The development of confocal microscopy in the 1980s and two-photon microscopy in the 1990s were milestones that made it possible to obtain high-contrast fluorescence images hundreds of micrometers below the surface of biological tissue. Subsequently, the green-fluorescent protein (GFP) revolution allowed for labeling of specific proteins and organelles inside living cells, and increases in computing power helped deal with large sets of imaging data.

The Nobel Prize was awarded for the work on GFP in 2008, and it would not be a surprise if two-photon microscopy, which has been widely adopted by biomedical research laboratories, will also receive such a distinction sooner or later.

Despite relentless progress in many areas, for a long time it looked as if fluorescence microscopy was facing a brick wall regarding efforts to improve its spatial resolution. The received wisdom was that the spatial resolution of light microscopy is fundamentally limited by the diffraction of light, and that the smallest structures that could be faithfully resolved were on the order of half the wavelength of the light used in the microscope (~200–300 nm). This limit had been enshrined as a physical law for over a hundred years; in fact Ernst Abbe's famous resolution formula (Eq. 2.1) is chiseled in stone at Zeiss in Jena, Germany, where he used to work.

If the diffraction barrier was indeed a hard limit, the study of cell biological structures and processes occurring on the "mesoscale" of 10–200 nm would essentially remain out of reach for fluorescence microscopy, including macromolecular complexes, cellular organelles, and subcellular signaling domains.

Electron microscopy, with its spatial resolution of a few nanometers, has essentially been the only way to access structural information of cells and organelles as well as protein localization on this spatial scale. However, it requires tissue fixation, which is problematic and which impedes understanding dynamic events. In addition, labeling of multiple proteins for electron microscopy is difficult and sampling of cellular volumes (3D reconstruction) is extremely labor intensive.

This chapter will focus on the basic optical principles of STED microscopy and its practical implementation, while only referring briefly to other recently developed super-resolution microscopy techniques. In addition, we will illustrate the potential of STED microscopy by reviewing recent proof-of-principle applications and highlighting some interesting new technical developments.

A New Wave of Imaging

Because of its wave nature, it is impossible to focus light to an infinitesimally small spot. Rather, the smallest spot size that can be achieved by focusing lenses is limited by diffraction, which refers to the phenomenon whereby a wave tends to spread out as it travels through a small opening (Born and Wolf, 1999). The actual size of the focal spot can be accurately predicted by optical physics, varying linearly with the wavelength of the light and inversely with the numerical aperture (NA) of the lens. This relationship is at the origin of Abbe's spatial resolution formula:

$$\Delta r = \frac{\lambda}{2\text{NA}} \tag{2.1}$$

where Δr defines the minimal distance between two objects for them to be detected as separate, λ is the wavelength of light, and NA denotes the numerical aperture of the lens.

If an image is "diffraction-limited," it is considered an achievement because it means that all other problem sources that can degrade image resolution, such as aberrations, light scattering, or a poor signal-to-noise ratio, have been reduced to a point where their effects on image resolution are negligible compared with the effect of diffraction, which places the final—or so it was thought—limit on spatial resolution of the optical system.

Whereas STED microscopy was the first concrete concept to break the diffraction limit (Hell and Wichmann, 1994; Klar et al., 2000), other powerful techniques, relying on different principles, have also been developed for nano-imaging of fluorescent samples. These include photo-activated localization microscopy (PALM; Betzig et al., 2006; Hess et al., 2006), stochastic optical reconstruction microscopy (STORM; Bates et al., 2007), and nonlinear structured illumination microscopy (SIM; Heintzmann et al., 2002; Gustafsson, 2005).

These new super-resolution techniques fall into two main categories: those based on stochastic switching and computational localization of single molecules (PALM/STORM) and those based on imaging ensembles of molecules using patterned illumination (STED/nonlinear SIM) where super-resolution is achieved optically. Because of differences in optical principles and implementation, the techniques have specific strengths and weaknesses in terms of temporal resolution, depth penetration, multicolor imaging, instrumentation requirements, practical handling, and so on.

They all have in common that theoretically there is no longer a hard resolution limit, and it is possible to achieve a resolution as high as a few nanometers under ideal conditions. However, in practice they are limited by signal noise (from drift inherent in samples, particularly in living biological samples, detector noise, etc.) to tens of nanometers.

STED Microscopy: The Basic Concept

In conventional (confocal or two-photon) laser-scanning microscopy the excitation light is focused by the microscope's objective to a small focal spot that is systematically moved across the specimen, typically in two or three spatial dimensions. Thus, images are constructed one pixel at a time by measuring the fluorescence signal at each position along the trajectory of the laser focal spot using a photodetector.

In general, the spatial resolution of an optical system is defined by the minimal distance between two fluorescent objects at which they can still be detected as separate objects. For a laser-scanning microscope this distance is determined by the size of the focal spot, because fluorescent objects that fall within the focal spot volume get excited at the same time and thus cannot be distinguished from each other at the level of the detector. Therefore, to increase the spatial resolution of a laser-scanning microscope, the focal spot should be made as small as possible.

However, because of the diffraction of light, any microscope objective will always produce an intensity distribution of finite size, no matter how jitter-free the scanner or how well aligned the laser beam is. The spatial extent of this blurry spot is called the point spread function (PSF), which is typically >250 nm for confocal microscopy and even wider for two-photon microscopy (>350 nm), which uses longer wavelength light (Eq. 2.1).

The basic idea of STED microscopy is to reduce the size of the focal spot, and hence improve spatial resolution, by actively inhibiting the fluorescence on the edge of the focal spot. This is achieved by routing in a second laser beam (called the STED laser) tuned to a longer wavelength, which has a doughnut-shaped or annular intensity distribution in the focal plane. The intense light of the doughnut quenches the fluorescence on the outer edge of the PSF by the process of stimulated emission at a wavelength that is longer than the fluorescence, allowing the fluorescence signal to be separately detected using standard optical filters.

The on/off switching of the fluorophores that underlies the engineering of the PSF, and hence the gain in spatial resolution, can be easily understood in terms of the energy diagram and transitions between ground (S_0) and excited (S_1) states of the fluorescent molecules, each of which have vibrational sublevels (Figure 2.1A).

Absorption of high-energy photons by the fluorescent molecules sends them into the first excited state (S_1), from which they rapidly pass (within picoseconds) to the lowest vibrational sublevel of the S_1 state. From there, they normally return to a vibrational sublevel of the S_0 state via spontaneous emission of a photon (=fluorescence) over

Figure 2.1 (A) Jablonksi diagram of the transitions of a fluorescent molecule. S_0 is the ground state, and S_1 is the first excited singlet state, each with multiple vibrational sublevels. (B) STED principle: the blue spot indicates the size of the excitation PSF, that is, the diffraction-limited intensity distribution of the blue excitation laser; the orange annulus is the intensity distribution of the STED laser, shaped like a doughnut, and the green spot denotes the size of the fluorescent spot after the suppression of the fluorescence induced by the STED doughnut. (C) Normalized intensity profiles of the blue excitation laser, the STED laser at low (dotted line) and high (solid line) intensity, and the fluorescence before (dotted line) and after (solid line) strong STED suppression. Please see color plates at the back of the book.

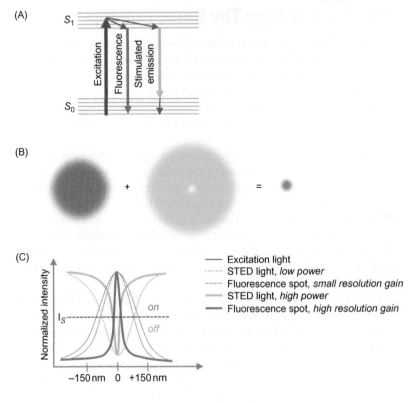

an average period determined by the fluorescence lifetime τ_{fl}, which is typically a few nanoseconds for organic molecules and fluorescent proteins, for example, ~4.1 ns for the Alexa Fluor 488 dye and ~2.8 ns for GFP.

In a STED microscope, the STED beam prevents fluorescence by forcing the molecules from the excited state to the ground state S_0 before fluorescence emission can happen. This occurs via stimulated emission, where an incident photon coming from the STED laser "harvests" a photon from the excited molecule that has the exact same energy and momentum. Like fluorescence, this process is also radiative, but the wavelength of the STED laser (λ_{STED}) is at the far right end of the fluorescence emission spectrum where it can be effectively filtered out without discarding too much of the fluorescence signal.

To achieve effective de-excitation and to prevent re-excitation by the STED laser, the molecules must be pumped into a higher vibrational sublevel of the S_0 state from which they rapidly escape to the final ground state S_0. This ensures that de-excitation by the STED laser predominates and re-excitation is negligible. In this case, the switching off can be approximated by the following exponential relationship:

$$N \approx e^{-I_{\mathrm{STED}}/I_s} \tag{2.2}$$

where N is the number of molecules that can still fluoresce, I_{STED} is the intensity of the STED laser, and I_S is a constant defined by the fluorophore (Dyba and Hell, 2002). Hence, if I_{STED} is three times larger than I_S the fluorophore is essentially switched off.

If the quenching effect is sufficiently strong ($I_{\mathrm{STED}} \gg I_S$), fluorescence can occur only from the center of the doughnut beam, where the intensity of the STED light vanishes due to destructive interference. The center area of the doughnut beam, which is called the null, can be made—in theory—arbitrarily small simply by increasing the power of the STED laser beam, I_{STED}. By saturating the quenching process on the rim of the doughnut, a very steep spatial gradient for molecules that are either "on" or "off" is created, which constricts the area where fluorescence is still permitted to occur (Figure 2.1B).

In this way, the size of the effective focal spot is reduced, and hence the spatial resolution of the optical system is improved. The current record in reducing the spot size is at 5.8 nm using diamond crystals, which is more than two orders of magnitude smaller than the wavelength of light that was used in the experiments (Rittweger et al., 2009).

Abbe's famous resolution formula picks up a second term in the denominator, which indicates that the spatial resolution barrier is not just shifted by a set amount, but is in fact broken (Hell, 2007) as follows:

$$\Delta r = \frac{\lambda}{2\mathrm{NA}\sqrt{1 + I_{STEP}/I_S}} \tag{2.3}$$

I_s is on the order of tens to hundreds of MW/cm^2 (peak power) when using STED pulses of 100–300 ps, which corresponds to an average focal power of a few milliwatts. Whereas this is about three orders of magnitude greater than the average intensity of the excitation laser beam (which typically is just a few microwatts), the peak power of the STED beam is still substantially less than the one typically used in two-photon microscopy, which is on the order of 200 GW/cm^2.

Together, the diffraction barrier of light microscopy can be effectively overcome by cleverly combining two laser beams, even though both of them are still limited by diffraction.

This may seem a bit counterintuitive at first, but one way to think about it is to look at I_S as a threshold value for the STED laser power,

I_{STED}, above which the fluorescence is turned off. As I_{STED} can be jacked up on the rim of the doughnut without ruining the null in the center, the threshold value I_S thus is reached at shorter distances from the center of the doughnut (Figure 2.1C). In theory, the distance to the threshold, that is, where $I_S = I_{STED}$, can be made arbitrarily short, which means that the resolution barrier is fundamentally broken.

Implementation of STED Microscopy

Whereas the optical principle behind STED microscopy with its gain in spatial resolution is well established, its implementation regarding instrumentation and design is still evolving, as the technology (e.g., lasers, scanners, detectors, fluorophores, etc.) keeps getting better and cheaper. Nevertheless, here we describe what a STED microscope, based on pulsed excitation and pulsed STED lasers, currently looks like. To begin with, a STED microscope is very similar in design and operation to a regular laser-scanning confocal fluorescence microscope.

The Microscope Base

Figure 2.2A shows a functional home-built STED microscope used for live-cell imaging. It is built around an inverted microscope stand and equipped for electrophysiology and two-photon microscopy to carry out neurobiological experiments in brain slices. Figure 2.2B is a schematic of the system, which shows the important optical components. The DMI 6000 (inverted) from Leica or the BX51 from Olympus (upright) standard research microscopes with or without motorized control of the wide-field epifluorescence and z-focus, can be used for this purpose.

Laser Sources

For excitation a pulsed laser diode (80 MHz pulses of 90 ps duration at a wavelength of 485 nm; LDH-P-C-485) is used. For the STED light a Ti:Sapphire laser (MaiTai) is used in combination with an optical parametric oscillator, (OPO) where the output is tuned to a wavelength of around 595 nm, which is suitable for quenching the fluorescence of GFP and yellow-fluorescent protein (YFP).

The Doughnut

The light pulses of the STED beam must be stretched in time to reduce the chance of re-excitation of the fluorophore and two-photon absorption by the sample. This is accomplished by sending the light from the OPO through a glass rod and a long optical fiber (100 m, polarization preserving fiber). By prebroadening the pulses

(A)

(B)

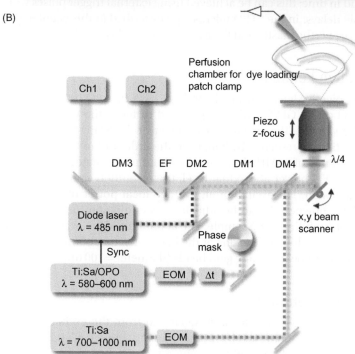

(C)

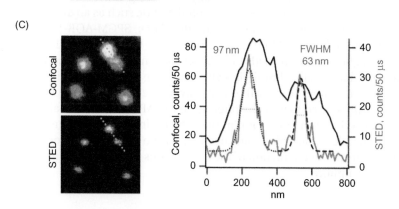

Figure 2.2 (A) Home-built STED microscope equipped for electrophysiological recordings and two-photon microscopy in living brain slices. (B) Schematic representation of the system in (A). (C) Imaging fluorescent beads to determine the spatial resolution of the STED microscope, direct comparison with confocal case. Line profiles across beads to determine their FWHM with the STED laser switched off (black trace) and on (red trace). Source: Reprinted from Tonnesen et al., 2011. Please see color plates at the back of the book.

from ~200 fs to a few picoseconds, the glass rod helps to protect its entry surface and avoid nonlinear effects inside the single-mode fiber. In the fiber the pulses are stretched to ~300 ps. The STED beam then passes through a polymeric phase plate imprinting a helical phase delay (vortex phase plate) on the wave front, which produces an annular intensity distribution in the focal plane of the lens with zero intensity at the center, hence the term doughnut.

Synchronization

The excitation and STED laser pulses need to be synchronized and overlaid in time; this can be achieved using external trigger pulses with variable delays. In the STED microscope described in this chapter, the STED laser is the master and the excitation diode laser is the slave.

Objective Lenses

An objective lens with a high NA (e.g., HCX PL APO 100×/1.40 OIL CS or 63×/1.30 GLYC CORR CS) is used to focus the incident light into the sample. The average focal power of the STED beam, which reduces the fluorescence by half, is on the order of 1 mW for YFP at a repetition rate of 80 MHz (measured without phase plate). By increasing the STED power above that level, the resolution can be continuously adjusted. For excitation an average focal power of 1–10 μW is sufficient; however, it can be adjusted depending on the brightness and photostability of the sample. The alignment of the excitation and STED lasers is done by looking at the back reflection of the excitation and doughnut beams using gold beads (diameter = 100 nm).

Emission Detection

The epifluorescence signal is separated from the excitation light (Figure 2.3) by a dichroic mirror. After passing through a bandpass filter to discard the STED light and any remaining excitation light, the signal photons are collected by a photodetector, such as an avalanche photodiode in photon-counting mode (APD; SPCM-AQR-13), but photomultipliers (PMT) in analog mode can be used as well. APDs have higher quantum yields (50–60% at 535 nm) than PMTs and virtually no dark counts, but their signal response becomes nonlinear at count rates larger than 2 MHz. For convenience, a multimode optical fiber can be used to pipe the photons to the APD, with the small fiber head serving as a confocal pinhole, which blocks out-of-focus light. In principle, a pinhole is not needed, as the STED resolution gain comes from the interaction of the laser beams in the focal plane with the fluorophore.

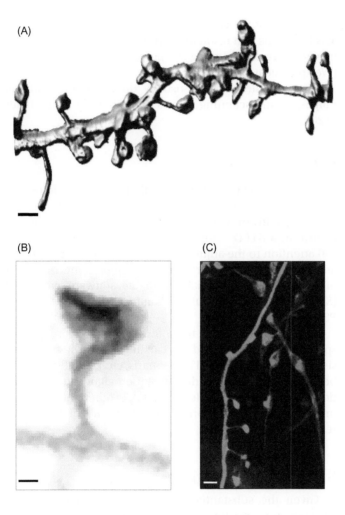

Figure 2.3 Examples of STED imaging of synapses in living brain slices. (A) Three-dimensional reconstruction of a stretch of dendrite of a pyramidal neuron filled with YFP based on a stack of STED images. Scale bar, 1 μm. Source: Reprinted from Nägerl et al., 2008. (B) Distribution of the cytoskeletal protein actin inside a dendritic spine imaged by STED using YFP-Lifeact as label for actin. Scale bar, 250 nm. Source: Reprinted from Urban et al., 2011. (C) Two-color STED microscopy of axons and dendrites using GFP and YFP as volume markers. Scale bar, 1 μm. Source: Reprinted from Tonnesen et al., 2011. Please see color plates at the back of the book.

Scanning Schemes

Super-resolved STED images can be acquired by superimposing the excitation and STED laser beams in all spatial dimensions and in time and moving them together across the sample using one of the following schemes for scanning, each of which has specific strengths and weaknesses: (1) beam scanning using galvanometer-driven mirrors, which allows for versatile and fast scanning over large fields of view; (2) stage scanning, where it is the sample that moves using a piezo stage, which is very precise, optically more simple and stable, and on the downside is relatively slow and makes it difficult to include micromanipulators for electrophysiological recordings, which must not move relative to the sample during scanning; or

(3) a hybrid of the two, where the beam is rapidly moved along one axis by means of a resonant mirror (e.g., 15 kHz), while the sample is moved more slowly along the perpendicular axis by the piezo stage (Westphal et al., 2008).

Whereas piezo stages usually do not come with large travel ranges, beam scanning allows for relatively large fields of view (>200 μm).

Software

As scanning and image acquisition in STED microscopy is nearly identical to the one used for traditional scanning microscopy, any scanning software can be used for data acquisition and instrument control of a STED setup. However, STED microscopy requires special attention to the adjustment of the excitation and the STED beam and conventional software packages can be somewhat clumsy for this purpose. The commercially available software package Imspector (Schonle, 2011) is a useful solution for this issue, as it facilitates rapid PSF measurements.

In contrast, regarding the software to analyze images, there are no special requirements, and one can fully rely on standard image analysis software (ImageJ, MATLAB, Imaris, etc.). Finally, STED images can be further enhanced by deconvolution algorithms, just like conventional confocal images. However, unlike other nanoscopy techniques, image processing, in principle, is not needed to obtain super-resolution images.

Stability Considerations

Given the substantial improvement in spatial resolution in a STED microscope, the hardware requirements are accordingly more stringent. For instance, the electronic circuit boards that drive the galvanometers controlling the mirror position in a scan head are constructed such that the position noise of the mirror is substantially below the resolution limit of a two-photon or confocal microscope; hence, this noise is not noticeable.

However, with an improvement in spatial resolution of about one order of magnitude these noise specs may not be good enough anymore, degrading the STED image appreciably. Taken together, whereas in general the architecture of the hardware can stay the same, the requirements in terms of electronic noise, thermal and mechanical stability, and so forth, are substantially more stringent in a STED microscope, so that in many instances components have to be upgraded and improved.

Checking the Resolution

The performance of the optical system is checked by imaging subdiffraction-sized fluorescent beads (e.g., diameter = 40 nm). After switching the STED laser on, the beads appear substantially smaller (Figure 2.2C), as confirmed by measuring their full-width-at-half-maximum (FWHM).

Sine Qua Non: Speed, Color, Depth, Live Imaging

STED microscopy certainly beats confocal/two-photon microscopy by a long way in terms of spatial resolution, but it still loses out when it comes to other aspects, which are also very important, such as choice of suitable dyes, multicolor imaging, signal-to-noise ratio, temporal resolution, compatibility with live-cell imaging, imaging deep inside complex biological samples, and so forth.

These drawbacks add on top of the relatively high cost and complexity to set up and maintain a commercial or a home-built STED microscope. Not surprisingly, its use in biological research is still fairly limited, despite its groundbreaking potential and all the fanfare. Its adoption has been hampered by the view of many biologists that its design is prohibitively complex and its scope too limited. However, this view is bound to change as technical improvements and commercial developments make STED microscopy more affordable, easier, and versatile.

Temporal Resolution and Imaging Speed

The speed of STED imaging is defined by the number of pixels and the pixel dwell time, just like for any single point-scanning technique. Usually, low signal counts are rate-limiting. Only if the sample is extremely bright does the hardware, that is, the speed of the scanner, become the limiting factor. Much like for confocal and two-photon microscopy, there is a trade-off between temporal and spatial resolution, such as acquisition speeds up to a few kilohertz can be achieved in line-scan mode, whereas larger images can take up to several seconds.

Given the enhancement in spatial resolution, pixel sizes should be reduced to comply with the Nyquist sampling theorem, which says that images should be acquired with a pixel size of less than half the optical resolution (typically 30 nm or less). This means that for an increase in lateral spatial resolution of a factor of 5, the acquisition time for a field of view of same size should increase by a factor of $5 \times 5 = 25$. In practice, the field of view is reduced substantially to keep the frame rate reasonable.

The pixel dwell time is longer for STED than for confocal or two-photon imaging to account for the fact the signal intensity in the center of the doughnut is typically cut in half, which means the frame acquisition time is twice as long.

Using a hybrid scanning system, moderate acquisition times of around 10 s per frame were used for imaging volume-labeled neuronal morphology over 15 µm fields of view (Nägerl et al., 2008). Using a faster scanning system based on a resonant mirror, STED imaging of synaptic vesicles at video rate could be performed on small scan areas (2.5×1.8 µm) using a bright organic dye (Westphal et al., 2008). More recently, multispot scanning with parallelized beams was used to increase the speed of STED microscopy by a factor of four (Bingen et al., 2011).

Labeling Strategies

An important challenge for STED microscopy lies in labeling and finding dyes that have suitable photo-physical properties and that can be targeted to specific organelles or proteins of interest inside cells. Several strategies exist, based on physical loading, transgenics, and/or chemistry, each having distinct weak and strong points (live-cell compatibility, specificity, choice of dyes, generalizability, etc.)

The classic way to achieve strong and specific labeling is based on immunohistochemistry using particularly photostable and bright organic dyes (such as Atto647N and Chromeo494) conjugated to secondary antibodies. This approach requires tissue fixation and membrane permeabilization, and thus does not permit investigations in living cells.

More recently, other labeling strategies were successfully used that are compatible with STED imaging in live preparations.

Transgenic expression of fluorescent proteins (such as GFP, YFP, and variants) was used to image subcellular organelles (Hein et al., 2008) or neuronal morphology (Nägerl et al., 2008) with a resolution better than 100 nm in live cells.

In addition, organic dyes can be introduced directly into single cells via patch pipettes, which was done for STED imaging of volume-labeled dendritic spines in brain slices using popular Alexa dyes (Ding et al., 2009; Tonnesen et al., 2011).

Finally, SNAP-tag and related approaches were recently used to label genetically encoded fusion-proteins with exogenous organic dyes for STED imaging in live cells (Hein et al., 2010; Pellett et al., 2011).

Multicolor Imaging

There is a clear need for multicolor solutions that integrate live-cell super-resolution laser-scanning microscopy with existing fluorescent labeling strategies and markers. For example, synapse biology would benefit greatly from multicolor STED imaging, as synapses are composed of presynaptic and postsynaptic compartments ensheathed by glial processes, which are typically too small to be properly resolved by conventional microscopic techniques.

As STED microscopy uses two separate laser beams, one for fluorescence excitation and another for fluorescence quenching, it is more difficult to incorporate multicolor imaging than conventional light microscopy.

However, several solutions exist for two-color imaging with STED microscopy, again each bearing particular strengths and weaknesses, regarding cost and complexity of the STED setup and the choice of fluorophores.

The first viable scheme for two-color STED microscopy relied on separate excitation and STED lasers for combinations of green and red fluorophores (Donnert et al., 2007; Meyer et al., 2008). This approach requires duplicating the lasers, ensuring their alignment, and compensating chromatic aberrations. In addition, it runs the risk that the STED laser for the green dye ends up exciting the red dye, causing excessive bleaching and phototoxicity.

Subsequently, a simpler approach was developed, which is also based on two excitation lasers, but only a single STED laser (Schmidt et al., 2008). It uses a pair of dyes with differential Stokes shifts, where the green dye undergoes a much larger shift than the red dye, allowing both dyes to be quenched by a single STED laser beam. This approach also forgoes the problem of excitation by the second STED laser. It was recently combined with SNAP- and CLIP-tag labeling to achieve two-color STED imaging in live cells (Pellett et al., 2011). The idea to use two dyes with different Stokes shifts for two-color nanoscopy is presently implemented in the commercial STED system from Leica using Chromeo494 and Atto647N as the dye pair.

Additional two-color nanoscopy schemes have recently been developed that only need a single pair of excitation and STED lasers. These approaches achieve spectral separation by (1) using dyes with differential fluorescence lifetimes (Buckers et al., 2011); (2) using a pair of photo-switchable fluorescent proteins (Padron and Dronpa) (Willig et al., 2011) that are spectrally similar, but can be specifically turned on or off by flashes of blue laser light (405 nm), which allows for sequential readout of their signals; or (3) simultaneously imaging two spectrally similar fluorophores followed by simple linear unmixing of the fluorescence intensity signal (Tonnesen et al., 2011). The last approach works readily for pairs of popular green-yellow fluorescent labels such as GFP, YFP, and Alexa Fluor 488, and is particularly easy to implement because it merely involves the addition of a second detector.

Importantly, since the same pair of lasers is used for different dyes, the measurements are effectively monochromatic, avoiding chromatic aberrations that could otherwise affect colocalization accuracy.

Depth Penetration and Spatial Resolution

Many important questions in biology require complex, thick tissue preparations such as brain slices or even intact brains. Confocal microscopy provides a way to optically section 3D fluorescent samples and acquire images with high signal-to-noise ratios several cell layers below tissue surface. Two-photon microscopy further improves on depth penetration and optical sectioning by the use of infrared light for two-photon fluorescence excitation.

However, regarding the new nanoscopy techniques, the gain in spatial resolution has so far been demonstrated mostly for very thin or surface preparations. The achievable spatial resolution usually deteriorates quickly with imaging depth because of scattering, absorption, and aberrations induced along the optical path and by the sample. While this is true for any laser-scanning light microscopy technique, these effects are particularly damaging for super-resolution microscopy.

In general, spherical aberrations are caused by a mismatch of refractive indices between the immersion medium of the objective lens and the tissue sample (Hell et al., 1993). Several strategies exist to reduce both effects, and thus (should) extend the ability to acquire super-resolved images deeper into living tissue.

The use of two-photon light for excitation reduces scattering of the excitation spot, and it was recently shown to be compatible with STED microscopy (Ding et al., 2009; Li et al., 2009; Moneron and Hell, 2009).

Adaptive optics can be used to reduce aberrations, and thus to ensure a high-quality STED doughnut. Using a glycerin immersion objective with an aberration-reducing correction ring, it was recently demonstrated that it is possible to achieve a lateral spatial resolution of 60 nm at imaging depths of ~100 μm below the surface (Urban et al., 2011) of living biological tissue.

Spatial Resolution in z

Despite the great improvement in lateral resolution brought about by the STED doughnut, the spatial resolution in z is typically only

500 nm or worse, depending on the objective, as expected for a confocal microscope, where the resolution formula for the optical axis is as follows:

$$\Delta z \approx \frac{n\lambda}{NA^2} \qquad (2.4)$$

where λ is the wavelength of light, NA denotes the numerical aperture of the lens, and n the refractive index.

However, the spatial resolution can also be improved along the optical axis by using either 4pi microscopy (Schmidt et al., 2008) or another type of phase mask that delivers STED laser power above and below the focal spot (Wildanger et al., 2009). Whereas the 4pi approach requires "sandwiching" the sample with a second objective to produce the desired effect, the latter approach is based on a single objective lens system and was reported to provide 3D resolution down to $45 \times 45 \times 108$ nm in a cell.

Live-cell Imaging

Much of the appeal of fluorescence microscopy stems from its ability to study dynamic cell biological processes inside living cells, setting it apart from electron microscopy. Therefore, there is a huge interest in getting the new super-resolution techniques to work in living cells.

A few years ago the idea of STED microscopy in living cells was met with quite a bit of skepticism due to difficulties with labeling and concerns over photodynamic damage induced by the STED laser. Fortunately, these and other challenges and fears have largely been overcome.

Concerning photodynamic damage that may be caused by the STED laser, it must be kept in mind that (1) the STED beam intensity ($I = 400$ MW/cm^2) is much lower than the intensity typically used for two-photon microscopy (200 GW/cm^2) and (2) that the absorption by water is much lower at the wavelength of 598 nm than in the near-infrared range, resulting in negligible temperature rises in the focal region (<0.2 K) (Schonle and Hell, 1998).

Whereas most early studies using STED have been carried out on fixed tissue preparations, several papers over the last couple of years have used it with living samples, such as cell lines (Hein et al., 2008; Pellett et al., 2011), dissociated neurons (Westphal et al., 2008), and brain slices (Nägerl et al., 2008; Ding et al., 2009; Tonnesen et al., 2011; Urban et al., 2011) (Figure 2.3A–C).

Summary and Outlook

Progress in expanding the scope and refining the use of STED microscopy for the biosciences is taking place at a rapid clip, which is going to accelerate as more laboratories join the effort, which is mainly driven by Stefan Hell in Göttingen and a few other labs.

Reversible Saturable Optical Fluorescence Transitions: A More General Principle for Nanoscopy

De-excitation by stimulated emission is not the only process by which a fluorophore can be deactivated to achieve super-resolution. It was shown that other switching mechanisms than quenching, such as photo-switching between excitable and nonexcitable states, can also achieve spatial resolution beyond the diffraction limit (Schwentker et al., 2007; Grotjohann et al., 2011). The more general concept behind STED microscopy is referred to as reversible saturable optical fluorescence transitions (RESOLFT) (Hofmann et al., 2005).

Other Areas of Development

Promising technical developments include "iso-STED," where the gain in resolution is extended into the z-axis (Schmidt et al., 2008; Wildanger et al., 2009); "gated-STED," which refers to time-resolved schemes to improve the spatial resolution of STED microscopy even further (Vicidomini et al., 2011); and "*in vivo* STED," which is intravital super-resolution imaging, such as in intact brains of anesthetized animals (Berning et al., 2012).

Because of the development of super-resolution microscopy, it is now possible to resolve details at the nanoscale in living biological specimens using fluorescence microscopy, which was considered a pipe dream just a few years ago. Therefore, STED microscopy and other super-resolution techniques are poised to transform bioscience research in the near future.

References

Bates, M., Huang, B., Dempsey, G.T., Zhuang, X., 2007. Multicolor super-resolution imaging with photo-switchable fluorescent probes. Science 317, 1749–1753.

Berning, S., Willig, K.I., Steffens, H., Dibaj, P., Hell, S.W., 2012. Nanoscopy in a living mouse brain. Science 335, 551.

Betzig, E., Patterson, G.H., Sougrat, R., Lindwasser, O.W., Olenych, S., Bonifacino, J.S., et al., 2006. Imaging intracellular fluorescent proteins at nanometer resolution. Science 313, 1642–1645.

Bingen, P., Reuss, M., Engelhardt, J., Hell, S.W., 2011. Parallelized STED fluorescence nanoscopy. Opt Express 19, 23716–23726.

Born, M., Wolf, E., 1999. Principles of Optic: Electromagnetic Theory of Propagation, Interference and Diffraction of Light. Cambridge University Press, New York.

Buckers, J., Wildanger, D., Vicidomini, G., Kastrup, L., Hell, S.W., 2011. Simultaneous multi-lifetime multi-color STED imaging for colocalization analyses. Opt Express 19, 3130–3143.

Ding, J.B., Takasaki, K.T., Sabatini, B.L., 2009. Supraresolution imaging in brain slices using stimulated-emission depletion two-photon laser scanning microscopy. Neuron 63, 429–437.

Donnert, G., Keller, J., Wurm, C.A., Rizzoli, S.O., Westphal, V., Schonle, A., et al., 2007. Two-color far-field fluorescence nanoscopy. Biophys J 92, L67–L69.

Dyba, M., Hell, S.W., 2002. Focal spots of size lambda/23 open up far-field fluorescence microscopy at 33 nm axial resolution. Phys Rev Lett 88, 163901.

Grotjohann, T., Testa, I., Leutenegger, M., Bock, H., Urban, N.T., Lavoie-Cardinal, F., et al., 2011. Diffraction-unlimited all-optical imaging and writing with a photochromic GFP. Nature 478, 204–208.

Gustafsson, M.G., 2005. Nonlinear structured-illumination microscopy: wide-field fluorescence imaging with theoretically unlimited resolution. Proc Natl Acad Sci USA 102, 13081–13086.

Hein, B., Willig, K.I., Hell, S.W., 2008. Stimulated emission depletion (STED) nanoscopy of a fluorescent protein-labeled organelle inside a living cell. Proc Natl Acad Sci USA 105, 14271–14276.

Hein, B., Willig, K.I., Wurm, C.A., Westphal, V., Jakobs, S., Hell, S.W., 2010. Stimulated emission depletion nanoscopy of living cells using SNAP-tag fusion proteins. Biophys J 98, 158–163.

Heintzmann, R., Jovin, T.M., Cremer, C., 2002. Saturated patterned excitation microscopy—a concept for optical resolution improvement. J Opt Soc Am A Opt Image Sci Vis 19, 1599–1609.

Hell, S.W., 2007. Far-field optical nanoscopy. Science 316, 1153–1158.

Hell, S.W., Wichmann, J., 1994. Breaking the diffraction resolution limit by stimulated emission: stimulated-emission-depletion fluorescence microscopy. Opt Lett 19, 780–782.

Hell, S.W., Reiner, G., Cremer, C., Stelzer, E.H., 1993. Aberrations in confocal fluorescence microscopy induced by mismatches in refractive index. J Microsc 169, 64.

Hess, S.T., Girirajan, T.P., Mason, M.D., 2006. Ultra-high resolution imaging by fluorescence photoactivation localization microscopy. Biophys J 91, 4258–4272.

Hofmann, M., Eggeling, C., Jakobs, S., Hell, S.W., 2005. Breaking the diffraction barrier in fluorescence microscopy at low light intensities by using reversibly photoswitchable proteins. Proc Natl Acad Sci USA 102, 17565–17569.

Klar, T.A., Hell, S.W., 1999. Subdiffraction resolution in far-field fluorescence microscopy. Opt Lett 24, 954–956.

Klar, T.A., Jakobs, S., Dyba, M., Egner, A., Hell, S.W., 2000. Fluorescence microscopy with diffraction resolution barrier broken by stimulated emission. Proc Natl Acad Sci USA 97, 8206–8210.

Li, Q., Wu, S.S., Chou, K.C., 2009. Subdiffraction-limit two-photon fluorescence microscopy for GFP-tagged cell imaging. Biophys J 97, 3224–3228.

Meyer, L., Wildanger, D., Medda, R., Punge, A., Rizzoli, S.O., Donnert, G., et al., 2008. Dual-color STED microscopy at 30-nm focal-plane resolution. Small 4, 1095–1100.

Moneron, G., Hell, S.W., 2009. Two-photon excitation STED microscopy. Opt Express 17, 14567–14573.

Nägerl, U.V., Willig, K.I., Hein, B., Hell, S.W., Bonhoeffer, T., 2008. Live-cell imaging of dendritic spines by STED microscopy. Proc Natl Acad Sci USA 105, 18982–18987.

Pellett, P.A., Sun, X., Gould, T.J., Rothman, J.E., Xu, M.Q., Correa Jr., I.R., et al., 2011. Two-color STED microscopy in living cells. Biomed Opt Express 2, 2364–2371.

Rittweger, E., Han, K.Y., Irvine, S.E., Eggeling, C., Hell, S.W., 2009. STED microscopy reveals crystal colour xcentres with nanometric resolution. Nat Photonics 3, 144–147.

Schmidt, R., Wurm, C.A., Jakobs, S., Engelhardt, J., Egner, A., Hell, S.W., 2008. Spherical nanosized focal spot unravels the interior of cells. Nat Methods 5, 539–544.

Schonle, A. (2011). Imspector. <www.max-planck-innovation.de/de/industrie/technologieangebote/software>.

Schonle, A., Hell, S.W., 1998. Heating by absorption in the focus of an objective lens. Opt Lett 23, 325–327.

Schwentker, M.A., Bock, H., Hofmann, M., Jakobs, S., Bewersdorf, J., Eggeling, C., et al., 2007. Wide-field subdiffraction RESOLFT microscopy using fluorescent protein photoswitching. Microsc Res Tech 70, 269–280.

Tonnesen, J., Nadrigny, F., Willig, K.I., Wedlich-Soldner, R., Nägerl, U.V., 2011. Two-color STED microscopy of living synapses using a single laser-beam pair. Biophys J 101, 2545–2552.

Urban, N.T., Willig, K.I., Hell, S.W., Nägerl, U.V., 2011. STED nanoscopy of actin dynamics in synapses deep inside living brain slices. Biophys J 101, 1277–1284.

Vicidomini, G., Moneron, G., Han, K.Y., Westphal, V., Ta, H., Reuss, M., et al., 2011. Sharper low-power STED nanoscopy by time gating. Nat Methods 8, 571–573.

Westphal, V., Rizzoli, S.O., Lauterbach, M.A., Kamin, D., Jahn, R., Hell, S.W., 2008. Video-rate far-field optical nanoscopy dissects synaptic vesicle movement. Science 320, 246–249.

Wildanger, D., Medda, R., Kastrup, L., Hell, S.W., 2009. A compact STED microscope providing 3D nanoscale resolution. J Microsc 236, 35–43.

Willig, K.I., Stiel, A.C., Brakemann, T., Jakobs, S., Hell, S.W., 2011. Dual-label STED nanoscopy of living cells using photochromism. Nano Lett 11, 3970–3973.

3

ENHANCEMENT OF OPTICAL RESOLUTION BY 4pi SINGLE AND MULTIPHOTON CONFOCAL FLUORESCENCE MICROSCOPY

W.A. van Cappellen[1,2,3], A. Nigg[2,3], and A.B. Houtsmuller[2,3]
[1]Department of Reproduction and Development, [2]Department of Pathology, [3]Erasmus Optical Imaging Centre (OIC), Erasmus Medical Centre, Rotterdam, The Netherlands

CHAPTER OUTLINE
Introduction   56
The 4pi Principle and Setup   56
Microscope Alignment   58
    Initial Alignment of the Excitation by Eye   58
    Optimizing the Coverslip Correction Ring   59
    Alignment of the Bottom Lens (*XYZ*)   59
    Alignment of the Second Mirror (*XYZ*)   60
4pi Imaging   60
4pi Deconvolution   61
Sample Preparation   62
    Fixation   63
    Selection of Fluorescent Dyes   64
Microtubule and Microtubule Plus End Imaging   65
Visualization of DNA   66
    4pi Imaging of Muntjac Chromosomes   67
**Single-photon Excitation (Measurement of the Redox
    State in Dopamine Neurons)   68**
**SYCP3 Axis as a Marker for Chromatin Organization
    in Mouse Spermatocytes   70**
Microbubbles with Medicine   75
Future of 4pi Imaging   77
Acknowledgment   77
References   77

Cellular Imaging Techniques for Neuroscience and Beyond.
DOI: http://dx.doi.org/10.1016/B978-0-12-385872-6.00003-9

Introduction

The idea to improve the axial (Z) resolution of an optical microscope by using two opposing lenses was first published by Stefan Hell and Ernst Stelzer in 1992 (Hell and Stelzer, 1992). In 1994 the first proof of principle was realized (Hell et al., 1994). It took ten years (2004) before the first commercial 4pi microscope was launched by Leica Microsystems. The commercial version of the 4pi microscope consists of a 4pi head attached to an advanced confocal microscope (SP2). Until today the 4pi principle is one of the best strategies for increasing the axial resolution of an optical system. In this chapter we will show how 4pi microscopy can be used in biomedical research and what advantages can be expected.

The 4pi Principle and Setup

Commercial 4pi microscopes make use of a diffraction-limited laser beam that is split and sent through two opposing lenses (Figure 3.1). When the two opposing beams meet each other an interference

Figure 3.1 The 4pi head with the two opposing lenses.

pattern is created in the axial direction. This pattern can be characterized as a point spread function (PSF) with a main lobe and several lower intensity axial side lobes. The distance of the side lobes from the main lobe is dependent on the numerical aperture (NA) of the objective lenses and the wavelength of the laser excitation light used. For excitation at 800 nm this distance is around 300 nm. The lateral resolution is slightly influenced by the opposing lenses (Sheppard et al., 2010). The number and intensity of the diffraction axial side lobes can be considerably reduced by using two-photon excitation instead of single photon excitation. For this reason most 4pi imaging has been performed with two-photon excitation.

The top lens in a 4pi objective lens configuration is fixed (Figure 3.2, 1), the bottom piezo-driven lens can be moved in X, Y, and Z for a proper beam path alignment (Figure 3.2, 2). Also the piezo-driven lower mirror can be moved in X, Y, and Z (Figure 3.2, 3) to compensate for phase differences and to control the angle of the lower beam. The microscope table is also piezo driven in the X-, Y-, and Z-direction (Figure 3.2, 4).

In the Leica 4pi microscope, an assembly of two quartz coverslips with the object sandwiched in between is inserted in the 4pi objective head. As objective lenses, glycerin immersion lenses with an NA of 1.3 are used. These lenses are optimized for a coverslip thickness of 230 μm, which makes cleaning of the expensive coverslips more feasible. The maximum available space between the two coverslips is 50 μm, which is a limitation imposed by the working distances of the lenses and the phase shift of the diffraction pattern. In practice this space ranges between 20 and 40 μm. Exchange of the lenses can only be done by taking the 4pi head apart. The signal coming from the small 4pi excitation spot is very clean with very little out-of-focus signal, and for this reason it is also quite low. In the 4pi head the signal

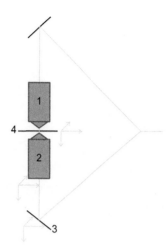

Figure 3.2 Schematic 4pi setup.

Figure 3.3 Output of the 4pi camera. (A) X and Y of excitation beams are misaligned, (B) slight misalignment, and (C) good alignment.

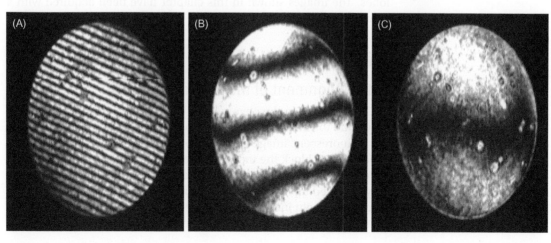

(A) (B) (C)

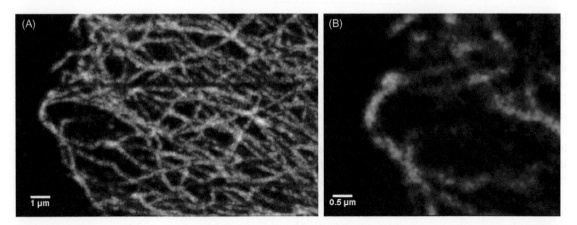

Figure 3.4 4pi image of β-tubulin Alexa Fluor 488 staining in a Hela cell. (A) Maximum projection and (B) magnification of one single plane showing inhomogeneous staining, which are probably artifacts.

is severely reduced. For this reason the setup is equipped with avalanche photo diodes (APDs) as detectors. APDs have a much higher quantum yield than photomultipliers (PMTs), but their dynamic range is much lower. In front of the two APDs the emission light can be split with a secondary dichroic mirror, which makes it possible to conduct two-color experiments.

There are different types of 4pi setup. In the 4pi type A setup, alignment of the two beams is only achieved in the excitation path. Detection is only performed on one lens or as an incoherent combination of both lenses. In the type B setup the two beams are aligned on the emission side of the signal. Type C combines type A and B, so the signal is aligned both at the excitation and at the emission side (http://www.mpibpc.mpg.de/groups/hell/4Pi.htm). The commercial Leica 4pi microscope is a type A microscope with detection from both lenses. Some systems can be upgraded to type C. Type C will have a somewhat higher resolution and lower refractive pattern side lobes. The images shown in this chapter have been acquired with a type A system.

Microscope Alignment

Initial Alignment of the Excitation by Eye

The first step in the alignment of the Leica 4pi microscope is very easy. Due to the nature of the setup it is possible to simultaneously see a fluorescent image through both lenses. By closing shutters it is possible to block the individual beam paths. With a big object, for example, a cell nucleus, it is easy to do coarse initial alignment. First, only the fixed top lens is activated and with the table's Z-stepper the object is brought into the focal plane. Then the object is focused with the bottom lens with the Z-stepper of the bottom lens. Switching on both lenses will show a distorted projection of a misaligned object. With the X and Y controllers of the bottom lens both images can be aligned. The camera built into the 4pi head will show an ellipse of light with a stripe pattern in it (Figure 3.3A).

Optimizing the Coverslip Correction Ring

In general the coverslip correction ring is optimized by maximization of the reflection of the excitation laser on the edge of the coverslip and the mounting medium. This can be done by intensity measurements of an XZ scan. The beam path in the 4pi microscope does not have a refractive index mismatch between coverslip and mounting medium, which results in an absence of reflection on this intermediate. For this reason one of the coverslips has a mirror damped on the quartz surface with a hole in the center for imaging.

A beam path is set up in which the excitation laser light is reflected into the spectral photomultiplier. In the case of the Leica SP2 confocal instrument this means that the acousto-optical beam splitter setting has to be optimized for reflection. The two-photon laser needs to be tuned down to 710 nm or a shorter wavelength, because the detector module blocks all wavelengths longer than 720 nm. Different excitation wavelengths (single-photon and two-photon combinations) can be used as readout at the same time. Continuous XZ scans are made with a pinhole at 0.2 Airy units to block most of the intense reflection light. With an optimal coverslip correction ring setting the reflection band is the smallest in area and the highest in intensity. Side bands of the refraction pattern should be as low as possible and the distribution should be symmetrical around the highest intensity band. This is done for both objective lenses. With high-quality lenses the focal planes for the different wavelengths will be the same. With the Z-stepper of the bottom lens the focal planes of the top and bottom lenses can be combined into a single plane.

Alignment of the Bottom Lens (XYZ)

Alignment of the bottom objective lens can be done with 100 nm fluorescent beads. The first step is the XY alignment by using the built-in camera. If the alignment is not correct the camera will show an ellipsoid spot with a striped diffraction pattern. By optimizing the X and Y positions of the bottom lens the spot will become mainly dark (Figure 3.3C). To align the Z position of the lower lens, XZ scans are made with the same 100 nm fluorescent bead showing the typical 4pi diffraction pattern. This pattern is measured and the Z-position of the lens is adjusted. When the main lobe of the refraction pattern has the highest intensity and the side lobes are symmetrical to the main lobe the lower lens is in the right position.

Alignment of the Second Mirror (*XYZ*)

The bottom mirror (Figure 3.2, 3) is used to make two opposing beams parallel and to change the distance between the two beams to get the opposing beams in phase. For this purpose the 100 nm fluorescent beads are used again. In this case it is important to again optimize the intensity of the main lobe of the refractive pattern, but now the signal of the refractive pattern side lobes should be reduced as much as possible and the signal between the lobes should be minimized. This can be done by adjusting the two angles and the axial (phase) direction of the mirror.

The complete alignment of the system takes 30–60 minutes and is always necessary at the beginning of an imaging session. The coverslip correction ring has to be readjusted after every sample change. The adjustment of the lower lens (Figure 3.2, 2) and mirror (Figure 3.2, 3) remains stable between sessions but must be readjusted after changing wavelengths. Depending on the accuracy of the temperature control of the room where the 4pi microscope is installed, small realignments of the bottom lens *XY* position have to take place every 0.5–3 hours. This alignment usually takes less than one minute.

The alignment can only be done with one excitation wavelength. For two-photon excitation this is not a problem, because its spectra are rather broad, but for a combination of two single-photon wavelengths or a combination of a single- and two-photon excitation experiments, a compromise has to be found.

4pi Imaging

The stage of the Leica 4pi microscope setup has a stepping limitation of 90 nm in the *Z*-direction, when making *XYZ* stacks. With an axial resolution in the order of 120 nm the Nyquist sampling rate is approximately 45 nm. These *Z* steps are too big to reach the highest axial resolution possible. Using the stage in *XZY*-mode, step sizes in the order of 10 nm are possible in the *X*- and *Z*-direction, but the step size in *Y* is limited to a minimum of 30 nm, increasing in 30 nm steps. For this reason most 4pi images are made in the *XZY*-mode with isotropic voxels of 30 nm in all three dimensions. Imaging a complete nucleus with this small voxel size results in a stack of 500 slices for a 15 μm field of view. To reduce the Schott noise, pixel intensities are often averaged, and if the signal is very low then accumulation of the signals is necessary, which means every slice is imaged four to eight times. With a typical image size of 512 × 512 pixels taking one second, making a complete stack will take 15 (average 2) to 60 (average 4, accumulation 2) minutes. This makes 4pi microscopy, like many other high-resolution techniques, less suitable for bulk scanning of cells. In practice it means that it is advisable to make a very good selection of the cells prior to microscopy.

4pi Deconvolution

Due to the diffraction pattern between the opposing excitation beams, diffraction side lobes will show up in 4pi images. The distances of the side lobes from the main lobe will be constant, depending on the used excitation wavelength. The constant intensity of the diffraction side lobes makes it relatively simple to correct the images for these side lobes. The Leica software offers the option to deconvolve the images with a three- or five-point Gaussian deconvolution algorithm. This algorithm was later improved to work also with an asymmetric distribution of the intensity of the diffraction side lobes. This deconvolution method is further referred to in this chapter as a 3-point deconvolution algorithm. In general, the signal-to-noise ratio (SNR) is quite low in raw 4pi images (SNR 2–7). When deconvolution calculations are done on these images the results are rather poor. For this reason the images are first filtered with a Gaussian 3D filter with a low kernel size (3 pixels, 90 nm) before deconvolution is performed. All information necessary for this deconvolution is the relative intensity of the diffraction side lobes and the distance between the diffraction main and side lobes, which can be derived from measuring a 100 nm bead. Although this type of deconvolution can be performed with the standard Leica 4pi package, a more flexible solution is available in a software package developed by the lab of Stefan Hell called Imspector (Max-Planck Innovation, http://www.max-planck-innovation.de/en/, Andreas Schönle and John Robbins). Recently an automated deconvolution method for 4pi microscopy was published by the same group (Vicidomini et al., 2010).

The above described deconvolution methods deconvolve the data in the axial direction, but not in the lateral direction. In an earlier publication (Hell et al., 1997) a more sophisticated deconvolution was performed; however, this method was never commercialized due to the low popularity of the 4pi microscope. A good thing about the latter method is that the Gaussian filter, which reduces the axial and lateral resolution, is not necessary and that both axial and lateral resolutions increase.

With all high-resolution techniques in which pixel precise calculations have to be done on noisy (SNR < 5) images, a prefiltering is needed before calculations can be made. Filtering often leads to the formation of artifacts in the images. This can be observed in 4pi images, but also in images obtained with structured illumination methods and in deconvolved images, especially if the SNR is not set at the appropriate value. These artifacts consist of small structures showing up in the background and in signals just above the background. Similar artifacts can also be observed in inhomogeneous structures like nuclear Hoechst/DAPI staining. As long as these artifacts can be recognized as such they will not pose a serious problem. With the higher resolutions of new microscopy techniques like 4pi and structured illumination microscopy or single molecule techniques like photoactivated localization microscopy and stochastic optical reconstruction microscopy, it will probably become clear that structures that presently appear smooth, for example, microtubules, are in reality much more punctuated (Figure 3.4). This will make the recognition of artifacts more difficult; that is, more effort will be needed to remove these artifacts from the images.

It is also important to check the Gaussian filter used. When using an 8-bit environment for filtering, round-off errors may severely influence the end result for low signals. A good Gaussian filter should work in a floating point environment.

Sample Preparation

In 4pi microscopy, as in most high-resolution techniques, it is very important to optimize the refractive index of the sample as much as possible. The fact that two opposing excitation laser beams must be aligned tends to make this optimization of the refractive index even more important. Optimizing the refractive index in this case means keeping the refractive index at a constant value for the complete beam path outside the objective lens. This can be done in different ways. The standard way is to use oil immersion ($\eta = 1.518$), a glass coverslip ($\eta = 1.526$), and a mounting medium that often has a slightly lower refractive index (e.g., a popular medium like VectaShield has a refractive index of 1.44; Florijn et al., 1995), which is not ideal. The cell also has a refractive index of its own, or actually a series of slightly different refractive index values depending on the location in the cell, but this is a property that cannot be changed. In the case of 4pi microscopy most experiments are conducted with a refractive index around 1.47 and using glycerin immersion lenses with glycerin as the immersion medium, quartz coverslips, and with a glycerin or glycerin-like mounting medium. To correct for the coverslip thickness, the glycerin immersion lenses are equipped with coverslip correction rings.

Fixation

Due to the nature of 4pi imaging most samples need to be fixed. In the case of 4pi microscopy it is important to leave the 3D structure intact as much as possible. The best results have been obtained with formaldehyde fixation; however, the protocol for this fixation is rather important, especially when fluorescent proteins are used as fluorescent dye. For example, when green fluorescent protein (GFP)-transfected cells are fixed with 4% buffered formaldehyde for 10 minutes, most of the signal will be gone. In the first few hours of the fixation movement of molecules will occur as shown by fluorescent redistribution after photobleaching (FRAP) experiments. The best results were obtained with a prolonged fixation of at least 3 hours with 4% buffered formaldehyde solution. A decrease of about 50% of the GFP signals is observed directly after replacement of the medium with formaldehyde (Figure 3.5 A and B). This low signal remains stable until 2 hours after the start of the fixation. At this point the GFP signal increases again to 30%. To study this effect in more detail GFP-transfected cells were fixed on the microscope stage. Propidium iodide (PI) was added to the cells to observe the degradation of the cell membrane. PI does not penetrate the limiting membranes of healthy cells. Directly after fixation with buffered 4% formaldehyde PI was still not able to penetrate the cells. After 2 to 3 hours two events took place: one by one the cells became stained, which meant that the PI was now penetrating the outer cell membranes and was binding to RNA, and there was an increase in the GFP signal (Figure 3.5 C and D). The pH of the formaldehyde was causing the initial decrease in GFP signal, but in all formaldehyde preparations whose pH ranged between 4 and 9 we observed a decrease of the GFP signal. However pH 8.6 gave the best end results. We found a similar effect with yellow fluorescent protein (YFP), but the decrease of signal after fixation was much less with cyan fluorescent protein (CFP). The tests were done with the enhanced versions of the fluorescent proteins.

Figure 3.5 Fixation of free GFP-transfected cells with the addition of PI. (A) Before fixation, (B) directly after the addition of 4% paraformaldehyde (PFA), (C) 2.5 hours after the addition of PFA, and (D) 3 hours after addition of paraformaldehyde. Please see color plates at the back of the book.

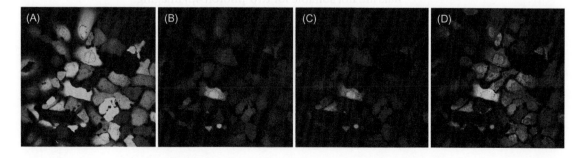

Selection of Fluorescent Dyes

The best 4pi results are obtained by two-photon excitation and with APDs as detectors. This opens the possibility of two-color imaging, but it limits the dyes, especially the combination of dyes that can be used. In general, dyes that can be excited with two-photon light perform well in the 4pi setup. For example, dyes like DAPI or Hoechst are particularly well suited for the study of DNA in a cell. Combinations of dyes with DAPI or Hoechst are difficult to use because of their broad emission spectra. In this respect PI is better suited for two-color 4pi imaging. The setup with the APDs is not as flexible as that with the standard spectral PMTs of the Leica SP2 where emission bands can be freely selected. In front of the APDs there is a filter cube with a secondary dichroic mirror and two emission filters. The filter cube can be exchanged with other cubes. Optimal two-color 4pi images should be made by excitation with one two-photon wavelength and with dyes with separate emission spectra. In this respect quantum dots could be perfect dyes for 4pi two-color imaging. In practice, however, the usage of quantum dots is very limited due to labeling difficulties. The 4pi setup with its incredible axial resolution is the perfect instrument for high-resolution 3D imaging, and for this reason it is very important to keep the 3D morphology of the biological substrate intact (e.g., by the application of a mild fixation with buffered formaldehyde). However, with this type of fixation it is not always possible to label structures of interest in the cell nucleus with quantum dots. The size of the quantum dots (5–15 nm, depending on the emission) is probably too large to penetrate incompletely broken down membranes in the cell. In this respect labeling of cytoplasmic components is somewhat easier. Popular dyes like DAPI, Hoechst, Alexa Fluor 488, FITC, Alexa Fluor 555, and TRITC, next to quantum dots, perform quite well. Fluorescent proteins like ECGP, EGFP, and EYFP were also imaged with good results. With mCherry the results were variable, but never as good as EGFP or EYFP. For all fluorescent dyes it is important to have a reasonable signal for success in the 4pi setup. Very weak signals are not suitable for two-photon 4pi imaging.

When single photon excitation is used, there are no restrictions to the dyes other than the laser lines available on the microscope. As high-resolution 4pi images are normally made with the instrument's APDs, there is less flexibility in the selection of combinations of dyes. As discussed before, the alignment with two different excitation lines is less optimal but quite often usable. Sequential scanning is normally the way to reduce bleed-through of the emission signals. It is also possible to combine two-photon excitation with single-photon excitation. This can be useful if the protein of interest can be imaged with two-photon laser scanning while a less important (counter) stain is used with one-photon excitation; for example, we have applied PI-stained DNA as a counterstain. With single-photon excitation it is possible to acquire images of quite weak signals. With very robust signals it is even possible to use standard PMTs with the flexibility of spectral detection and a combination of dyes.

Microtubule and Microtubule Plus End Imaging

One of our first 4pi microscopy projects was the visualization of microtubules in mammalian (Hela) cells. Microtubules are, as known from electron microscopic observations, 25 nm thick polymer fibers composed of α- and β-tubulin dimers. These fibers originate in the microtubule organizing center and spread out to the outer cell membrane. Because of their high amount of tubulin protein, a nice, robust staining of the fibers can be achieved in Hela cells, especially if the cells are fixed with methanol, which allows good penetration of antibodies against tubulin. The 25 nm thick microtubules create very nice diffraction pattern side lobes as can be seen in Figure 3.6. With 4pi microscopy, microtubules with a spatial separation in the axial direction of only 400 nm can be resolved. In addition we observed twisting patterns between microtubules. A drawback of methanol fixation is shrinkage of the cell, which results in a very flat cell and nucleus. The nucleus of the cell is nicely outlined by the microtubules and the microtubule organizing center is clearly visible in a maximum projection *XY* view (Figure 3.7). Microtubules nicely demonstrate the 4pi diffraction pattern side lobes and show the ability to dispose of these side lobes by three-point deconvolution of the images. Microtubule images are also great for making a good estimation of the axial resolution of the 4pi microscope. From a more biological point of view it is interesting to see whether the

Figure 3.6 *XZ* image of a single slice of a β-tubulin staining in a COS cell. (A) Raw 4pi data, (B) Gauss-filtered data, and (C) three-point Gaussian deconvolution.

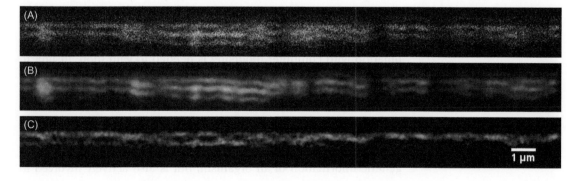

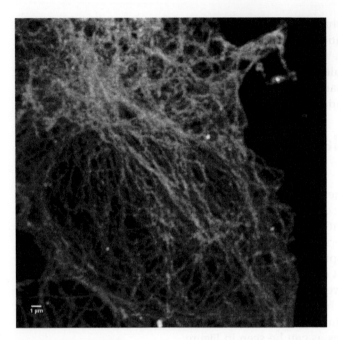

Figure 3.7 *XY* maximum projection of a β-tubulin staining of a COS cell.

4pi microscope is capable of showing how microtubules cross into different *Z*-planes (Figure 3.6). Very nice microtubule images can be acquired with a standard confocal microscope. However, as the dimensions of microtubules (i.e., diameter smaller than 25 nm) are smaller than the limited (Abbe) axial resolution (i.e., 800 nm) of the confocal microscope, all microtubules will be recorded as if they lie in the same confocal optical plane while their details in the axial direction, such as orientation and approaching and crossing each other, will be beyond recording. Deconvolution of the 4Pi images can be done quite successfully, and the axial resolution is then increased to 120–140 nm. Due to imperfection of the beam path by refractive index variations of the interior of cells, a small phase shift might happen between the lower and upper parts of the cell, which leads to a slow change of the PSF (mainly the intensity of the two diffraction side lobes). A flexible three-point Gaussian deconvolution would gain better results; however, this algorithm is not yet available. Phase shift is not really a problem for most biological applications. With an axial resolution of approximately 130 nm it is currently possible to distinguish different layers of the microtubule network in Hela cells (Figure 3.6).

Microtubules have a shrinking phase called a catastrophe (Galjart, 2010). During this phase the end points of microtubules become unstable and they start to fall apart, which is characterized by the occurrence of curved protofilaments in electron microscopic images. Attempts to resolve these "banana-shaped" protofilaments with the 4pi microscope were not successful. A still higher resolution is needed to resolve these details. Ultimately, visualization of this process in living cells with a higher resolution and higher speed in combination with computer modeling is needed to gain more insight in the process of growth and shrinkage of microtubules.

Visualization of DNA

DNA is a very intriguing molecule that not only contains all the genetic information of an organism, but is also actively involved in regulatory processes like gene silencing and gene activation. Studying this kind of behavior is severely restricted by the limitations of the available imaging tools. The high amount of DNA inside cells and the high speed of processes in which DNA molecules are involved make it a complex and challenging structure to study. Specific staining of parts of the DNA in living cells is still in the early stage of development. Very high resolution in combination with a specific staining is needed for imaging. With general DNA staining, 4pi microscope images were made to discover how much extra information could be gained with its high axial resolution. With fluorescent probes it is quite easy to stain the complete DNA, and a dye like Hoechst or DAPI is perfect for two-

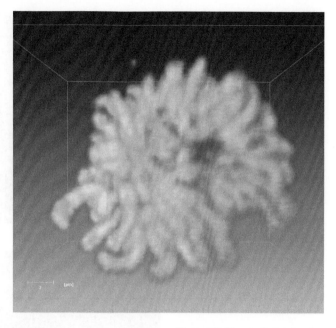

Figure 3.8 Three-dimensional volume rendering of a Hela cell in mitosis stained with H2B-GFP.

photon excitation. Visualization of the histone protein H2b coupled to GFP can also be used to gain information about the DNA morphology at light microscopic resolution. With the 4pi setup it is almost possible to observe the individual chromosomes on the equatorial plate during mitosis (Figure 3.8). The study of chromosomes and chromosome distribution can be simplified by using a cell system with a low number of chromosomes.

4pi Imaging of Muntjac Chromosomes

The Muntjac (Muntiacinae, Cervidae) is one of the phylogenetically oldest species of deer and can be found in many (Asian) countries. The cells of the Indian Muntjac, *Muntiacus muntjak*, possess only six diploid chromosomes (males have seven chromosomes). A subspecies of the Asian Muntjac living in China, *M. reveesi*, has 46 chromosomes (Wurster and Benirschke, 1970). During the prophase when the DNA starts to compact it is possible to distinguish the individual chromosomes with 4pi microscopy (Figure 3.9). However, this can also be done with a classical confocal microscope. The real advantage of high-resolution microscopy should be to reveal the structure of DNA in an interphase nucleus, mainly focusing on the question: Do proteins bind to DNA or to other structures in the nucleus? The interphase nucleus of a Muntjac cell reveals a very inhomogeneous DNA distribution with most of the DNA close to the nuclear periphery (Figure 3.10).

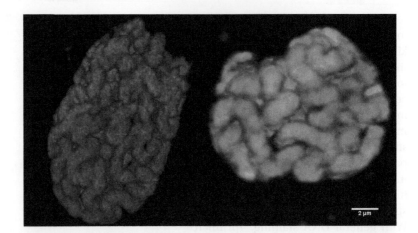

Figure 3.9 Three-dimensional volume rendering of two prophase Muntjac cell nuclei stained with Hoechst.

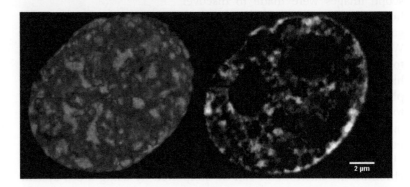

Figure 3.10 Three-dimensional volume rendering (left) and a single slice of an interphase Muntjac cell.

Single-photon Excitation (Measurement of the Redox State in Dopamine Neurons)

Although two-photon excitation is the preferred method in 4pi microscopy, single-photon excitation can also be used. With the right lenses and a 4pi C setup (with matched interference patterns both in the excitation and the emission beam paths) it is even possible to gain diffraction side lobes as low as in standard 4pi two-photon microscopy (Lang et al., 2007a,b).

Two-photon 4pi microscopy needs good fluorescence signals. In some cases it is not possible to increase the fluorescence signal without disturbing the morphology of the cell too much, and in some cases single-photon imaging can produce quite good results. A type A 4pi setup with standard glycerin immersion objective lenses still gives *XY* planes with much less out-of-focus light compared to a confocal microscope due to the "smaller" PSF of the 4pi microscope.

The 4pi microscope with single-photon excitation was used in a study of the intracellular oxidoreductive (redox) state of dopamine

Figure 3.11 Two different 3D reconstructions of oxidation (green) in endoplasmic reticulum-labeled (red) cells. In these images empty portions represent reduced areas. Please see color plates at the back of the book.

(DA) neurons of the substantia nigra pars compacta (SNc) in rat brain for the purpose of measuring the oxidation signal and its possible overlap with the endoplasmic reticulum of the dopaminergic neurons. DA cells possess a distinct physiology intrinsically associated with elevated reactive oxygen species (ROS) production, and they selectively degenerate in Parkinson's disease under oxidative stress conditions (Horowitz et al., 2011). Understanding how ROS modulates the intracellular thiol redox state in health and disease is critical to unraveling the responses activated by DA cells to preserve normal function and prevent pathogenesis under conditions of redox imbalance. DA neurons in the SNc are part of the basal ganglia motor circuitry (Rinne, 1993; Björklund and Dunnett, 2007). The study of redox homeostasis in SNc DA neurons is of extreme interest for understanding the biology of these cells, as their particular physiology is intrinsically associated with elevated ROS production. Redox immunohistochemistry (RHC) detects areas of higher oxidation at the subcellular level. 4pi microscopy was performed on cells expressing endoplasmic reticulum-targeted red fluorescent protein (RFP; red) and prepared for RHC to label only the oxidized disulfides (green). Yellow (colocalization) indicates endoplasmic reticulum areas with higher oxidation. Figure 3.11 depicts two different 3D reconstructions of oxidation (green) in endoplasmic reticulum-labeled (red) cells. In these images empty portions represent reduced areas.

Moreover, when used in combination with higher resolution imaging techniques such as 4pi microscopy the method can detect differences in oxidation at the subcellular level. As expected, the endoplasmic reticulum appears as an area of higher oxidation (Figure 3.11). It is intriguing that higher magnification images generated with 4pi microscopy reveal that the oxidation signal seems to emerge from a heterogeneous network. This evidence, which was undetectable by conventional confocal microscopy, is consistent with the notion that the intracellular redox state differs substantially between subcellular compartments (Figure 3.11).

SYCP3 Axis as a Marker for Chromatin Organization in Mouse Spermatocytes

Sexual reproduction is an important step in evolution because it facilitates the exchange of chromatin between different individuals. The exchange process takes place during the meiosis of the oocytes and spermatocytes. During this process the paternal and maternal chromosomes initially have to find each other in a limited space of about $1500 \mu m^3$. This process may be improved if there is good organization or a good process available to untie entangled DNA or a combination of both. The key feature here is the formation and repair of DNA double-strand breaks by the topoisomerase-like enzyme SPO11 (Inagaki et al., 2010). The enzyme induces several hundreds of DNA breaks throughout the genome. The homologous chromosome is used as a template to repair the breaks, and then the formation of the synaptonemal complex, the protein structure that synapses both chromosomes, is started.

During the leptotene and zygotene phase of meiosis the synaptonemal complex is further established, and during the pachytene phase there is full synapsis. Next to DNA repair another process called meiotic bouquet formation is involved in the process of chromosome alignment (Ding et al., 2007; Bhalla and Dernburg, 2008). During early leptotene, just before SPO11 induces the double-strand breaks, the telomeres of the (mouse) chromosomes associate with the outside of the nucleus and the chromosome ends are moved to each other. Thereafter the telomeres spread out over the nuclear surface again. Most studies of synaptonemal axis formation have been conducted in spread preparations by which the cells are spread out on a microscope slide. In this procedure the 3D morphology is completely disrupted. With a synaptonemal complex protein 3 (SYCP3) staining it is possible to identify the meiotic stage in both oocytes and spermatocytes. SYCP3 is a protein found in the lateral part of the synaptonemal complex. The two lateral components binding to the paternal and maternal chromosomes are kept together with the axial components of the synaptonemal complex. To visualize the DNA a DAPI staining may be applied; however, in this case the broad DAPI emission in combination with the good excitation of DAPI with the two-photon laser is not an option. A better DNA stain in this case is PI. As a reference point to find any organization in the chromosome distribution in the spermatocyte we used an antibody against γH2Ax to identify the *XY*-body. Because the detection is limited to two channels we used the SYCP3/PI combination or the SYCP3/γH2Ax combination.

Mouse testes were dissected, and the seminiferous tubules were isolated and cut in small pieces. The pieces were put on quartz coverslips and put in an incubator for 24 hours. The sample was fixed with 2% formaldehyde in PBS for 15 minutes. Synaptonemal complexes were detected with a rabbit anti-SYCP3 antibody (de Vries et al., 2005) combined with an Alexa Fluor 488-tagged secondary antibody. Nuclei were counterstained with PI or a γH2Ax mouse monoclonal antibody and a TRITC-conjugated secondary antibody to mark the *XY*-body. Images were made with the 4pi microscope in *XZ*-mode with steps in the *Y*-direction. The diameter of a spermatocyte is in the order of 10–20 µm, which means that imaging a typical nucleus with a diameter of 15 µm with 30 nm isotropic voxels results in a stack of 500 frames. For this reason only a limited amount of cells could be analyzed. The thickness of the synaptonemal complex is between 120 and 400 nm, which makes it very interesting to study with 4pi microscopy. Electron microscopic images do show the synaptonemal complex as the lateral and central elements with a distance between the lateral elements of about 100 nm (Yanagibashi and Kusanagi, 1973). It is not always clear if this distance is always the same and constant during spermatogenesis. Due to the occurrence of diffraction side lobes in the 4pi images it is difficult to judge if there are two different axes when looking in the axial direction. After 3D Gaussian blurring (with a $3 \times 3 \times 3$ kernel) and deconvolution the signal is no longer homogeneous and there could be two axes (Figure 3.12). It should be noted that putting a filter like a Gaussian filter over an image to remove noise quite often results in artifactual structures showing up in the images. Low-intensity (high Schott noise) nuclear signals are especially liable to this effect. Images are made of spermatocytes in the pachytene stage of meiosis when the synapsis is complete. Images were acquired from a total of 34 cells, but the *XY*-body could only be successfully identified in 13 cells. Three-dimensional reconstruction of the first nuclei immediately revealed that during pachytene most synaptonemal axes still attach on both sides of the nuclear envelope (Figure 3.13). All axes were measured and the length of each axis normalized to the total length of the axis (Figure 3.14). The position of axis compared to the center of mass of the complete nucleus

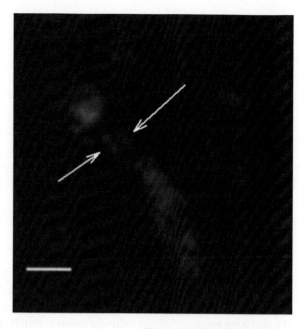

Figure 3.12 SYCP3 staining faintly showing two individual axes between the arrows. Scale bar, 500 nm.

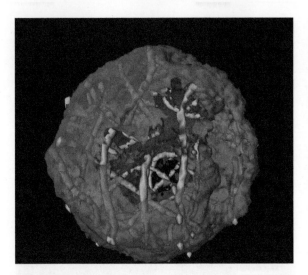

Figure 3.13 Surface-rendered SYCP3 Alexa Fluor 488 (green) and DNA PI (red) staining of a pachytene mouse spermatocyte. Please see color plates at the back of the book.

was also measured. An attempt to identify the individual chromosomes on the basis of the length of the synaptonemal complexes was unsuccessful. The individual differences and the amount of variation in axis length were too great to positively identify all axes (chromosomes). Also, the curvature of the axis in combination with its length did not lead to a positive identification of the axes. It is not clear yet if the curvature of the axes is dependent on the stage of meiotic development, or if this feature must be interpreted as individual variation. Studies in living samples are needed to explain this phenomenon.

The observation in spread preparations that the *XY*-body is always at the periphery of the nucleus can be explained after 3D examination. In the 3D reconstructed nucleus almost all synaptonemal complexes, and all chromosomes, touch the nuclear envelope on both sides quite often in a horseshoe-like curvature. This organization seems to be different from the organization found in interphase nuclei (Cremer and Cremer, 2001).

Synaptonemal axes do not end on the nuclear periphery in all nuclei. In some cells three axis ends were seen to be close to each other somewhere in the nucleus, probably in a nucleolus. This is clearly visible in a rotating 3D model of this particular cell nucleus, but it is difficult to see on a printed 2D image. It can also be shown by quantification of the distance of every axis end point to the center of mass of the nucleus where three axes do have an end point less than 2 μm from the center of the nucleus (Figure 3.15). The curvature of the axis can be quantified by dividing the distance of the end points by the length of the axis. A completely stretched axis has a curvature value of 1 while a strongly curved axis will have a curvature value <0.5. The combination of the axis curvature and the axis end points makes it possible to distinguish three different groups (morphologies). In group one all axis ends are on the nuclear envelope and most of the axes have a strong curvature, with the middle of the axis pointing toward the center of the nucleus (Figure 3.16). This morphology often resembles the bouquet stadium. In group 2 all axes still end on the nuclear envelope, but most of the axes are stretched (Figure 3.17). Group 3 consists of cells with three axes ending somewhere in the center of the nucleus, probably in a nucleolus (Figure 3.15). The different groups probably represent different stages of development during the pachytene stage. This can only be resolved by a time-lapse study on living material and for that another microscope and

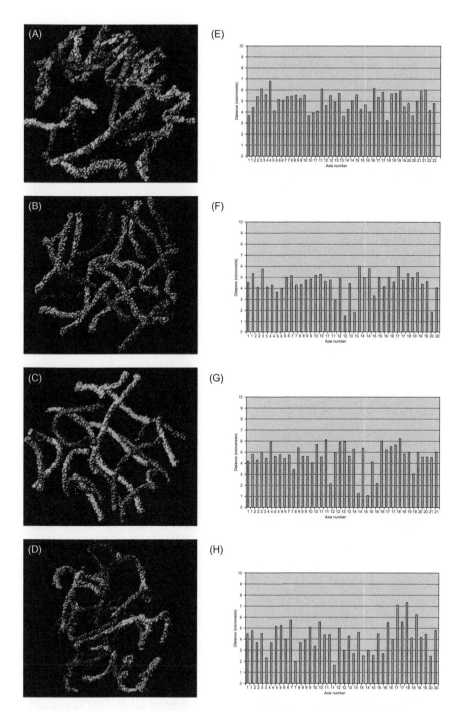

Figure 3.14 Surface rendering, segmentation (A, B, C, D), and analysis (E, F, G, H) of the SYCP3 Alexa Fluor 488 staining in mouse spermatocytes. Please see color plates at the back of the book.

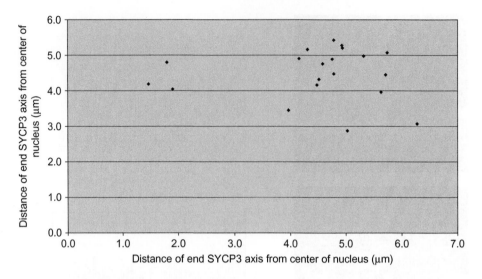

Figure 3.15 Distance from SYCP3 axis end points to the center of mass of the spermatocyte nucleus.

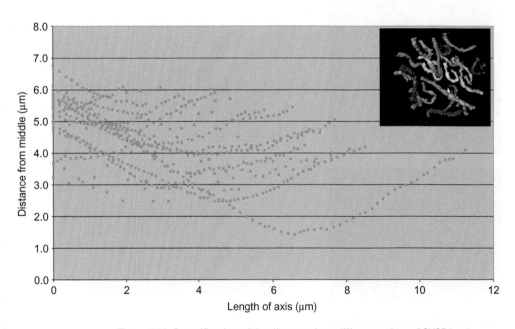

Figure 3.16 Quantification of the distance from different points of SYCP3 axis toward the center of the mouse spermatocyte. Almost all axes from this one cell are curved. Please see color plates at the back of the book.

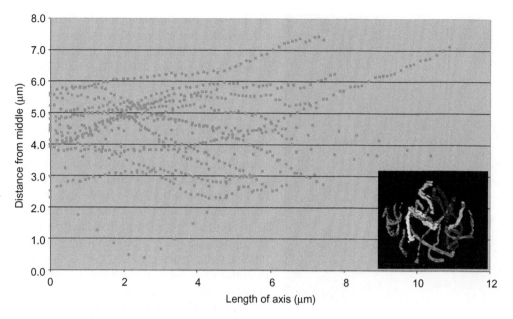

Figure 3.17 Quantification of the distance from different points of the SYCP3 axis toward the centre of the mouse spermatocyte. Almost all axes from this one cell are stretched. Please see color plates at the back of the book.

transgenic mice are needed, because 4pi microscopy is not compatible with living material.

Microbubbles with Medicine

Ultrasound not only images anatomical structures, it can also locally distribute agents in the human body. Ultrasound contrast agents consist of gas-filled–coated microbubbles with diameters between 1 and 10 μm. Microbubbles are often coated with a lipid layer. Besides imaging, ultrasound can also be used to blow up microbubbles to locally release medicinal drugs. Strong differences in the behavior of similar sized microbubbles have been reported. Heterogeneous coating of the microbubbles has been suggested as the underlying cause. Most studies were performed dynamically using a setup of vibrating microbubbles in an ultrasound field. Very little is known about the distribution of the lipid coating of the outer layer of microbubbles, which might be an indication for the pressure inside the microbubbles (Kooiman 2011). One component of the lipid layer was fluorescently tagged to study the lipid distribution in a confocal microscope. For microbubbles with diameters between 15 and 75 μm it was found in previous studies that the

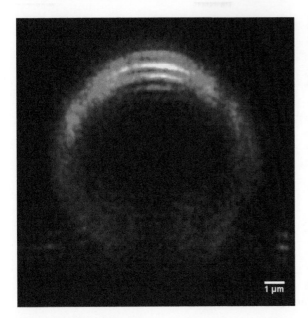

Figure 3.18 Typical 4pi side lobe showing up on *XZ* images of a lipid-labeled microbubble. Please see color plates at the back of the book.

Figure 3.19 Lipid distribution of microbubble coating with different sizes. Please see color plates at the back of the book.

lipids were not homogeneously distributed throughout the microbubble coating layer (Kim et al., 2003; Borden et al., 2004, 2006; Lozano and Longo, 2009). The lipid distribution for clinically relevant microbubble sizes in the 1–10 μm diameter range has never been investigated. Probably because of the high refractive mismatch between the interior of the microbubble and the surrounding environment (water) it was not possible to image more than the lower half of the microbubbles in a confocal microscope. The 4pi setup was used in this case as a "double" microscope to visualize both the upper and lower halves of microbubbles labeled with fluorescent DSPE-PEG2000. The microbubbles were embedded in glycerin, similar to all other 4pi samples. Surprisingly, the typical axial 4pi diffraction side lobes do still show up on the thin lipid layer of the microbubbles (Figure 3.18). All microbubbles studied displayed an inhomogeneous distribution of the fluorescent label that was independent of the size of the microbubble (Figure 3.19). Also, areas without fluorescence were detected, indicating local exclusion of the DSPE-PEG2000 lipid and hence a high concentration of one or both of the other coating components, namely DSPC and PEG-40 stearate. In microbubbles of the same size, different distributions were detected in contrast to what has been previously reported in larger bubbles (Kim et al., 2003). To calculate the surface shear viscosity of the coating, the diffusion of the lipids, D, was determined using FRAP (Kooiman, 2011).

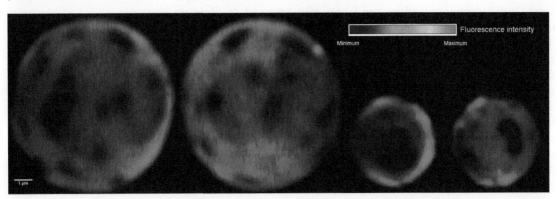

Future of 4pi Imaging

4pi microscopy is still one of the best ways to improve the axial resolution of the optical microscope. For this reason the technique is showing up in all kinds of new developments. One interesting development is the attempt to simplify the setup by replacing one of the lenses with a mirror (Mudry et al., 2010a,b; Le Moal et al., 2011). It would be interesting to see if this setup is really usable with biological samples.

The optical techniques with the highest resolution at the moment are the stochastic single molecule localization techniques. Most of these techniques, however, lack good localization in the axial direction. Improvements have been made with optical astigmatism (Huang et al., 2008), but the combination of 4pi microscopy with techniques like STED with 4pi (Schmidt et al., 2008) and, more recently, single molecule imaging in a 4pi setup (Aquino et al., 2011) prove that 4pi remains a very interesting technique.

The 4pi microscope has not developed into a commercial success, probably due to the expected difficulties with the alignment of the two beams. Automation of this process similar to the automatic alignment of the STED beam may improve acceptance. Further development of single-photon 4pi microscopy (Martinez-Corral et al., 2003; Lang et al., 2007b) in combination with automated deconvolution (Vicidomini et al., 2009, 2010) and automated alignment will further stimulate the acceptance of the 4pi microscope.

Acknowledgment

We want to thank Pier Mastroberardino for the information about the "redox state in dopamine neurons," Klazina Kooiman for the information and the images of the microbubbles, and José Tielen for critical reading of the manuscript.

References

Aquino, D., Schonle, A., Geisler, C., Middendorff, C.V., Wurm, C.A., Okamura, Y., et al., 2011. Two-color nanoscopy of three-dimensional volumes by 4Pi detection of stochastically switched fluorophores. Nat Methods 8, 353–359.

Bhalla, N., Dernburg, A.F., 2008. Prelude to a division. Annu Rev Cell Dev Biol 24, 397–424.

Björklund, A., Dunnett, S.B., 2007. Dopamine neuron systems in the brain: an update. Trends Neurosci 30, 194–202.

Borden, M.A., Martinez, G.V., Ricker, J., Tsvetkova, N., Longo, M., Gillies, R.J., et al., 2006. Lateral phase separation in lipid-coated microbubbles. Langmuir 22, 4291–4297.

Borden, M.A., Pu, G., Runner, G.J., Longo, M.L., 2004. Surface phase behavior and microstructure of lipid/PEG-emulsifier monolayer-coated microbubbles. Colloids Surf B Biointerfaces 35, 209–223.

Cremer, T., Cremer, C., 2001. Chromosome territories, nuclear architecture and gene regulation in mammalian cells. Nat Rev Genet 2, 292–301.

de Vries, F.A., de Boer, E., van den Bosch, M., Baarends, W.M., Ooms, M., Yuan, L., et al., 2005. Mouse Sycp1 functions in synaptonemal complex assembly, meiotic recombination, and XY body formation. Genes Dev 19, 1376–1389.

Ding, X., Xu, R., Yu, J., Xu, T., Zhuang, Y., Han, M., 2007. SUN1 is required for telomere attachment to nuclear envelope and gametogenesis in mice. Dev Cell 12, 863–872.

Florijn, R.J., Slats, J., Tanke, H.J., Raap, A.K., 1995. Analysis of antifading reagents for fluorescence microscopy. Cytometry 19, 177–182.

Galjart, N., 2010. Plus-end-tracking proteins and their interactions at microtubule ends. Curr Biol 20, R528–R537.

Hell, S.W., Schrader, M., van der Voort, H.T., 1997. Far-field fluorescence microscopy with three-dimensional resolution in the 100-nm range. J Microsc 187, 1–7.

Hell, S.W., Stelzer, E.H., 1992. Fundamental improvement of resolution with a 4Pi-confocal fluorescence microscope using two-photon excitation. Optics Communications 93, 227–282.

Hell, S.W., Stelzer, E.H., Lindek, S., Cremer, C., 1994. Confocal microscopy with an increased detection aperture: type-B 4Pi confocal microscopy. Opt Lett 19, 222.

Horowitz, M.P., Milanese, C., Di Maio, R., Hu, X., Montero, L.M., Sanders, L.H., et al., 2011. Single-cell redox imaging demonstrates a distinctive response of dopaminergic neurons to oxidative insults. Antioxid Redox Signal 15, 855–871.

Huang, B., Wang, W., Bates, M., Zhuang, X., 2008. Three-dimensional super-resolution imaging by stochastic optical reconstruction microscopy. Science 319, 810–813.

Inagaki, A., Schoenmakers, S., Baarends, W.M., 2010. DNA double strand break repair, chromosome synapsis and transcriptional silencing in meiosis. Epigenetics 5, 255–266.

Kim, D.H., Costello, M.J., Duncan, P.B., Needham, D., 2003. Mechanical properties and microstructure of polycrystalline phospholipid monolayer shells: Novel solid microparticles. Langmuir 19, 8455–8466.

Kooiman, K., 2011. Therapeutic bubbles. In: Biomedical Engineering. Erasmus University Medical Center, Rotterdam.

Lang, M., Muller, T., Engelhardt, J., Hell, S.W., 2007a. 4Pi microscopy of type A with 1-photon excitation in biological fluorescence imaging. Opt Express 15, 2459–2467.

Lang, M.C., Engelhardt, J., Hell, S.W., 2007b. 4Pi microscopy with linear fluorescence excitation. Opt Lett 32, 259–261.

Le Moal, E., Mudry, E., Chaumet, P.C., Ferrand, P., Sentenac, A., 2011. Isotropic single-objective microscopy: theory and experiment. J Opt Soc Am A Opt Image Sci Vis 28, 1586–1594.

Lozano, M.M., Longo, M.L., 2009. Microbubbles coated with disaturated lipids and DSPE-PEG2000: phase behavior, collapse transitions, and permeability. Langmuir 25, 3705–3712.

Martinez-Corral, M., Caballero, M.T., Pons, A., Andres, P., 2003. Sidelobe decline in single-photon 4Pi microscopy by Toraldo rings. Micron 34, 319–325.

Mudry, E., Chaumet, P.C., Belkebir, K., Maire, G., Sentenac, A., 2010a. Mirror-assisted tomographic diffractive microscopy with isotropic resolution. Opt Lett 35, 1857–1859.

Mudry, E., Le Moal, E., Ferrand, P., Chaumet, P.C., Sentenac, A., 2010b. Isotropic diffraction-limited focusing using a single objective lens. Phys Rev Lett 105, 203903.

Rinne, J.O., 1993. Nigral degeneration in Parkinson's disease. Mov Disord 8 (Suppl. 1), S31–S35.

Schmidt, R., Wurm, C.A., Jakobs, S., Engelhardt, J., Egner, A., Hell, S.W., 2008. Spherical nanosized focal spot unravels the interior of cells. Nat Methods 5, 539–544.

Sheppard, C.J., Gong, W., Si, K., 2010. Polarization effects in 4Pi microscopy. Micron (Epub ahead of print).

Vicidomini, G., Hell, S.W., Schonle, A., 2009. Automatic deconvolution of 4Pi-microscopy data with arbitrary phase. Opt Lett 34, 3583–3585.

Vicidomini, G., Schmidt, R., Egner, A., Hell, S., Schonle, A., 2010. Automatic deconvolution in 4Pi-microscopy with variable phase. Opt Express 18, 10154–10167.

Wurster, D.H., Benirschke, K., 1970. Indian muntjac, Muntiacus muntjak: a deer with a low diploid chromosome number. Science 168, 1364–1366.

Yanagibashi, K., Kusanagi, A., 1973. Electronmicroscopic ammoniacal silver reaction for the synaptonemal complex of the mouse. Exp Cell Res 78, 228–230.

4

NANO RESOLUTION OPTICAL IMAGING THROUGH LOCALIZATION MICROSCOPY

Helge Ewers

Institute of Biochemistry, ETH Zurich, Zürich, Switzerland

CHAPTER OUTLINE

Introduction   82
Superresolution Microscopy Techniques   82
    The Concept Behind Localization Microscopy   83
    Matters of Concern   85
The Main Approaches to Single-molecule Localization-based
  Superresolution Microscopy   87
    Using Fluorescent Proteins: PALM and FPALM   87
    Using a Pair of Interacting Cyanine Dyes: STORM   88
    Using Blinking Dyes: dSTORM   88
    Using High-intensity Light: GSDIM   88
    Using Targeted Molecules: BALM   88
Fluorescent Probes   89
Fluorescent Proteins   89
    Organic Fluorophores   90
    Labeling Techniques: 90
Multicolor Localization Microscopy   90
    Single Particle Tracking   91
    Approaches to 3D PALM   92
    Application in Neurons   92
Outlook   93
    Technical Development   93
    Biological Questions   96
Conclusion   96
Acknowledgments   96
References   97

Cellular Imaging Techniques for Neuroscience and Beyond.
DOI: http://dx.doi.org/10.1016/B978-0-12-385872-6.00004-0

Introduction

Fluorescence microscopy has become one of the most important techniques in cell biology and (cellular) neurobiology. The labeling of specific cellular components via antibodies coupled to organic fluorescent dyes made it possible to determine the subcellular localization of proteins in the context of the cell. When different proteins were labeled with fluorophores that could be spectrally separated, the extent of overlap of the fluorescence signal from both molecules could support the conclusion that they localize to the same cellular compartment or to a specific structure. The possibility of genetically tagging proteins via green fluorescent protein (GFP) then made it possible to follow cellular components, even in live cells, over time and with ever more efficient methods used to deliver recombinant DNA into neurons. By now, entire genomes are available as GFP-tagged libraries and allow the study of virtually any protein by fluorescence microscopy in neurons. However, fluorescence suffers from one fundamental limit, the diffraction limit of light (Abbe, 1882), which makes it impossible to spatially resolve two molecules that are separated by a distance less than about half the wavelength of light (for GFP that would be around 250 nm). In a cellular context this is a large distance, about ten times the diameter of a microtubule, and two proteins that fall within this distance, while they are not distinguishable by fluorescence microscopy, do not necessarily interact.

Neuroscientists are interested in the events taking place at synapses, for example, where thousands of molecules are localized within an area similar in size to one diffraction-limited spot. A detailed investigation of the organization of these important structures by light microscopy has thus far been impossible, and the development of novel fluorescence microscopy techniques that circumvent this problem has been received with great enthusiasm. In this chapter I will discuss techniques based on the assembly of high-resolution images from the accumulated positions of thousands of individual molecules, summarized under the term localization microscopy. The special appeal of this technique, in contrast to other approaches, is that along with the higher resolution of a structure of interest it also yields at the same time the positions of thousands of individual molecules located to this structure. Here, I will review the technical requirements and pitfalls of this technique, the first studies that have made use of this technique in neurons, and briefly discuss the technical challenges that lie ahead.

Superresolution Microscopy Techniques

Besides localization microscopy, other approaches to circumvent the Abbe limit have been developed; the most prominent approaches are stimulated emission depletion (STED; Klar et al., 2000) and structured illumination microscopy (SIM; Gustafsson, 2005). In contrast to localization microscopy, both of these approaches are ensemble methods that yield images of structures rather than molecular positions. STED is a scanning technique in which the effective size of the point spread function (PSF) of the illuminating laser beam is reduced by depleting the excited state in a circular pattern around the center of the excitation beam by a second beam (detailed in Chapter 2) The other approach, SIM, is based on sequential excitation of the sample with different illumination patterns and subsequent computational reconstruction of a super-resolved image from the frames resulting from several different patterns.

One advantage of these techniques is that theoretically they are less restricted in the choice of fluorophore labels, whereas one disadvantage is that they may lead to significant photobleaching. In contrast, the main advantage of localization microscopy is that, along with the resolution of structures, it also yields the localization of the molecules that make up the structure.

The novel superresolution techniques have been extensively covered in several reviews (Bates et al., 2008; Ji et al., 2008; Lippincott-Schwartz and Manley, 2009; Huang et al., 2010; Schermelleh et al., 2010) and, especially for localization microscopy, several detailed and carefully described protocols for the execution of experiments have been published that deal with every aspect from microscope assembly over experiments to data processing (Ji et al., 2008; Shroff et al., 2008; Manley et al., 2011; van de Linde et al., 2011b).

The Concept Behind Localization Microscopy

While all relying on the same principle (Figure 4.1), many different acronyms have been coined for what we will call localization microscopy methods. These include photoactivated localization microscopy (PALM; Betzig et al., 2006), fluorescence photoactivated localization microscopy (FPALM; Hess et al., 2006), stochastic optical reconstruction microscopy (STORM; Rust et al., 2006), direct stochastic optical reconstruction microscopy (dSTORM; Heilemann et al., 2008), ground-state depletion and single molecule return (GDSIM; Fölling et al., 2008), binding-activated localization microscopy (BALM; Schoen et al., 2011), complementation-activated localization microscopy (CALM; Pinaud and Dahan, 2011), point accumulation imaging of nanoscale topology (PAINT; Sharonov and Hochstrasser, 2006), and universal point accumulation imaging of nanoscale topology (uPAINT; Giannone et al., 2010). While extensive, this is probably an incomplete list. All of these techniques are based on the same idea: when many fluorophores

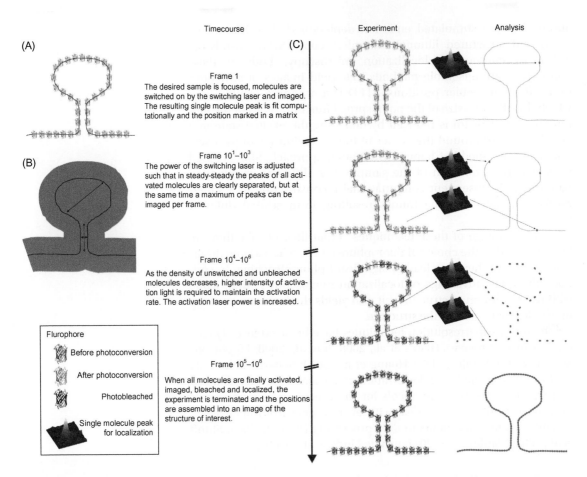

Figure 4.1 Principle of localization microscopy. Please see color plates at the back of the book.

within the area of a diffraction-limited spot cannot be resolved due to the diffraction limit of light, conditions must be created where only one fluorophore at a time emits photons in one detection channel. In this way, individual molecules can then be localized with subdiffraction accuracy by using fitting algorithms similar to those in single particle tracking. The position of this molecule is noted, and when it is bleached another molecule will be brought into an emitting state and imaged. There are several different ways in which most molecules are kept undetectable, while individual ones are stochastically moved to a detectable state. The previously mentioned approaches differ in how this is achieved. The specific advantages and disadvantages of the different approaches will be discussed in a later section. First, several parameters that need to be considered for a single molecule localization-based superresolution experiment will be discussed.

Matters of Concern

Single molecule localization microscopy is still being established and the first commercial systems have entered the market. Various approaches that differ mostly in the way the molecules of interest are labeled with fluorophores are now available, but most of them still involve a trade-off between different parameters that need to be optimized for the specific research question addressed. These parameters include localization precision, linkage error, specificity, labeled fraction, and labeling density. No single approach to localization microscopy is optimal for all of these parameters (Table 4.1), and the approach needs to be tailored to the specific problem under investigation.

Localization Precision

The first concern when planning a single-molecule localization microscopy experiment is the localization precision and the goal becomes to achieve high localization precision. Achieving this goal depends directly on the number of photons collected from an activated molecule compared to the background.

Proteins that can be photoswitched, photoactivated, and photoconverted typically yield from 10 to 500 photons (~20–50 nm accuracy) and organic fluorophores can yield up to ~10,000 photons (~2–5 nm accuracy). In this type of experiment the organic dyes are localized and not the target molecule, and the way the fluorophore is targeted to the molecule of interest can result in considerable error.

Table 4.1 Properties of the Most Common Labeling Approaches

	Fluorescent Protein	Organic Dye on Antibody	Organic Dye on Nanobody	Enzymatic Labeling	Specific Organic Dye
Specificity	+++	+	++	++	++
Brightness	++	+++	+++	+++	+++
Linkage error	+++	+	+++	++	+++
Labeled fraction	+++	+	++	?	?
Background labeling	+++	+	+	+	++
Intracellular labeling	+++	++	+	+	++
Membrane labeling	+++	+++	+++	+++	+++
Live cell labeling	+++	+	+	++	+
Labeling density	+++	+	+++	+++	++

Linkage Error

In localization microscopy the fluorophore is localized accurately on the scale of the size of many proteins, and the exact localization of the fluorophore in respect to the target molecule is of considerable importance. Some molecules can extend over tens or even hundreds of nanometers, for example, many filamentous structural or linker molecules. Thus, it can make a difference where the fluorescent protein tag is incorporated into the construct; for example, the N- and C-termini of talin have been localized ~40 nm apart (Kanchanawong et al., 2010) using constructs tagged at the respective position.

When organic dyes are delivered to the target structure via immunofluorescence labeling, the fluorophore may still be localized with an accuracy of a few nanometers. However, this advantage is lost, because this is less than the size of an antibody (around 10 nm), and when the dye is attached to the secondary antibody in an immunofluorescence experiment, the significant linkage error may even lead to artifactual staining, for example, the detection of single microtubules as two parallel lines. The linkage error can be significantly reduced, when small single-domain antibodies are used for detection (Ries et al., 2012).

Specificity

The fusion of a photoswitchable protein to the target molecule theoretically leads to the highest possible specificity; however, as in conventional fluorescence microscopy, not all fusion constructs behave like an endogenous protein and overexpression may lead to artifacts. The same holds true for the enzymatic labeling schemes such as the SNAP-tag (Gautier et al., 2008; Testa et al., 2010; Banala et al., 2011; Klein et al., 2011), and since here organic dyes need to be added exogenously, they may bind unspecifically to the cells and lead to significant background.

In immunofluorescence labeling unspecific binding is also a problem that depends on the quality of the primary and secondary antibody, the labeling protocol, and the cell type.

Labeled Fraction

When a molecule of interest is studied, ideally all molecules should be labeled, such that all possible cellular functions and localizations of the molecule will be sampled. Again a fusion construct with a fluorescent protein will ideally lead to 100% labeling, but this depends on the folding characteristics of the fluorescent protein used, which is not completely understood for most molecules and will typically be well below 100%. Furthermore, unless overexpression of tagged molecules is combined with RNAi against the protein of interest or knock-out animals, endogenous pools of the protein and maybe even different splice isoforms, will be present in the cell as well, which should be considered. This also holds true for enzymatic labeling methods, but here the fact that likely less than 100% of molecules will react with the exogenously added fluorescent dye is an additional handicap.

Specific antibodies, on the other hand, have the advantage that there is no overexpression and they could detect all molecules of the species of interest. However, some epitopes may be masked, especially in dense multiprotein complexes or when working with monoclonal antibodies.

Labeling Density

When molecules can be localized with very high accuracy, say a few nanometers, the resolution of the structure of interest depends on the labeling density. When individual molecules are localized with an accuracy of 5 nm, it is not sufficient to have one fluorophore every 50 nm to resolve the outline of the structure of interest. Ideally, if molecules can be localized with 10 nm precision, a sampling rate of 5 nm is desirable, which would be 50 molecules in one dimension within one PSF and hundreds in two dimensions. The labeling density, indeed, can quickly become the limiting factor in well-established assays.

The Main Approaches to Single-molecule Localization-based Superresolution Microscopy

While localization microscopy is used here as a general term for all techniques that render individual molecules stochastically detectable in sparse populations to assemble a superresolved image over time, different acronyms are still mostly connected to different approaches to how this is achieved. These are outlined in the following sections.

Using Fluorescent Proteins: PALM and FPALM

The first published method that used sequential localization of single molecules to assemble the nanoscopic outline of a microscopic structure in 2006 made use of fluorescent proteins that could be switched from dark or green states to bright or spectrally shifted states (Betzig et al., 2006, Hess et al., 2006). Although photo-uncaging of organic dyes was shown in this work as well, PALM is now generally referred to as localization microscopy with fluorescent proteins.

Using a Pair of Interacting Cyanine Dyes: STORM

Published in 2006 as well, STORM relies on the phenomenon previously described in the lab (Bates et al., 2005): specific dye pairs may act as switches where the emission of the longer wavelength fluorophore can be switched on by absorption through the shorter wavelength fluorophore. Special switchable fluorescent proteins and lasers are needed for the respective switching and emission wavelength. STORM requires a switching laser and specific buffers.

Using Blinking Dyes: dSTORM

Since many fluorescent dyes can be rendered photoswitchable through near-UV light under specific buffer conditions, they can be used for localization microscopy as well. The combination of regular organic dyes with reducing and oxygen-scavenging buffer systems is often referred to as dSTORM. Again switching and excitation lasers and a specific buffer are required (Heilemann et al., 2008).

Using High-intensity Light: GSDIM

Some organic dyes have fairly stable triplet states into which they can be pumped by intense illumination without too much photobleaching. They will stochastically return to the ground state and then be detectable again. If the excitation laser power imaging of such dyes is kept high enough, most dyes will soon be in the triplet state, and only the ones dropping back to the ground state will be imaged. This makes the use of a switching laser obsolete, but requires very high illumination intensities (Fölling et al., 2008).

Using Targeted Molecules: BALM

When many photons can be collected from a fluorophore and it can thus be localized more precisely, the error introduced by the targeting of the fluorophore to the structure of interest becomes more important. For a system in which a primary antibody is used to bind to the target structure and is then detected by a secondary antibody, the dye on the secondary antibody could be located as far as 20 nm away from the target structure. This problem can be avoided when the fluorophore binds directly to the target. For several biological structures, specific dye conjugates or specific binding dyes exist. The most common are fluorescent conjugates of the actin-binding phallotoxin phalloidin, which can be readily used in commercially available form with chemically induced photoblinking (dSTORM) (Heilemann et al., 2008; Testa et al., 2010; Xu et al., 2012) and DNA-binding dyes such as YOYO-1 and Pico-green

(Flors et al., 2009; Schoen et al., 2011). The latter case is especially appealing since the dyes are virtually nonfluorescent when not bound to DNA. This becomes an advantage because several rounds of binding and unbinding of dyes to a DNA structure can be imaged, thus an extremely high density of labeling can be achieved (up to 1/nm; Schoen et al., 2011). This concept of BALM could also be used for other dyes that exhibit marked shifts in their fluorescence spectrum or an increase in brightness when binding to a target structure such as the lipophilic dye Nile red when it inserts into membranes (Sharonov and Hochstrasser, 2006) or the split-GFP (Pinaud and Dahan, 2011). However, due to its intrinsic specificity, it can only be used for the respective target structures, which are hardly ever proteins. This approach requires no switching laser or specific buffer conditions, but it is very limited in its targets.

Fluorescent Probes

There are several excellent reviews on the choice of fluorophores in cell biology in general and superresolution microscopy in particular (Fernández-Suárez and Ting, 2008; Patterson et al., 2010; Dempsey et al., 2011). The present range of fluorophores will not be comprehensively covered in this chapter.

Fluorescent Proteins

The range of fluorescent proteins suitable for PALM has been reviewed previously (Lippincott-Schwartz and Patterson, 2009) and the number of fluorescent proteins used for PALM is growing steadily (Fuchs et al., 2010; Hoi et al., 2010; Subach et al., 2010; Brakemann et al., 2011). Briefly, mEOS2 (McKinney et al., 2009), PAmCherry (Subach et al., 2009), and PSCFP2 (Chudakov et al., 2004) are established molecules with good photon yield and widespread use. The recent additions to the ever-expanding pool that are of special interest include the reliable mClavGR (Hoi et al., 2010), which is similar to mEOS2 in many aspects; PSOrange, which switches to the far red part of the spectrum (Subach et al., 2011); the Dreiklang fluorescent protein, where the excitation and off-switching wavelength are decoupled (Brakemann et al., 2011); and a novel, reversibly photoswitchable GFP that may be useful for repeated imaging in live cells (Grotjohann et al., 2011). A special case is the mIrisFP, which can be switched on and off in a green state and after conversion to a red state can also be switched on and off. This may allow the use of localization microscopy for the observation of protein turnover or exchange (Fuchs et al., 2010).

Organic Fluorophores

A wide range of bright organic fluorophores is available from several commercial providers covering the entire spectrum of visible light and beyond. The discovery that the intrinsic photoblinking of dyes can be used for localization microscopy has lead to a deeper investigation of the mechanism of photoblinking. By now the mechanism for cyanine dyes (Dempsey et al., 2009) and Alexa dyes (Vogelsang et al., 2008; Endesfelder et al., 2011; van de Linde et al., 2011a) are well understood and the community has contributed valuable comparative analyses of dyes (Dempsey et al., 2011; van de Linde et al., 2011b). Conveniently, many dyes that have been used for years in fluorescence microscopy such as AF647 turned out to be ideal molecules for localization microscopy, such that a vast number of reagents labeled with these molecules is available. Most photoblinking techniques, however, rely on the use of reducing buffer systems and the absence of oxygen in the imaging buffer, which is not necessarily compatible with live-cell imaging. An alternative may be the recently developed photoswitchable and photocaged organic dyes (Bossi et al., 2008; Belov et al., 2010; Banala et al., 2011).

Labeling Techniques

While fluorescent proteins are delivered to their target by genetically engineering fusion constructs that are then transiently or stably expressed in neurons or other cells, the delivery of organic dyes to their target requires more invasive techniques. For most approaches, cells need to be fixed and permeabilized (see Table 4.2). So can organic dyes be delivered to their target structure via antibodies, nanobodies, enzymatic labeling, or directly by specific dyes?

Multicolor Localization Microscopy

Several approaches to localization microscopy with more than one color have been proposed using fluorescent proteins and organic fluorophores. In general, it is possible to use localization microscopy in two colors. In practice this is not trivial, especially when using genetically encoded fluorophores. While several bright photoswitchable fluorophores exist that can be used in the red emission channel, the options in the green channel remain poor. While PA-GFP and DRONPA yield few photons (Betzig et al., 2006), PSCFP and rsGFP solutions may have been found and several assays using fluorescent proteins have been published (Shroff

Table 4.2 Mechanisms of Photoactivation

Mode of activation	Example	Reference
Fluorescent proteins		
Photoactivation	PAGFP	Betzig et al., 2006; Hess et al., 2006
Photoswitching	Dronpa	Betzig et al., 2006
Photoconversion	EOS	Betzig et al., 2006
Assembly of fluorophore	Split-GFP	Pinaud and Dahan, 2011
Organic dyes:		
Photoactivation	Photoactivatable Rhodamine	Fölling et al., 2007
Photo-uncaging of dye		Betzig et al., 2006
Dark state pumping plus switching	AF647	Heilemann et al., 2008
Photoblinking in proximity of other dye	Cy3–Cy5	Rust et al., 2006
Dark state pumping	Atto 532, citrine	Fölling et al., 2008
Environmental sensitivity	BALM, PAINT	Sharonov and Hochstrasser, 2006; Schoen et al., 2011
Specific binding at low concentration	uPAINT	Giannone et al., 2010
Other probes		
Quantum dots	Blinking	Lidke et al., 2005

et al., 2007; Andresen et al., 2008; Gunewardene et al., 2011). The wide variety of organic fluorophores makes multicolor labeling using commercial dyes much easier and multicolor labeling with up to six colors has been reported (Bates et al., 2012). One appealing approach is the use of dye pairs in STORM, because several combinations of activator dyes with one emission dye are possible, such that all emissions can be detected in one channel, minimizing chromatic aberration.

Single Particle Tracking

Since localization microscopy is based on the sequential localization of many molecules, the data gathered are indistinguishable from single-molecule tracking experiments. Thus the most obvious expansion of this technique is the application to single-particle tracking and this was quickly realized in sptPALM (Manley et al., 2008). The advantage over traditional methods is that over time hundreds and thousands of particles are followed (instead of single particles), and a huge amount of diffusion data can be collected on a single cell in a short time.

Approaches to 3D PALM

One shortcoming of the original PALM method is the confinement to two dimensions and its preferred application in total internal fluorescence microscopy. Cells are 3D structures, so they fall in the nanoscopic range. A resolution of 20 nm in xy can only lead to limited conclusions if the resolution in z is 500 μm. Several approaches have been published: the simultaneous imaging of two slightly shifted image planes and the correlation of the PSFs in the two planes (Juette et al., 2008) or the distortion of the PSF according to its z-position by the addition of a cylindrical lens in the beam path (Huang et al., 2008).

The most powerful, but at the same time most technically demanding, approaches for resolving the localization of single molecules in 3D are based on the interference of the emission light collected from the sample by two objectives on both sides of the sample (Shtengel et al., 2009; Xu et al., 2012). This approach allows for an axial resolution of a few nanometers (Shtengel et al., 2009) and has dramatically demonstrated the true potential of localization microscopy to resolve the assembly of multiprotein complexes in cells (Kanchanawong et al., 2010). In this eye-opening study of focal adhesions, the authors localized thousands of molecules from several different species and could resolve the relative organization of the molecules in functional layers or modules within the complex. While the individual protein–protein interactions within this complex have long been identified by biochemical and structural biology methods, this study allows insight into how thousands of proteins are organized in the molecular machines that execute cellular function on the molecular level.

Application in Neurons

Many structures in neurons such as dendritic filopodia (Figure 4.2A) and spines (Figure 4.2B) and the synaptic domains are too small to be quantitatively studied without superresolution microscopy methods, but they are some of the most heavily investigated structures in biology. Consequently, among the first publications that applied localization microscopy to biological questions were several on neurons and specifically on dendritic spines and synapses (Tatavarty et al., 2009; Dani et al., 2010; Frost et al., 2010; Izeddin et al., 2011).

Single-molecule localization-based microscopy techniques are still not mainstream methods. This is reflected in the literature, where most publications using PALM have been collaborations with groups involved in the establishment and further development of PALM or STORM (Dani et al., 2010; Frost et al., 2010; Izeddin et al., 2011). Nevertheless, these studies have shown the potential of these techniques for the neurosciences. In a fascinating study using 3D STORM via astigmatism imaging, Dani et al. (2010) have resolved the relative positions of 10 different presynaptic and postsynaptic proteins of glutamatergic central nervous system synapses around the synaptic cleft (Figure 4.2C).

Localization microscopy in neurons builds on a long tradition of single-particle tracking studies investigating the interaction of membrane receptors with cytoskeletal scaffold molecules (Winckler et al., 1999; Borgdorff and Choquet, 2002; Nakada et al., 2003; Boiko et al., 2007; Newpher and Ehlers, 2008; Triller and Choquet, 2008; Opazo et al., 2010), and this combination is likely to yield novel insight in the future.

Furthermore, sptPALM can be used in combination with quickly diffusing molecules to resolve the nanoscopic structure of dendritic filopodia or spines over time (Figure 4.3). If the photobleached molecule can be exchanged quickly with unbleached molecules, several superresolution images can be composed over time from the same structure (see also Grotjohann et al., 2011).

Outlook

Technical Development

In the five years since the publication of the PALM method, substantial progress has been made in terms of fluorescent probes and labeling approaches, fitting and primary data analysis, and expansion of the method into the third dimension. However, several important problems remain in all of these areas.

First, while there are a number of very good photoswitchable proteins in the red, a photoswitchable GFP with a similar photon yield is still lacking, let alone the far ends of the fluorescent protein color palette. This limits work with live cells.

Second, while the localization software has matured over the last years, there is room for improvement in the analysis of the relative positions of the tens of thousands to millions of localizations in respect to one another. While first steps have been made in that direction (Owen et al., 2010; Sengupta et al., 2011), the postprocessing needs to be developed further to get the most out of the information contained in the vast amounts of data produced in single molecule localization measurements.

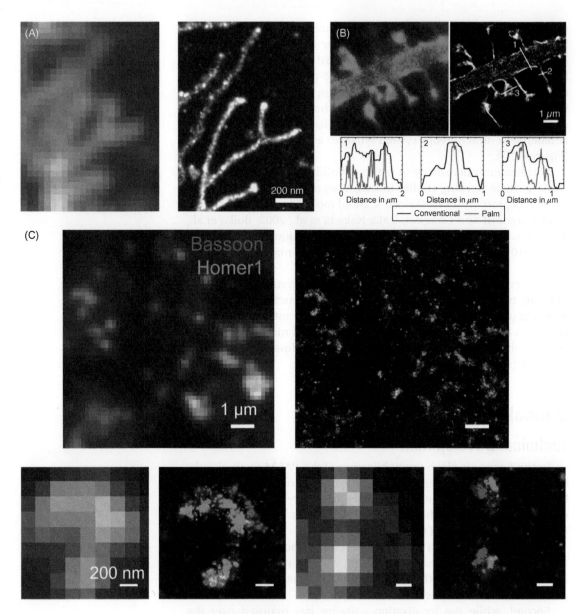

Figure 4.2 Applications of localization microscopy techniques in neurons. (A) Superresolution microscopy of dendritic filopodia. Rat hippocampal neurons were transfected with Myr/Palm mEOS2 and imaged in PALM (Ries and Ewers, unpublished). (B) Conventional (left) and localization microscopy (right) images of a rat hippocampal neuron expressing ABP-tdEosFP and fixed at day *in vitro* 25. Bottom, Profiles across the dendritic shaft (1), the spine neck (2), and the spine head (3), as indicated. *From* Izeddin et al., 2011. (C) Presynaptic protein Bassoon and postsynaptic protein Homer1 in the mouse MOB glomeruli were identified by immunohistochemistry using Cy3-A647- and A405-A647-conjugated antibodies,

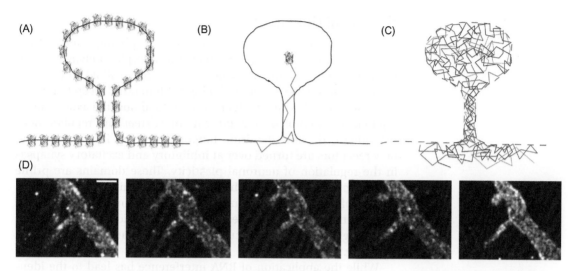

Thirdly, while the very complete paper from Betzig et al. (2006) already demonstrated correlative data of localization microscopy with electron microscopy, surprisingly this very important technical challenge has not been further developed. The embedding of the information gathered from thousands of individual molecules in specific structures into their structural environment will yield invaluable new insights.

Finally, the holy grail of single molecule microscopy would be quantitative imaging of single molecules, but this is complicated by the fact that organic dyes switch on more than one time during imaging (Dempsey et al., 2011), and it is unclear how quantitatively they can be delivered to their target. While fluorescent proteins may be more reliable in quantitative labeling, they unfortunately often blink as well, which can lead to clustering artifacts that prohibit quantitative imaging (Annibale et al., 2011). These issues require further investigation. The development of novel labeling methods such as fluorescent proteins that can be switched twice in different colors (Fuchs et al., 2010) and new image analysis approaches may make it possible to quantify synaptic receptor numbers in a temporally resolved manner.

Figure 4.3 Live-cell imaging with PALM. (A) If a membrane-associated fusion construct with a photoswitchable fluorescent protein is expressed in neuronal cells, individual activated molecules can be localized several times as they move around in on the cell surface (B). Since the molecules can only move within the boundaries of the cell membrane, the outline of the neuron can be assembled from the localizations (C). (D) Experimental data from a neuron expressing Myr/Palm mEOS2 (Ries and Ewers, unpublished). For every image, 5000 consecutive frames have been accumulated. Scale bar, 300 nm. Please see color plates at the back of the book.

Figure 4.2 (Continued) respectively. The conventional fluorescence image (top left) shows punctate patterns that are partially overlapping, whereas the STORM image (top right) of the same area clearly resolves distinct synaptic structures. Further zoom-in of the conventional images does not reveal detailed structure of the synapses, whereas the corresponding STORM images distinguish the presynaptic Bassoon and postsynaptic Homer1 clusters. *From* Dani et al., 2010. Please see color plates at the back of the book.

Biological Questions

Neurons are cells of very complex morphology and exhibit many prominent domains in which the interplay between membrane proteins and the cytoskeleton via motor, scaffold, and signaling molecules govern defined biological functions. Despite intense investigation, it is still not clearly understood how the axonal initial segment is assembled, how growth cones steer and regulate neurite outgrowth, how dendritic filopodia develop into spines, and how receptors are turned over at inhibitory and excitatory synapses in the regulation of neuronal plasticity. These domains are prime targets for a technique such as localization microscopy, and the tracing of hundreds of membrane molecules in relation to interacting intracellular molecules will help to elucidate how neuronal physiology is controlled by cues from the cellular environment and signaling.

While the application of RNA interference has lead to the identification of many new factors involved in these processes, what is needed now is a detailed understanding of how these factors are involved in the regulation of the nanoscopic organization of the complex neuronal cytoskeleton. With localization microscopy an ideal tool is now available to take a closer look at these structures by combining the specificity of fluorescence microscopy with molecular resolution. Undoubtedly, soon we will see a number of exciting studies published on these problems.

Conclusion

Localization microscopy techniques provide access to a wealth of information on how molecules interact in multiprotein complexes with unprecedented statistics and resolution. This is especially valuable in the cellular neurosciences, where many cellular functions are executed by highly complex and dynamic multiprotein complexes that contain hundreds of molecules and are thus especially well suited for localization microscopy.

Acknowledgments

I apologize to all colleagues whose work could not be cited due to the overwhelming wealth of literature on this topic or literature that has been missed. This work was supported by the NCCR Neural Plasticity and Repair and the Holcim Foundation for the Advancement of Science.

References

Abbe, E., 1882. The relation of aperture and power in the microscope. J R Microsc Soc 2, 300–309.

Andresen, M., Stiel, A.C., Fölling, J., Wenzel, D., Schönle, A., Egner, A., et al., 2008. Photoswitchable fluorescent proteins enable monochromatic multilabel imaging and dual color fluorescence nanoscopy. Nat Biotechnol 26, 1035–1040.

Annibale, P., Vanni, S., Scarselli, M., Rothlisberger, U., Radenovic, A., 2011. Identification of clustering artifacts in photoactivated localization microscopy. Nat Methods 8, 527–528.

Banala, S., Maurel, D., Manley, S., Johnsson, K., 2011. A caged, localizable rhodamine derivative for superresolution microscopy. ACS Chem Biol 7, 289–293.

Bates, M., Blosser, T.R., Zhuang, X., 2005. Short-range spectroscopic ruler based on a single-molecule optical switch. Phys Rev Lett 94, 108101.

Bates, M., Dempsey, G.T., Chen, K.H., Zhuang, X., 2012. Multicolor super-resolution fluorescence imaging via multi-parameter fluorophore detection. Chemphyschem 13, 99–107.

Bates, M., Huang, B., Zhuang, X., 2008. Super-resolution microscopy by nanoscale localization of photo-switchable fluorescent probes. Curr Opin Chem Biol 12, 505–514.

Belov, V.N., Wurm, C.A., Boyarskiy, V.P., Jakobs, S., Hell, S.W., 2010. Rhodamines NN: a novel class of caged fluorescent dyes. Angew Chem Int Ed Engl 49, 3520–3523.

Betzig, E., Patterson, G.H., Sougrat, R., Lindwasser, O.W., Olenych, S., Bonifacino, J.S., et al., 2006. Imaging intracellular fluorescent proteins at nanometer resolution. Science 313, 1642–1645.

Boiko, T., Vakulenko, M., Ewers, H., Yap, C.C., Norden, C., Winckler, B., 2007. Ankyrin-dependent and -independent mechanisms orchestrate axonal compartmentalization of L1 family members neurofascin and L1/neuron-glia cell adhesion molecule. J Neurosci 27, 590–603.

Borgdorff, A.J., Choquet, D., 2002. Regulation of AMPA receptor lateral movements. Nature 417, 649–653.

Bossi, M., Fölling, J., Belov, V.N., Boyarskiy, V.P., Medda, R., Egner, A., et al., 2008. Multicolor far-field fluorescence nanoscopy through isolated detection of distinct molecular species. Nano Lett 8, 2463–2468.

Brakemann, T., Stiel, A.C., Weber, G., Andresen, M., Testa, I., Grotjohann, T., et al., 2011. A reversibly photoswitchable GFP-like protein with fluorescence excitation decoupled from switching. Nat Biotechnol 29, 942–947.

Chudakov, D.M., Verkhusha, V.V., Staroverov, D.B., Souslova, E.A., Lukyanov, S., Lukyanov, K.A., 2004. Photoswitchable cyan fluorescent protein for protein tracking. Nat Biotechnol 22, 1435–1439.

Dani, A., Huang, B., Bergan, J., Dulac, C., Zhuang, X., 2010. Superresolution imaging of chemical synapses in the brain. Neuron 68, 843–856.

Dempsey, G.T., Bates, M., Kowtoniuk, W.E., Liu, D.R., Tsien, R.Y., Zhuang, X., 2009. Photoswitching mechanism of cyanine dyes. J Am Chem Soc 131, 18192–18193.

Dempsey, G.T., Vaughan, J.C., Chen, K.H., Bates, M., Zhuang, X., 2011. Evaluation of fluorophores for optimal performance in localization-based super-resolution imaging. Nat Methods 8, 1027–1036.

Endesfelder, U., Malkusch, S., Flottmann, B., Mondry, J., Liguzinski, P., Verveer, P.J., Heilemann, M., 2011. Chemically induced photoswitching of fluorescent probes—a general concept for super-resolution microscopy. Molecules 16, 3106–3118.

Fernández-Suárez, M., Ting, A.Y., 2008. Fluorescent probes for super-resolution imaging in living cells. Nat Rev Mol Cell Biol 9, 929–943.

Flors, C., Ravarani, C.N.J., Dryden, D.T.F., 2009. Super-resolution imaging of DNA labelled with intercalating dyes. Chemphyschem 10, 2201–2204.

Fölling, J., Belov, V., Kunetsky, R., Medda, R., Schönle, A., Egner, A., et al., 2007. Photochromic rhodamines provide nanoscopy with optical sectioning. Angew Chem Int Ed Engl 46, 6266–6270.

Fölling, J., Bossi, M., Bock, H., Medda, R., Wurm, C.A., Hein, B., et al., 2008. Fluorescence nanoscopy by ground-state depletion and single-molecule return. Nat Methods 5, 943–945.

Frost, N.A., Shroff, H., Kong, H., Betzig, E., Blanpied, T.A., 2010. Single-molecule discrimination of discrete perisynaptic and distributed sites of actin filament assembly within dendritic spines. Neuron 67, 86–99.

Fuchs, J., Böhme, S., Oswald, F., Hedde, P.N., Krause, M., Wiedenmann, J., et al., 2010. A photoactivatable marker protein for pulse-chase imaging with superresolution. Nat Methods 7, 627–630.

Gautier, A., Juillerat, A., Heinis, C., Corrêa, I.R., Kindermann, M., Beaufils, F., Johnsson, K., 2008. An engineered protein tag for multiprotein labeling in living cells. Chem Biol 15, 128–136.

Giannone, G., Hosy, E., Levet, F., Constals, A., Schulze, K., Sobolevsky, A.I., et al., 2010. Dynamic superresolution imaging of endogenous proteins on living cells at ultra-high density. Biophys J 99, 1303–1310.

Grotjohann, T., Testa, I., Leutenegger, M., Bock, H., Urban, N.T., Lavoie-Cardinal, F., et al., 2011. Diffraction-unlimited all-optical imaging and writing with a photochromic GFP. Nature 478, 204–208.

Gunewardene, M.S., Subach, F.V., Gould, T.J., Penoncello, G.P., Gudheti, M.V., Verkhusha, V.V., et al., 2011. Superresolution imaging of multiple fluorescent proteins with highly overlapping emission spectra in living cells. Biophys J 101, 1522–1528.

Gustafsson, M.G.L., 2005. Nonlinear structured-illumination microscopy: wide-field fluorescence imaging with theoretically unlimited resolution. Proc Natl Acad Sci USA 102, 13081–13086.

Heilemann, M., van de Linde, S., Schüttpelz, M., Kasper, R., Seefeldt, B., Mukherjee, A., et al., 2008. Subdiffraction-resolution fluorescence imaging with conventional fluorescent probes. Angew Chem Int Ed Engl 47, 6172–6176.

Hess, S.T., Girirajan, T.P.K., Mason, M.D., 2006. Ultra-high resolution imaging by fluorescence photoactivation localization microscopy. Biophys J 91, 4258–4272.

Hoi, H., Shaner, N.C., Davidson, M.W., Cairo, C.W., Wang, J., Campbell, R.E., 2010. A monomeric photoconvertible fluorescent protein for imaging of dynamic protein localization. J Mol Biol 401, 776–791.

Huang, B., Babcock, H., Zhuang, X., 2010. Breaking the diffraction barrier: super-resolution imaging of cells. Cell 143, 1047–1058.

Huang, B., Wang, W., Bates, M., Zhuang, X., 2008. Three-dimensional super-resolution imaging by stochastic optical reconstruction microscopy. Science 319, 810–813.

Izeddin, I., Specht, C.G., Lelek, M., Darzacq, X., Triller, A., Zimmer, C., et al., 2011. Super-resolution dynamic imaging of dendritic spines using a low-affinity photoconvertible actin probe. PLoS One 6, e15611.

Ji, N., Shroff, H., Zhong, H., Betzig, E., 2008. Advances in the speed and resolution of light microscopy. Curr Opin Neurobiol 18, 605–616.

Juette, M.F., Gould, T.J., Lessard, M.D., Mlodzianoski, M.J., Nagpure, B.S., Bennett, B.T., et al., 2008. Three-dimensional sub-100 nm resolution fluorescence microscopy of thick samples. Nat Methods 5, 527–529.

Kanchanawong, P., Shtengel, G., Pasapera, A.M., Ramko, E.B., Davidson, M.W., Hess, H.F., Waterman, C.M., 2010. Nanoscale architecture of integrin-based cell adhesions. Nature 468, 580–584.

Klar, T.A., Jakobs, S., Dyba, M., Egner, A., Hell, S.W., 2000. Fluorescence microscopy with diffraction resolution barrier broken by stimulated emission. Proc Natl Acad Sci USA 97, 8206–8210.

Klein, T., Löschberger, A., Proppert, S., Wolter, S., van de Linde, S., Sauer, M., 2011. Live-cell dSTORM with SNAP-tag fusion proteins. Nat Methods 8, 7–9.

Lidke, K., Rieger, B., Jovin, T., Heintzmann, R., 2005. Superresolution by localization of quantum dots using blinking statistics. Opt Express 13, 7052–7062.

Lippincott-Schwartz, J., Manley, S., 2009. Putting super-resolution fluorescence microscopy to work. Nat Methods 6, 21–23.

Lippincott-Schwartz, J., Patterson, G.H., 2009. Photoactivatable fluorescent proteins for diffraction-limited and super-resolution imaging. Trends Cell Biol 19, 555–565.

Manley, S., Gillette, J.M., Patterson, G.H., Shroff, H., Hess, H.F., Betzig, E., Lippincott-Schwartz, J., 2008. High-density mapping of single-molecule trajectories with photoactivated localization microscopy. Nat Methods 5, 155–157.

Manley, S., Gunzenhäuser, J., Olivier, N., 2011. A starter kit for point-localization super-resolution imaging. Curr Opin Chem Biol 15, 813–821.

McKinney, S.A., Murphy, C.S., Hazelwood, K.L., Davidson, M.W., Looger, L.L., 2009. A bright and photostable photoconvertible fluorescent protein. Nat Methods 6, 131–133.

Nakada, C., Ritchie, K., Oba, Y., Nakamura, M., Hotta, Y., Iino, R., Kasai, R.S., Yamaguchi, K., Fujiwara, T., Kusumi, A., 2003. Accumulation of anchored proteins forms membrane diffusion barriers during neuronal polarization. Nat Cell Biol 5, 626–632.

Newpher, T.M., Ehlers, M.D., 2008. Glutamate receptor dynamics in dendritic microdomains. Neuron 58, 472–497.

Opazo, P., Labrecque, S., Tigaret, C.M., Frouin, A., Wiseman, P.W., De Koninck, P., Choquet, D., 2010. CaMKII triggers the diffusional trapping of surface AMPARs through phosphorylation of stargazin. Neuron 67, 239–252.

Owen, D.M., Rentero, C., Rossy, J., Magenau, A., Williamson, D., Rodriguez, M., Gaus, K., 2010. PALM imaging and cluster analysis of protein heterogeneity at the cell surface. J Biophotonics 3, 446–454.

Patterson, G., Davidson, M., Manley, S., Lippincott-Schwartz, J., 2010. Superresolution imaging using single-molecule localization. Annu Rev Phys Chem 61, 345–367.

Pinaud, F., Dahan, M., 2011. PNAS Plus: Targeting and imaging single biomolecules in living cells by complementation-activated light microscopy with split-fluorescent proteins. Proc Natl Acad Sci USA 108, E201–210.

Ries, J., Kaplan, C., Platonova, E., Eghlidi, H., Ewers, H., 2012. A simple, versatile method for GFP-based superresolution microscopy via nanobodies. Nat Methods 10.1038/NMETH.1991.

Rust, M.J., Bates, M., Zhuang, X., 2006. Sub-diffraction-limit imaging by stochastic optical reconstruction microscopy (STORM). Nat Methods 3, 793–795.

Schermelleh, L., Heintzmann, R., Leonhardt, H., 2010. A guide to super-resolution fluorescence microscopy. J Cell Biol 190, 165–175.

Schoen, I., Ries, J., Klotzsch, E., Ewers, H., Vogel, V., 2011. Binding-activated localization microscopy of DNA structures. Nano Lett 11, 4008–4011.

Sengupta, P., Jovanovic-Talisman, T., Skoko, D., Renz, M., Veatch, S.L., Lippincott-Schwartz, J., 2011. Probing protein heterogeneity in the plasma membrane using PALM and pair correlation analysis. Nat Methods 8, 969–975.

Sharonov, A., Hochstrasser, R.M., 2006. Wide-field subdiffraction imaging by accumulated binding of diffusing probes. Proc Natl Acad Sci USA 103, 18911–18916.

Shroff, H., Galbraith, C.G., Galbraith, J.A., White, H., Gillette, J., Olenych, S., Davidson, M.W., Betzig, E., 2007. Dual-color superresolution imaging of genetically expressed probes within individual adhesion complexes. Proc Natl Acad Sci USA 104, 20308–20313.

Shroff, H., White, H., Betzig, E., 2008. Photoactivated localization microscopy (PALM) of adhesion complexes. Curr Protoc Cell Biol (Chapter 4, Unit 4.21).

Shtengel, G., Galbraith, J.A., Galbraith, C.G., Lippincott-Schwartz, J., Gillette, J.M., Manley, S., et al., 2009. Interferometric fluorescent super-resolution microscopy resolves 3D cellular ultrastructure. Proc Natl Acad Sci USA 106, 3125–3130.

Subach, F.V., Patterson, G.H., Manley, S., Gillette, J.M., Lippincott-Schwartz, J., Verkhusha, V.V., 2009. Photoactivatable mCherry for high-resolution two-color fluorescence microscopy. Nat Methods 6, 153–159.

Subach, F.V., Patterson, G.H., Renz, M., Lippincott-Schwartz, J., Verkhusha, V.V., 2010. Bright monomeric photoactivatable red fluorescent protein for two-color super-resolution sptPALM of live cells. J Am Chem Soc 132, 6481–6491.

Subach, O.M., Patterson, G.H., Ting, L.-M., Wang, Y., Condeelis, J.S., Verkhusha, V.V., 2011. A photoswitchable orange-to-far-red fluorescent protein, PSmOrange. Nat Methods 8, 771–777.

Tatavarty, V., Kim, E.-J., Rodionov, V., Yu, J., 2009. Investigating sub-spine actin dynamics in rat hippocampal neurons with super-resolution optical imaging. PLoS one 4, e7724.

Testa, I., Wurm, C.A., Medda, R., Rothermel, E., Middendorf von, C., Fölling, J., et al., 2010. Multicolor fluorescence nanoscopy in fixed and living cells by exciting conventional fluorophores with a single wavelength. Biophys J 99, 2686–2694.

Triller, A., Choquet, D., 2008. New concepts in synaptic biology derived from single-molecule imaging. Neuron 59, 359–374.

van de Linde, S., Krstić, I., Prisner, T., Doose, S., Heilemann, M., Sauer, M., 2011a. Photoinduced formation of reversible dye radicals and their impact on super-resolution imaging. Photochem Photobiol Sci 10, 499–506.

van de Linde, S., Löschberger, A., Klein, T., Heidbreder, M., Wolter, S., Heilemann, M., Sauer, M., 2011b. Direct stochastic optical reconstruction microscopy with standard fluorescent probes. Nat Protoc 6, 991–1009.

Vogelsang, J., Kasper, R., Steinhauer, C., Person, B., Heilemann, M., Sauer, M., Tinnefeld, P., 2008. A reducing and oxidizing system minimizes photobleaching and blinking of fluorescent dyes. Angew Chem Int Ed Engl 47, 5465–5469.

Winckler, B., Forscher, P., Mellman, I., 1999. A diffusion barrier maintains distribution of membrane proteins in polarized neurons. Nature 397, 698–701.

Xu, K., Babcock, H.P., Zhuang, X., 2012. Dual-objective STORM reveals three-dimensional filament organization in the actin cytoskeleton. Nat Methods 9, 185–188.

5

OPTICAL INVESTIGATION OF BRAIN NETWORKS USING STRUCTURED ILLUMINATION

Marco Dal Maschio, Francesco Difato, Riccardo Beltramo, Angela Michela De Stasi, Axel Blau, and Tommaso Fellin
Department of Neuroscience and Brain Technologies, Istituto Italiano di Tecnologia, Genova, Italy

CHAPTER OUTLINE
Introduction   101
Structuring Light by Phase Modulation Using SLMs   103
Wavefront Engineering Using SLMs: The Optical Setup   104
    Computational Aspects of Phase Modulation Using SLMs   104
    Phase Modulation and Temporal Focusing   107
    Phase Modulation and Generalized Phase Contrast   108
    Coupling the SLM with Scanning Systems: Extending
        the Optical Performance of the Microscope   109
    Phase Modulation and Optical Resolution   110
Light-sensitive Molecular Tools for the Investigation
    of the Central Nervous System   111
SLM-based Approaches for the Optical Dissection
    of Brain Microcircuits   112
    Photostimulation of Neuronal Cells with Complex Light Patterns   114
    Activation of Light-sensitive Proteins Expressed Over Large Areas   114
    Illuminating Neurons with 3D Light Patterns   115
Conclusions   116
Acknowledgments   116
References   116

Introduction

The term structured, tailored, or patterned light usually refers to the ability of illuminating an object with complex patterns of light. It is thereby just one aspect of a more general concept, the spatiotemporal engineering of wavefronts and of the spectral characteristics of

Cellular Imaging Techniques for Neuroscience and Beyond.
DOI: http://dx.doi.org/10.1016/B978-0-12-385872-6.00005-2

electromagnetic fields (Mias and Camon, 2008; Weiner, 2011). This technique serves many different purposes, including light and video projection in ubiquitous entertainment devices and 3D reconstruction of macroscopic surface features (Gong and Zhang, 2010). In life sciences, one example is structured illumination microscopy (SIM). The primary goal in SIM is to enhance spatial resolution beyond the diffraction limit. While different SIM configurations exploit different physical phenomena (e.g., the photophysics of the fluorophores), their common working principle is the confinement of light to generate predefined static patterns. Another important application of structured light is adaptive optics (AO). Spatial inhomogeneity in the refractive index, which is typical in biological samples, generates distortions in the light wavefront, also known as aberrations. Aberrations lead to a reduced signal-to-noise ratio (SNR), introduce distortions in the image, and limit the imaging depth in thick biological samples, such as the brain (due to decreased excitation efficiency). These sample-associated aberrations and artifacts can be eliminated using AO by applying dynamic optical elements to introduce corrections to the wavefront that compensate for the aberrations. Other applications of structured light modulation include, but are not limited to, optical traps (Dholakia and Cizmar, 2011) for noninvasive object manipulation and pico-Newton force spectroscopy (Cojoc et al., 2007; Difato et al., 2011b), laser ablation or surgery (Jayasinghe et al., 2011; Difato et al., 2011b), actuation of photoswitchable molecules (Lutz et al., 2008; Nikolenko et al., 2008) or proteins (Papagiakoumou et al., 2010; Stirman et al., 2011), activation of photosensitive polymers (Ghezzi et al., 2011), and functional fluorescence microscopy (Nikolenko et al., 2008; Dal Maschio et al., 2011), which is discussed in more detail in the section "SLM-based Approaches for the Optical Dissection of Brain Microcircuits" in this chapter.

There are two principal strategies for generating structured light. The first is based on designing a structured light source. Any type of device with individually addressable emitters, such as monochromatic LED arrays, can generate arbitrary optical excitation patterns on a sample with micrometer and sub-millisecond resolutions (Choi et al., 2004; Poher et al., 2007; Grossman et al., 2010). The second strategy shapes light into a desired profile by distributing it using various interference schemes or optical elements, such as fibers, lenses, masks, grids, filters, prisms, and mirrors, or a combination thereof. These elements can be passive and static or actively adjustable, such as irises, shutters, diaphragms, and liquid (Berge and Peseux, 2000; Tsai et al., 2008) or polymer (Beadie et al., 2008) lenses. Lenses, masks, and mirrors can also be arranged in matrices to generate a parallel, yet spatially partitioned, light distribution. Two-dimensional pattern devices for the generation of arbitrary and parallel permissive or reflective light paths include liquid crystal or electrophoretic displays, microlens arrays (Ren et al., 2004), micro electromechanical system (MEMS)-based (Miles, 1999) or liquid (Brown et al., 2009) interferometric modulator displays, single- or multi-axis digital micromirror devices (Boysel, 1991; Monk and Gale, 1995; Jerome et al., 2011; Liang et al., 2011), or cantilever-based thin-film micromirror arrays (Kim et al., 1999). However, only a few technologies allow for control over all spatiotemporal degrees of freedom for positioning an arbitrary number of independent light spots in 3D space at any given time. These 3D light shapers include vertically and analogically actuated, segmented or nonsegmented deformable mirrors (Bortolozzo et al., 2010; Bifano, 2011; Bonora, 2011) and a variety of liquid crystal on silicon spatial light modulators (SLMs; Efron, 1994; Maurer et al., 2011; Venediktov et al., 2011). Depending on their embodiments, SLMs can be addressed electrically or optically (Moddel et al., 1989; Stanley et al., 2000; Mathur et al., 2009) in reflection (Sanford et al., 1998) or transmission mode (Nikolenko et al., 2010). The various underlying design principles are compared and discussed in detail by Hornbeck (1998) and Collings et al. (2011).

In the remaining sections, we will focus on SLM-based technology in reflection mode. First, we will briefly summarize the basic principles of SLM operation and then provide a detailed description of the different hardware configurations in which SLMs can be integrated. Finally, we will present an overview of results, recently obtained using this approach, of imaging and photostimulation experiments in neuroscience.

Structuring Light by Phase Modulation Using SLMs

The spatial distribution of light illuminating a sample is the result of the propagation of a radiation wavefront through a compound system formed by the objective lens, immersion medium, and sample. By modulating the phase of light, it is possible to introduce a change into the propagating wavefront so that the light distribution at the sample plane can be controlled and optimized. According to the Huygens-Fresnel diffraction theory (Goldman, 2005), the objective lens performs an optical transformation of the light wavefront that can be mathematically represented as a Fourier integral. This operation transforms the initial complex field distribution, which is expressed at the back focal plane of the objective in terms of spatial frequency components, into a corresponding spatial distribution of light intensities at the sample plane. Thus, the complex spatial field distribution at the focal plane can be defined by a map of phase delays applied to the corresponding spatial frequency distribution in the back focal plane. These phase maps are generally referred to as diffractive optical elements (DOEs). SLMs allow the generation of different DOEs at the back focal plane of the objective through a dynamically programmable matrix of active pixels (Efron, 1994). Each pixel contains liquid crystals that are controlled by an electrical bias voltage. By changing the bias voltage, the orientation of the liquid crystal molecule with respect to the propagation direction of light is modified (Khoo, 2007). The birefringent properties of the liquid crystals are used to control the effective refraction index and, as a consequence, the phase delay that is experienced by light as it travels through the SLM (Vicari, 2003).

Wavefront Engineering Using SLMs: The Optical Setup

The simplest optical design for projecting complex patterns of light onto the sample plane with an SLM resembles a classical 4f configuration (Lee et al., 2007; Lutz et al., 2008; Golan et al., 2009) and is shown in Figure 5.1A. The setup includes an SLM, continuous-wave or pulsed laser source (S), intensity modulation unit (e.g., Pockels cell, P), and some coupling optics. The SLM is placed in a plane optically conjugated to the back focal plane of the objective lens. Two telescopes (L_1, L_2 and L_3, L_4, respectively) are used to couple the SLM to the laser source and to the objective (OBJ). L_1 and L_2 expand the laser beam to match the dimensions of the active window of the SLM, typically in the range of $1-2\times1-2\,cm^2$. The second telescope (L_3 and L_4) scales the beam diameter to fit the dimensions of the back aperture of the microscope objective. Given that different objectives have distinct pupil apertures, the choice of L_3 and L_4 depends on the type of objective used (Difato et al., 2011a). A field stop can be inserted between L_3 and L_4 to remove the zero-order nondiffracted light (Dal Maschio et al., 2010). It is important to remember that the SLM is sensitive to the polarization of the incident light (Dal Maschio et al., 2010) and acts as a phase-only modulator for light that is linearly polarized in the direction corresponding to the liquid crystal orientation. Thus, it is necessary to adjust a half-wave plate ($\lambda/2$ in Figure 5.1A), positioned in front of the SLM, to match the polarization of the incident beam with the working orientation of the SLM. A detailed step-by-step implementation of an SLM-based microscope is described in (Difato et al., 2011a).

Computational Aspects of Phase Modulation Using SLMs

A key aspect of SLMs is the identification of proper phase maps (DOEs) that result in the desired intensity distributions at the sample.

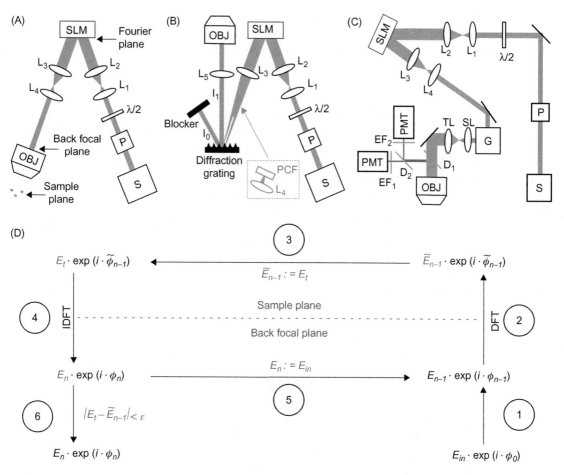

Figure 5.1 Optical configuration for wavefront engineering using SLMs. (A) A basic system consists of a laser source (S), Pockels cell (P), half-wave plate (λ/2), two telescopes (L₁, L₂ and L₃, L₄), spatial light modulator (SLM), and objective (OBJ). (B) To achieve temporal focusing, a diffraction grating is inserted between lenses L₃ and L₅ and a beam blocker is used to suppress the zero-order. By adding a phase distribution resembling that of a prism to the DOE, the first and zero diffracted orders can be spatially separated and further dispersed in different directions by the grating. The zero-order can then be eliminated by using an optical blocker, which results in a total power loss of approximately half of the initial power, while the first diffracted order is directed into the objective. To obtain generalized phase contrast, a PCF and a lens (L₄) are positioned along the optical path (blue inset, see also Papagiakoumou et al., 2010). (C) To perform simultaneous scanning imaging and inertia-free z-focusing, a more complex experimental setup is needed. G, galvanometric mirrors; SL, scan lens; TL, tube lens; D1, 660 nm long-pass dichroic mirror; D₂, 575 nm long-pass dichroic mirror; EF₁ and EF₂, emission filters; PMTs, photomultiplier tubes. Modified from Dal Maschio et al., 2011. (D) Schematic representation of the procedural steps used to iteratively generate DOEs. E_in, input intensity; E_t, target intensity; Φ₀, initial estimate for the phase map. Please see color plates at the back of the book.

To calculate these phase maps, several mathematical transformation algorithms, primarily iterative Fourier transform algorithms (Kim et al., 2004), have been designed on the basis of generalized discrete Fourier transforms (DFTs). These algorithms optimize the spatial light intensity distribution in terms of illumination efficiency and uniformity among different illuminated subregions. The simplest and computationally fastest approach is the so-called "gratings and lenses" algorithm (Leach et al., 2006). This algorithm is based on analytical expressions of optical transformations generated by basic elements, such as prisms and lenses. Here, a lens imposes a radial distribution of phase delays onto an incoming plane wave, modifying the convergence/divergence properties of its wavefront. Similarly, a prism steers the wavefront direction as a consequence of a linear phase delay gradient. In the case of a light spot placed at an arbitrary position (x_s, y_s, z_s) at the sample plane, the corresponding mathematical representation of these transformations returns a matrix of phase delays to be applied to the pixel matrix $(x_i, y_j; 1 < i < N, 1 < j < M)$ of the SLM. This basic approach can be extended to generate any arbitrary distribution of light spots in 3D by superimposing the phase matrices from separate holograms. However, this approach does not work for the generation of illumination patterns with extended regions of interest (ROIs) and, in general, suffers from poor uniformity across the generated pattern. To illuminate extended areas, different implementations of iterative algorithms have been adopted (Sinclair et al., 2004; Kuzmenko, 2008; Engstrom et al., 2009). Most of these algorithms are based on modifications of the procedure originally designed by Ralph Gerchberg and Owen Saxton (1971). Here, the goal is to iteratively converge onto a specific phase distribution that transforms an incoming wavefront, with some assumed profile, into the desired intensity distribution at the plane of interest in the sample. All of these algorithms share the following set of basic procedural steps (Leach et al., 2006; Lutz et al., 2008). First, a field, characterized by an input intensity distribution and a random phase (step #1, Figure 5.1D), is propagated from the back focal plane to the sample plane by means of a DFT (step #2, Figure 5.1D). The phase information of the resulting field at the sample plane is preserved while the intensity distribution is substituted with the desired one (step #3, Figure 5.1D). This modified complex field is then backpropagated to the back focal plane by an inverse DFT (step #4, Figure 5.1D), where the resulting field is modified again by retaining the phase information and substituting the intensity distribution with the initial input intensity distribution (step #5, Figure 5.1D). After a few iterations, the algorithm converges to the desired phase distribution (step #6, Figure 5.1D). At the sample plane, these phase distributions are characterized by spatial light artifacts, which lead to intensity distortions in the projected pattern (Palima and Gluckstad, 2008; Golan and Shoham, 2009). These distortions result from the algorithm substituting amplitude information, which is obtained from the propagation procedure, with given distributions at the SLM and sample, respectively. Specific algorithms (called "input-output" algorithms) have been developed to address this issue (Fienup, 1982; Georgiou et al., 2008). During each iteration, the difference between the amplitudes computed at the sample plane and target intensities is evaluated. A small feedback correction is then applied to the target intensity pattern to reduce noise contributions. If the image plane is limited to specific subregions of interest by a proper formulation of the feedback, this procedure results in an improved SNR in the projected pattern (Kim et al., 2004).

Phase Modulation and Temporal Focusing

The generation of extended 2D areas of illumination introduces unwanted effects on the optical features of the projected pattern. One of the most evident is the quasilinear dependency of the axial extension on the lateral dimension of the excitation profile (Lutz et al., 2008). For example, a shape of lateral dimension around 20 μm, reaches an axial extent of approximately 35 μm (Lutz et al., 2008). If a pulsed laser is used, this limitation can be overcome by the use of "temporal focusing." Using this technique, an effective axial confinement is achieved by temporally focusing spatially separated spectral components of a light pulse (Oron et al., 2005; Papagiakoumou et al., 2008, 2009). This spectral separation is generally obtained by projecting a laser beam onto a blazed grating (Figure 5.1B). The dispersed spectral components are collimated by a lens and focused by the objective at the sample. Here, the spectral components temporally overlap to produce a short pulse only at the focal plane and are temporally dispersed in the out-of-focus regions. This technique was originally introduced by Yaron Silberberg (Oron et al., 2005) as a method of performing scan-less two-photon imaging while preserving z-sectioning efficiency. The technique was then adopted for optimizing structured illumination as a method for reducing the axial extent of 2D illumination patterns (Papagiakoumou et al., 2008, 2009). With respect to the basic design described in the previous section (Figure 5.1A), the implementation of the temporal focusing requires the introduction of a blazed diffraction grating, which is placed in a plane optically conjugated to the sample plane (Figure 5.1B). Using temporal focusing, 2D extended patterns with a maximum axial extent of about 5 μm can be achieved, which is several times smaller than the value that can be obtained without temporal focusing (Papagiakoumou et al., 2008).

Phase Modulation and Generalized Phase Contrast

Aside from the aforementioned axial extent, another important aspect to consider when generating extended 2D coherent illumination patterns is the appearance of significant intensity fluctuations across the illuminated area, which is a phenomenon known as speckles. These intensity variations in the spatial pattern are mainly a result of cross talk between adjacent points of the band-limited frequency plane at the SLM, where large phase modulations are superimposed. These contributions sum up randomly at the sample plane, resulting in constructive/destructive interference (Palima and Gluckstad, 2008; Golan and Shoham, 2009). The discrete levels of phase delays and spatial frequency quantization due to the pixel size of the SLM also contribute to the generation of speckles. For many applications, speckles are an unwanted side effect, and different methods have been developed to minimize their visibility, including: (1) development of ad hoc algorithms (Golan and Shoham, 2009), (2) insertion of diffusing systems in the optical path (Papagiakoumou et al., 2009; Zahid et al., 2010), and (3) use of generalized phase contrast (GPC; Papagiakoumou et al., 2010). GPC is an extension of the phase contrast method originally introduced by Frederik Zernike for the visualization of transparent samples. It is based on the conversion of weak phase perturbations into more consistent and detectable intensity variations. The hardware implementation resembles that of a common path interferometer and differs from that of the basic holographic scheme shown in Figure 5.1A. For GPC, the SLM plane is optically conjugated to the sample plane through a telescope, and a phase contrast filter (PCF; Figure 5.1B, blue insert) is placed in a plane optically conjugated to the back focal plane of the objective (Papagiakoumou et al., 2010). This configuration implies that no calculation of Fourier transforms is needed, and the DOE applied to the SLM is simply the intensity pattern desired at the sample plane (in binary format). The PCF is designed with a central region that has a constant phase difference with respect to the surrounding area. Therefore, the PCF operates on the propagated beam as a spatial filter, introducing phase delays between the diffracted off-axis and nondiffracted on-axis components of the beam. The phase-shifted, nondiffracted on-axis component is transformed by the objective lens in a synthetic reference wave that interferes with the diffracted component at the sample plane (Gluckstad and Palima, 2010). In this way, the binary phase pattern imposed by the SLM is reproduced at the sample plane as a binary intensity pattern. The advantages of this configuration include (1) increased illumination uniformity due to spatial interference, (2) no need for complex algorithm development and reduced computation times, and (3) improved light efficiency.

Coupling the SLM with Scanning Systems: Extending the Optical Performance of the Microscope

The SLM can be inserted in the light path of a laser scanning microscope to improve the optical performance of the imaging system. For example, we recently combined a commercial two-photon scanning microscope (from Prairie Technologies) with an optical module containing an SLM (Dal Maschio et al., 2011). To couple the two systems together, before impinging on the galvanometric mirrors (Figure 5.1C, G), the laser beam must first be expanded using a telescope (Figure 5.1C, L_1 and L_2) and deflected onto the SLM, which is positioned in a plane optically conjugated to the galvano-

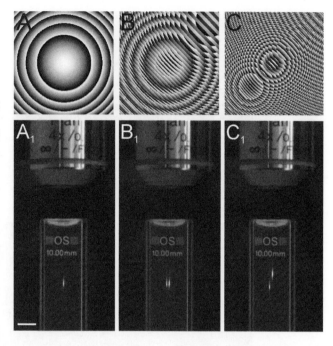

Figure 5.2 Generating 3D illumination patterns using SLMs. (A, A_1) The image in A_1 shows a cuvette of fluorescein excited by a two-photon ($\lambda = 800$ nm) illumination spot generated by the phase hologram shown in (A). Scale bar, 5 mm. B, B_1 and C, C_1 show the same as in A, A_1 for illumination patterns generating two spots in the same plane (B, B_1) and two simultaneous spots at different axial positions (C, C_1)

metric mirrors. A second telescope (Figure 5.1C, L_3 and L_4) is used to match the beam size to the dimensions of the galvanometric mirrors. In this configuration, the SLM was used to shift the axial position of the imaging spot without having to move the objective (Dal Maschio et al., 2011). By using DOEs, which radially distribute phase delays, an upward or downward offset was imposed onto the laser spot, shifting the image plane in the z-direction (Figure 5.2C, C_1). We demonstrated that, depending on the numerical aperture of the objective used, this approach achieves z-focusing in the range of tens to hundreds of microns with minimal variation in the point spread function (PSF; Dal Maschio et al., 2011). This system was used to perform inertia-free 3D imaging *in vivo* (Figure 5.3; see also below). Using DOEs that steer the laser spot in the x-y plane, the same optical configuration may be used to extend the lateral dimensions of the raster-scanned field of view, which is usually a subregion of the field of view of the objective (Hanes et al., 2009). Alternatively, the galvanometric mirrors can be used to shift the region where structured light can be projected in the x-y plane.

Aside from providing shifts in the x, y, and z directions, SLMs can be used to correct optical aberrations in scanning fluorescence imaging systems (Ji et al., 2008, 2010; see AO in the Introduction). This is particularly relevant for *in vivo* applications where the aberrations introduced by the objective lens sum up with those introduced by the

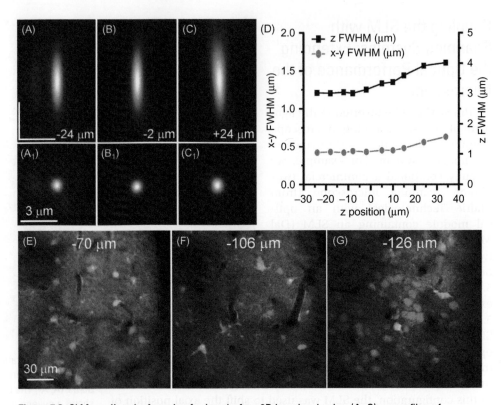

Figure 5.3 SLM-mediated z-focusing for inertia-free 3D imaging *in vivo*. (A–C) x-z profiles of illuminated fluorescent beads, 170 nm in diameter, obtained by moving the sample in the z-direction with a piezoelectric translator. Beads were dispersed in agarose and coverslipped. Profiles are obtained at different refocusing axial positions (+24 μm, −2 μm, and −24 μm) operated by the SLM control (see values in microns on the image). The corresponding x-y profiles are shown in A_1–C_1. Scale bars, 3 μm. (D) Plot of the x-y FWHM (gray circles) and z FWHM (black squares) as a function of the axial position of the laser beam. (E–G) Images of neocortical cells loaded with Oregon Green BAPTA and sulforhodamine in anesthetized mice at different z-positions (see values in microns on the images) obtained with the inertia-free SLM-based focus control. Modified with permission from Dal Maschio et al., 2011. Please see color plates at the back of the book.

highly scattering brain tissue (Rueckel et al., 2006). The main challenge in this approach is to infer wavefront changes of the laser beam as it travels through the sample. Direct measurements of the laser wavefront (Neil et al., 2000; Rueckel et al., 2006), as well as computational methods to optimize the fluorescence signal (Albert et al., 2000; Sherman et al., 2002; Debarre et al., 2009), have been demonstrated.

Phase Modulation and Optical Resolution

In terms of diffraction efficiency and light confinement, the optical performances of an SLM-based microscope may be different from those of a conventional optical microscope and may depend on the particular illumination profile adopted. The discretization of the Fourier plane due to the finite pixel size of the SLM array and the fact that distinct patterns may have different spatial frequency content are two of the prominent reasons for the reduction in lateral diffraction efficiency. This reduction leads to nonuniform lateral intensity distribution between points at the center and at the edges of the field of view (Yang et al., 2011). The finite pixel size of the SLM and the difference in spatial frequency content also affect the diffraction efficiency variation along the propagation direction. When out-of-focus illumination patterns are generated (Figure 5.2), relatively strong phase variations are projected onto the external ring of the pupil lens. These components experience strong aberrations resulting in decreased performance. Thus, when the SLM projects light to planes out of the objective focus, there is progressive loss in the axial and lateral resolution due to the reduced effective numerical aperture (NA). Considering a $60 \times$ WI 0.9 NA objective and light at 900 nm, the axial resolution varies from 3.02 to 4.02 µm (full-width at half-maximum, FWHM, values) while the lateral resolution varies from 0.42 to 0.63 µm within a defocus range of 60–70 µm (Dal Maschio et al., 2011). Using the same objective with light at 405 nm, an excitation volume $80 \times 80 \times 100$ µm^3 can be addressed with light efficiency greater than 50% and with resolution losses within 66%. In this latter case, zFWHM varies from 1.2 to $\sim$2 µm and the x-y FWHMs vary from 0.3 to $\sim$0.4 µm (Yang et al., 2011).

Light-sensitive Molecular Tools for the Investigation of the Central Nervous System

The mammalian brain is an extremely complex structure, containing billions of cells that form highly interconnected networks. Each neuron within this network displays an elaborate 3D structure characterized by prolonged and thin processes (dendrites and axons), which can span distances from hundreds of microns to several centimeters. Optical approaches have traditionally been fundamental in unraveling this complex 3D cellular network. Bright field microscopy of Golgi-stained *fixed* samples facilitated the understanding of the fine details of cell morphology (Golgi, 1885). More recently, the introduction of fluorescence markers (Heim et al., 1994; Heim and Tsien, 1996) and indicators (Miyawaki et al., 1997; Tsien, 1980, 1981), together with the use of fluorescence microscopy, has permitted thorough optical investigation of the structure and function of brain cells in *living* tissue. For example, the development of genetically modified fluorescent proteins, which can be expressed with cellular and subcellular specificity, has allowed monitoring of cellular dynamics during migration, differentiation, and so forth (Heim and Tsien, 1996). Similarly, electrical activity and biochemical signals in both cultured brain cells and intact brain tissue can be traced with unprecedented resolution by using fluorescent indicators that modify their fluorescent properties based on changes in voltage or concentrations of specific ions (Tsien, 1980, 1981; Tsutsui et al., 2001; Knopfel et al., 2003; Kerr et al., 2005; Garaschuk et al., 2006; Kerr and Denk, 2008). Parallel to the development of these fluorescent dyes, a class of molecules called caged compounds, has also been introduced (Adams and Tsien, 1993; Ellis-Davies, 2007). Caged compounds are formed using a core, a biologically relevant molecule (e.g., glutamate or GABA, which are two major neurotransmitters), which is rendered physiologically inactive by a photolabile link to a chemical group (the "cage"). Following the delivery of a light flash of an appropriate wavelength and intensity, the bond to the cage group is broken and the active molecule is released.

The broad applications of these light-sensitive molecular tools have created a need for new and advanced optical methodologies. In this context, we believe that structured light using phase modulation with SLMs is an extremely promising technique. In the following section, we will present an overview of recent studies that have applied SLM technology for the investigation of the central nervous system and focus on imaging and photostimulation applications.

SLM-based Approaches for the Optical Dissection of Brain Microcircuits

When combined with fluorescent indicators and caged compounds, SLMs represent a unique tool for investigating brain circuits. They have several advantages over more traditional approaches (e.g., wide field and sequential scanning microscopy) for a number of reasons (Watson et al., 2010). SLMs permit simultaneous light delivery to multiple ROIs in the field of view with high spatiotemporal control. This gives unprecedented precision in illuminating cellular networks, in particular when combined with two-photon excitation. Several applications of SLM-mediated illumination in combination with the use of light-sensitive molecules have been demonstrated. First, SLMs can be used for uncaging purposes (Lutz et al., 2008; Nikolenko et al., 2008; Dal Maschio et al., 2010). Diffraction limited spots as well as complex ROIs within the cellular processes of a neuron can be simultaneously illuminated to release caged glutamate, thus mimicking the complex patterns of synaptic inputs that neurons experience under physiological conditions

(Lutz et al., 2008; Yang et al., 2011). Second, when combined with fluorescence indicators and camera imaging (Nikolenko et al., 2008; Dal Maschio et al., 2010), SLM-mediated parallel illumination at different positions in the field of view allows faster functional imaging compared to sequential scanning approaches (Figure 5.4). Recently, the strengths of these two different applications were combined in a dual holographic optical setup for combined two-photon imaging and uncaging with structured light illumination (Figure 5.4) (Dal Maschio et al., 2010). Third, the SLM can be used to correct the aberrations that affect the excitation beam traveling through the highly scattering biological tissue. Using pupil segmentation to measure image shifts between sequentially and complementary illuminated subregions, correction of large aberrations and significant increases in image quality were demonstrated (Ji et al., 2008, 2010). Finally, when positioned in the fluorescence emission light path, SLMs have been used to capture high-resolution, nonscanning 3D images of biological samples without having to move the objective (Rosen and Brooker, 2008).

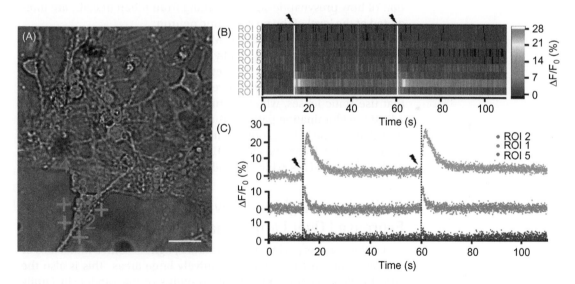

Figure 5.4 Fast holographic fluorescence imaging combined with galvo-steered uncaging. (A) Based on the image, ROIs corresponding to different cells in culture are identified (green dots numbered 1 to 9). DOEs enabling fluorescence imaging of calcium indicators over time in only those ROIs are then projected onto the SLM. Scale bar, 20 μm. (B) Values of $\Delta F/F_0$ for Fluo-4 fluorescence are shown as a function of time for the nine regions identified in A. The arrows indicate the time of delivery of the four photolysis stimuli (red crosses in A) to uncage the MNI-glutamate present in the bathing medium. (C) Time course of the fluorescence signal in ROIs 1, 2, and 5. ROI 1 and 2, but not 5, display a clear response to MNI-glutamate uncaging. Modified with permission from Dal Maschio et al., 2010. Please see color plates at the back of the book.

Photostimulation of Neuronal Cells with Complex Light Patterns

One fundamental advantage of SLMs is that the pattern of illumination can be precisely tailored to the structure of the cell or network under investigation. To this aim, the SLM has to be combined with high-resolution imaging techniques, such as high-frequency/low-frequency sequential acquisition microscopy (Zahid et al., 2010) or two-photon scanning microscopy (Dal Maschio et al., 2010). Using these approaches, different research groups were able to demonstrate simultaneous photorelease of caged glutamate at specific locations along cellular processes to study the integration of synaptic inputs at the dendritic tree (Lutz et al., 2008; Nikolenko et al., 2008). This approach can also potentially be useful for suprathreshold, cell type-specific stimulation in connectivity studies. For example, using a newly synthesized form of caged glutamate (RuBi-glutamate) and scanning two-photon excitation, the connectivity between a particular subtype of interneurons and glutamatergic neurons could be mapped (Fino and Yuste, 2011). In this context, SLM technology, with its ability to excite different cells in parallel, will allow the investigation of how presynaptic signals, coming from different cells, are integrated at the level of the postsynaptic neuron.

Moreover, by coupling the SLM to an imaging system, the illumination pattern can be dynamically adjusted according to the changes in experimental conditions (Lutz et al., 2008; Dal Maschio et al., 2010). This provides more flexibility over structured illumination systems that use static lenses, which require changes to the optical setup to adjust the illumination pattern. The rate of change of light projections in systems using SLMs is only limited by the relaxation time of the liquid crystals (Dal Maschio et al., 2011), and the refresh rate of commercially available SLMs, which typically is on the order of 60–100 Hz.

Activation of Light-sensitive Proteins Expressed Over Large Areas

Under many experimental conditions, light-sensitive molecules or indicators are distributed over relatively large areas. This is also the case for light-sensitive membrane proteins of the rhodopsin family (e.g., channelrhodopsins and halorhodopsin). Channelrhodopsin-2 (ChR2) and halorhodopsin (HR) are membrane proteins that mediate the transfer of ions across the cell membrane upon light absorption, thus leading to excitation (ChR2) or inhibition (HR) of the illuminated neuron (Zhang et al., 2007). Networks of rhodopsin-expressing neurons are generally stimulated using wide field illumination techniques (Zhang et al., 2010). By illuminating the whole field of view simultaneously, this approach precludes any cell specificity and leads to synchronous activation/inhibition of all rhodopsin-expressing cells. This simultaneous activation might not reflect the true physiological pattern of cellular activation. An alternative approach to overcome this issue is sequential scanning of a diffraction limited spot by using galvanometric mirrors or acousto-optic devices. However, this method results in suboptimal activation of rhodopsins. The diffraction-limited spot only illuminates a restricted number of molecules, and sequential scanning of the beam leads to a poor temporal summation of single activation/inhibition events. To achieve a sufficiently large depolarizing/hyperpolarizing effect, a large number of rhodopsins need to be activated simultaneously, which typically requires illuminating areas on the order of tens of square micrometers (Andrasfalvy et al., 2010). SLMs provide an ideal solution to this problem because they can simultaneously deliver light to independent areas of arbitrary shape distributed over a large region. This allows the activation of numerous receptors, while maintaining cell-type specificity within a network of different rhodopsin-expressing cells. The application of SLMs and two-photon excitation for ChR2 activation has been successfully demonstrated (Papagiakoumou et al., 2010).

Illuminating Neurons with 3D Light Patterns

The SLM permits the generation of complex illumination patterns in 2D and 3D (Emiliani et al., 2005; Figure 5.2C, C_1). The group of Valentina Emiliani has recently proven this approach feasible for the activation of the dendritic tree of hippocampal neurons in slice preparations for studying the integration properties of neuronal dendrites in 3D (Yang et al., 2011). We have taken advantage of the ability of SLMs to shape the light in the axial direction to perform inertia-free 3D imaging. By using photomultiplier-based detection and a scanning system, we used the SLM to remotely control the focus position and imaged fluorescent neurons and glial cells in the neocortex of anesthetized mice at multiple planes (Figure 5.3). The minimum time required to move the diffraction limited spot between two different positions was 21 ± 2 ms. Using electrically tunable lenses, similar speeds for z-positioning can be obtained without moving the objective (Grewe et al., 2011). These remote focusing approaches promise to be important for fast and inertia-free 3D functional imaging in living animals. These techniques overcome previously reported temporal (10 Hz) and perturbation limits imposed by mechanically moving parts, such as translating objectives, while imaging in living animals (Gobel et al., 2007). These advantages may become particularly relevant when imaging experiments are combined with microelectrode insertion for electrophysiological recordings. Therefore, *inertia-free* SLM-focus control is an indispensable tool for imaging living animals at frequencies higher than 10 Hz. It is also important to note that the use of SLMs for *z*-focusing may be useful for other important neuroscience applications. For example, when combined with two laser sources (one laser beam going through the SLM light path and one through scanning mirrors), this technology could allow for *simultaneous* imaging at one plane and photoactivation at different planes, which cannot be achieved by using a piezoelectric actuated objective lens.

Conclusions

In this chapter, we provided an overview of some major applications of SLMs for the study of mammalian brains. The application in neuroscience of SLMs is only beginning. The development of next-generation SLMs, with improved characteristics and faster and more user-friendly software, will provide better optical performance and allow this technology to be broadly applied in life sciences research. In combination with the ever-growing number of light-sensitive biological sensors and actuators, we foresee SLM-based approaches providing precise and noninvasive control of neuronal cells and networks in their natural environment in living animals. This will allow the hypothesis-driven testing of the contribution of specific cellular subpopulations to the generation and modulation of network activity driving behavior and thus promises to advance our understanding of the complex cellular dynamics underpinning brain function.

Acknowledgments

We wish to thank F. Benfenati and E. Ronzitti for critical reading of the manuscript. This work was supported by Telethon-Italy (GGP10138), San Paolo "Programma in Neuroscienze," FIRB (RBAP11×42L) to T.F. and FP7 FOCUS Project to M.D.M.

References

Adams, S.R., Tsien, R.Y., 1993. Controlling cell chemistry with caged compounds. Annu Rev Physiol 55, 755–784.

Albert, O., Sherman, L., Mourou, G., Norris, T.B., Vdovin, G., 2000. Smart microscope: an adaptive optics learning system for aberration correction in multiphoton confocal microscopy. Opt Lett 25, 52–54.

Andrasfalvy, B.K., Zemelman, B.V., Tang, J., Vaziri, A., 2010. Two-photon single-cell optogenetic control of neuronal activity by sculpted light. Proc Natl Acad Sci USA 107, 11981–11986.

Beadie, G., Sandrock, M.L., Wiggins, M.J., Lepkowicz, R.S., Shirk, J.S., Ponting, M., et al., 2008. Tunable polymer lens. Opt Express 16, 11847–11857.

Berge, B., Peseux, J., 2000. Variable focal lens controlled by an external voltage: an application of electrowetting. Eur Phys J E: Soft Matter Biol Physics 3, 159–163.

Bifano, T., 2011. Adaptive imaging: MEMS deformable mirrors. Nat Photon 5, 21–23.

Bonora, S., 2011. Distributed actuators deformable mirror for adaptive optics. Opt Commun 284, 3467–3473.

Bortolozzo, U., Bonora, S., Huignard, J.P., Residori, S., 2010. Continuous photocontrolled deformable membrane mirror. Appl Phys Lett 96, 251108.

Boysel, R.M., 1991. A 128 × 128 frame-addressed deformable mirror spatial light modulator. Opt Eng 30, 1422–1427.

Brown, C.V., Wells, G.G., Newton, M.I., McHale, G., 2009. Voltage-programmable liquid optical interface. Nat Photon 3, 403–405.

Choi, H.W., Jeon, C.W., Dawson, M.D., 2004. High-resolution 128 × 96 nitride microdisplay. Electron Device Lett IEEE 25, 277–279.

Cojoc, D., Difato, F., Ferrari, E., Shahapure, R.B., Laishram, J., Righi, M., et al., 2007. Properties of the force exerted by filopodia and lamellipodia and the involvement of cytoskeletal components. PLoS One 2, e1072.

Collings, N., Davey, T., Christmas, J., Chu, D.P., Crossland, B., 2011. The applications and technology of phase-only liquid crystal on silicon devices. J Display Technol 7, 112–119.

Dal Maschio, M., De Stasi, A.M., Benfenati, F., Fellin, T., 2011. Three-dimensional in vivo scanning microscopy with inertia-free focus control. Opt Lett 36, 3503–3505.

Dal Maschio, M., Difato, F., Beltramo, R., Blau, A., Benfenati, F., Fellin, T., 2010. Simultaneous two-photon imaging and photo-stimulation with structured light illumination. Opt Express 18, 18720–18731.

Debarre, D., Botcherby, E.J., Watanabe, T., Srinivas, S., Booth, M.J., Wilson, T., 2009. Image-based adaptive optics for two-photon microscopy. Opt Lett 34, 2495–2497.

Dholakia, K., Cizmar, T., 2011. Shaping the future of manipulation. Nat Photon 5, 335–342.

Difato, F., Dal Maschio, M., Beltramo, R., Blau, A., Benfenati, F., Fellin, T., 2011a. Spatial light modulators for complex spatio-temporal illumination of neuronal networks. In: Fellin, T., Halassa, M.M. (Eds.), Neuronal Network Analysis: Concepts and Experimental Approaches. Springer, New York.

Difato, F., Dal Maschio, M.D., Marconi, E., Ronzitti, G., Maccione, A., Fellin, T., et al., 2011b. Combined optical tweezers and laser dissector for controlled ablation of functional connections in neural networks. J Biomed Opt 16, 051306–051309.

Efron, U., 1994. Spatial Light Modulator Technology: Material, Devices and Applications. Marcel Dekker, New York.

Ellis-Davies, G.C., 2007. Caged compounds: photorelease technology for control of cellular chemistry and physiology. Nat Methods 4, 619–628.

Emiliani, V., Cojoc, D., Ferrari, E., Garbin, V., Durieux, C., Coppey-Moisan, M., et al., 2005. Wave front engineering for microscopy of living cells. Opt Express 13, 1395–1405.

Engstrom, D., Frank, A., Backsten, J., Goksor, M., Bengtsson, J., 2009. Grid-free 3D multiple spot generation with an efficient single-plane FFT-based algorithm. Opt Express 17, 9989–10000.

Fienup, J.R., 1982. Phase retrieval algorithms: a comparison. Appl Opt 21, 2758–2769.

Fino, E., Yuste, R., 2011. Dense inhibitory connectivity in neocortex. Neuron 69, 1188–1203.

Garaschuk, O., Milos, R.I., Konnerth, A., 2006. Targeted bulk-loading of fluorescent indicators for two-photon brain imaging in vivo. Nat Protoc 1, 380–386.

Georgiou, A., Christmas, J., Collings, N., Moore, J., Crossland, W.A., 2008. Aspects of hologram calculation for video frames. Appl Opt 10, 4793–4803.

Gerchberg, R.W., Saxton, W.O., 1971. Phase determination for image and diffraction plane pictures in the elctron microscope. Optik 34, 275–284.

Ghezzi, D., Antognazza, M.R., Dal Maschio, M., Lanzarini, E., Benfenati, F., Lanzani, G., 2011. A hybrid bioorganic interface for neuronal photoactivation. Nat Commun 2, 166.

Gluckstad, J., Palima, D., 2010. Generalized Phase Contrast. Springer in association with Canopus Academic Publishing Limited, Dordrecht, Netherlands.

Gobel, W., Kampa, B.M., Helmchen, F., 2007. Imaging cellular network dynamics in three dimensions using fast 3D laser scanning. Nat Methods 4, 73–79.

Golan, L., Reutsky, I., Farah, N., Shoham, S., 2009. Design and characteristics of holographic neural photo-stimulation systems. J Neural Eng 6, 066004.

Golan, L., Shoham, S., 2009. Speckle elimination using shift-averaging in high-rate holographic projection. Opt Express 17, 1330–1339.

Goldman, J.W., 2005. Introduction to Fourier Optics. Roberts & Company Publishers, Greenwood Village, CO.

Golgi, C., 1885. Sulla fina anatomia degli organi centrali del sistema nervoso. pp. 72–123.

Gong, Y., Zhang, S., 2010. Ultrafast 3-D shape measurement with an off-the-shelf DLP projector. Opt Express 18, 19743–19754.

Grewe, B.F., Voigt, F.F., van 't Hoff, M., Helmchen, F., 2011. Fast two-layer two-photon imaging of neuronal cell populations using an electrically tunable lens. Biomed Opt Express 2, 2035–2046.

Grossman, N., Poher, V., Grubb, M.S., Kennedy, G.T., Nikolic, K., McGovern, B., et al., 2010. Multi-site optical excitation using ChR2 and micro-LED array. J Neural Eng 7, 16004.

Hanes, R.D., Jenkins, M.C., Egelhaaf, S.U., 2009. Combined holographic-mechanical optical tweezers: construction, optimization, and calibration. Rev Sci Instrum 80, 083703.

Heim, R., Prasher, D.C., Tsien, R.Y., 1994. Wavelength mutations and posttranslational autoxidation of green fluorescent protein. Proc Natl Acad Sci USA 91, 12501–12504.

Heim, R., Tsien, R.Y., 1996. Engineering green fluorescent protein for improved brightness, longer wavelengths and fluorescence resonance energy transfer. Curr Biol 6, 178–182.

Hornbeck, L.J., 1998. From cathode rays to digital micromirrors: a history of electronic projection display technology. TI Techn J 15, 7–46.

Jayasinghe, A.K., Rohner, J., Hutson, M.S., 2011. Holographic UV laser microsurgery. Biomed Opt Express 2, 2590–2599.

Jerome, J., Foehring, R.C., Armstrong, W.E., Spain, W.J., Heck, D.H., 2011. Parallel optical control of spatiotemporal neuronal spike activity using high-speed digital light processing. Front Syst Neurosci 5, 70.

Ji, N., Milkie, D.E., Betzig, E., 2010. Adaptive optics via pupil segmentation for high-resolution imaging in biological tissues. Nat Methods 7, 141–147.

Ji, N., Shroff, H., Zhong, H., Betzig, E., 2008. Advances in the speed and resolution of light microscopy. Curr Opin Neurobiol 18, 605–616.

Kerr, J.N., Denk, W., 2008. Imaging in vivo: watching the brain in action. Nat Rev Neurosci 9, 195–205.

Kerr, J.N., Greenberg, D., Helmchen, F., 2005. Imaging input and output of neocortical networks in vivo. Proc Natl Acad Sci USA 102, 14063–14068.

Khoo, IA, 2007. Liquid Crystals. Wiley, Hoboken, NJ.

Kim, H., Yang, B., Lee, B., 2004. Iterative Fourier transform algorithm with regularization for the optimal design of diffractive optical elements. J Opt Soc Am A Opt Image Sci Vis 21, 2353–2365.

Kim, S.G., Hwang, K.H., Kim, J.S., Choi, Y.J., Kim, J.M., 1999. Thin-film micromirror array (TMA) for high luminance and cost-competitive information display systems. Proc SPIE 3634, 207–216.

Knopfel, T., Tomita, K., Shimazaki, R., Sakai, R., 2003. Optical recordings of membrane potential using genetically targeted voltage-sensitive fluorescent proteins. Methods 30, 42–48.

Kuzmenko, A.V., 2008. Weighting iterative Fourier transform algorithm of the kinoform synthesis. Opt Lett 33, 1147–1149.

Leach, J., Wulff, K., Sinclair, G., Jordan, P., Courtial, J., Thomson, L., et al., 2006. Interactive approach to optical tweezers control. Appl Opt 45, 897–903.

Lee, W.M., Reece, P.J., Marchington, R.F., Metzger, N.K., Dholakia, K., 2007. Construction and calibration of an optical trap on a fluorescence optical microscope. Nat Protoc 2, 3226–3238.

Liang, C.W., Mohammadi, M., Santos, M.D., Tang, C.M., 2011. Patterned photostimulation with digital micromirror devices to investigate dendritic integration across branch points. J Vis Exp pii 2003. doi:10.3791/2003.

Lutz, C., Otis, T.S., DeSars, V., Charpak, S., DiGregorio, D.A., Emiliani, V., 2008. Holographic photolysis of caged neurotransmitters. Nat Methods 5, 821–827.

Mathur, V., Vangala, S.R., Qian, X., Goodhue, W.D., Haji-Saeed, B., Khoury, J., 2009. All optically driven MEMS deformable mirrors via direct cascading with wafer bonded GaAs/GaP PIN photodetectors. Proc IEEE Photon Soc, 156–157.

Maurer, C., Jesacher, S., Bernet, M., Ritsch-Marte, M., 2011. What spatial light modulators can do for optical microscopy. Laser Photon Rev 5, 81–101.

Mias, S., Camon, H., 2008. A review of active optical devices: I. Amplitude modulation. J Micromechan Microeng 18, 083001.

Miles, M., 1999. MEMS-based interferometric modulator for display applications. Proc SPIE 3876, 20.

Miyawaki, A., Llopis, J., Heim, R., McCaffery, J.M., Adams, J.A., Ikura, M., et al., 1997. Fluorescent indicators for Ca^{2+} based on green fluorescent proteins and calmodulin [see comments]. Nature 388, 882–887.

Moddel, G., Johnson, K.M., Li, W., Rice, R.A., Pagano-Stauffer, L.A., Handschy, M.A., 1989. High-speed binary optically addressed spatial light modulator. Appl Phys Lett 55, 537–539.

Monk, D.W., Gale, R.O., 1995. The digital micromirror device for projection display. Microelectron Eng 27, 489–493.

Neil, M.A., Juskaitis, R., Booth, M.J., Wilson, T., Tanaka, T., Kawata, S., 2000. Adaptive aberration correction in a two-photon microscope. J Microsc 200, 105–108.

Nikolenko, V., Peterka, D.S., Yuste, R., 2010. A portable laser photostimulation and imaging microscope. J Neural Eng 7, 045001.

Nikolenko, V., Watson, B.O., Araya, R., Woodruff, A., Peterka, D.S., Yuste, R., 2008. SLM microscopy: scanless two-photon imaging and photostimulation with spatial light modulators. Front Neural Circuits 2, 5–19.

Oron, D., Tal, E., Silberberg, Y., 2005. Scanningless depth-resolved microscopy. Opt Express 13, 1468–1476.

Palima, D., Gluckstad, J., 2008. Comparison of generalized phase contrast and computer generated holography for laser image projection. Opt Express 16, 5338–5349.

Papagiakoumou, E., Anselmi, F., Begue, A., de Sars, V., Gluckstad, J., Isacoff, E.Y., et al., 2010. Scanless two-photon excitation of channelrhodopsin-2. Nat Methods 7, 848–854.

Papagiakoumou, E., de Sars, V., Emiliani, V., Oron, D., 2009. Temporal focusing with spatially modulated excitation. Opt Express 17, 5391–5401.

Papagiakoumou, E., de Sars, V., Oron, D., Emiliani, V., 2008. Patterned two-photon illumination by spatiotemporal shaping of ultrashort pulses. Opt Express 16, 22039–22047.

Poher, V., Zhang, H.X., Kennedy, G.T., Griffin, C., Oddos, S., Gu, E., et al., 2007. Optical sectioning microscopes with no moving parts using a micro-stripe array light emitting diode. Opt Express 15, 11196–11206.

Ren, H., Fan, Y.H., Gauza, S., Wu, S.T., 2004. Tunable microlens arrays using polymer network liquid crystal. Opt Commun 230, 267–271.

Rosen, J., Brooker, G., 2008. Non-scanning motionless fluorescence three-dimensional holographic microscopy. Nat Photon 2, 190–195.

Rueckel, M., Mack-Bucher, J.A., Denk, W., 2006. Adaptive wavefront correction in two-photon microscopy using coherence-gated wavefront sensing. Proc Natl Acad Sci USA 103, 17137–17142.

Sanford, J.L., Greier, P.F., Yang, K.H., Lu, M., Olyha, R.S., Narayan, C., et al., 1998. A one-megapixel reflective spatial light modulator system for holographic storage. IBM J Res Dev 42, 411–426.

Sherman, L., Ye, J.Y., Albert, O., Norris, T.B., 2002. Adaptive correction of depth-induced aberrations in multiphoton scanning microscopy using a deformable mirror. J Microsc 206, 65–71.

Sinclair, G., Leach, J., Jordan, P., Gibson, G., Yao, E., Laczik, Z., et al., 2004. Interactive application in holographic optical tweezers of a multi-plane Gerchberg-Saxton algorithm for three-dimensional light shaping. Opt Express 12, 1665–1670.

Stanley, M., Conway, P.B., Coomber, S.D., Jones, J.C., Scattergood, D.C., Slinger, C.W., et al., 2000. Novel electro-optic modulator system for the production of dynamic images from giga-pixel computer-generated holograms. Proc SPIE 3956, 13–22.

Stirman, J.N., Crane, M.M., Husson, S.J., Wabnig, S., Schultheis, C., Gottschalk, A., et al., 2011. Real-time multimodal optical control of neurons and muscles in freely behaving *Caenorhabditis elegans*. Nat Methods 8, 153–158.

Tsai, F.S., Cho, S.H., Lo, Y.H., Vasko, B., Vasko, J., 2008. Miniaturized universal imaging device using fluidic lens. Opt Lett 33, 291–293.

Tsien, R.Y., 1980. New calcium indicators and buffers with high selectivity against magnesium and protons: design, synthesis, and properties of prototype structures. Biochemistry 19, 2396–2404.

Tsien, R.Y., 1981. A non-disruptive technique for loading calcium buffers and indicators into cells. Nature 290, 527–528.

Tsutsui, H., Wolf, A.M., Knopfel, T., Oka, Y., 2001. Imaging postsynaptic activities of teleost thalamic neurons at single cell resolution using a voltage-sensitive dye. Neurosci Lett 312, 17–20.

Venediktov, V.Y., Nevskaya, G.E., Tomilin, M.G., 2011. Liquid crystals in dynamic holography (review). Opt Spectrosc (English translation of Optika i Spektroskopiya) 111, 113–133.

Vicari, L., 2003. Optical Applications of Liquid Crystals. Institute of Physics Publishing Ltd, London, UK.

Watson, B.O., Nikolenko, V., Araya, R., Peterka, D.S., Woodruff, A., Yuste, R., 2010. Two-photon microscopy with diffractive optical elements and spatial light modulators. Front Neurosci 4, pii 29.

Weiner, A.M., 2011. Ultrafast optical pulse shaping: a tutorial review. Opt Commun 284, 3669–3692.

Yang, S., Papagiakoumou, E., Guillon, M., de Sars, V., Tang, C.M., Emiliani, V., 2011. Three-dimensional holographic photostimulation of the dendritic arbor. J Neural Eng 8, 046002.

Zahid, M., Velez-Fort, M., Papagiakoumou, E., Ventalon, C., Angulo, M.C., Emiliani, V., 2010. Holographic photolysis for multiple cell stimulation in mouse hippocampal slices. PLoS One 5, e9431.

Zhang, F., Aravanis, A.M., Adamantidis, A., de Lecea, L., Deisseroth, K., 2007. Circuit-breakers: optical technologies for probing neural signals and systems. Nat Rev Neurosci 8, 577–581.

Zhang, F., Gradinaru, V., Adamantidis, A.R., Durand, R., Airan, R.D., de Lecea, L., et al., 2010. Optogenetic interrogation of neural circuits: technology for probing mammalian brain structures. Nat Protoc 5, 439–456.

6

MULTIPHOTON MICROSCOPY ADVANCES TOWARD SUPER RESOLUTION

Paolo Bianchini[1], Partha P. Mondal[2], Shilpa Dilipkumar[2], Francesca Cella Zanacchi[1], Emiliano Ronzitti[1], and Alberto Diaspro[1]

[1]*Deparment of Nanophysics, Istituto Italiano di Tecnologia, Genova, Italy,*
[2]*Department of Instrumentation and Applied Physics, Indian Institute of Science, Bangalore, India.*

CHAPTER OUTLINE
Introduction   121
Point Spread Function for Single- and Multiphoton Imaging   123
Super Resolution Techniques for Multiphoton Fluorescence
    Microscopy   126
    Two-photon 4pi Microscopy   126
    Multiple Excitation Spot Optical Microscopy   127
    Multiphoton Localization Techniques   129
    Multiphoton STED Microscopy   131
Conclusions   134
Acknowledgments   134
References   134

Introduction

Multiphoton imaging is one of the key imaging techniques for fields ranging from bioimaging to optical engineering. This technique has an intrinsic optical sectioning capability (Sheppard and Kompfner, 1978; Denk et al., 1990; Diaspro et al., 2005) and the ability of essay-based physiological study for deeper understanding of biological processes from the molecular to the tissue level. Multiphoton excitation (MPE)-based fluorescence imaging techniques have the additional advantage of exciting biomolecules classically excitable in the ultraviolet-visual spectroscopy (UV-VIS) range using visual

Cellular Imaging Techniques for Neuroscience and Beyond.
DOI: http://dx.doi.org/10.1016/B978-0-12-385872-6.00006-4

spectroscopy-infrared (VIS-IR) light (Mertz et al., 1995; Lippitz et al., 2002; Diaspro, 2004). While UV lasers are now available for direct excitation of fluorophores, the issues of phototoxicity (Patterson and Piston, 2000; Chirico et al., 2003; Mondal and Diaspro, 2007; Mondal et al., 2010) and scattering critically hamper the result.

The intrinsic optical sectioning property of the multiphoton system makes it a very attractive and unique technique for both *in vivo* and *in vitro* imaging. The unique extraction of focal plane information facilitates plane-by-plane scanning and the images can be stacked together to obtain a 3D map of biological specimens (Diaspro and Robello, 2000; Diaspro, 2004). Quantum yield, photostability and pulse width are key factors for high signal-to-noise ratio (SNR) quantitative images (Koester et al., 1999; Konig et al., 1999; Konig, 2000; Drobizhev et al., 2011). Emitted fluorescence falls off with time, which is attributed to photobleaching. Additionally, multiphoton imaging does not suffer from photobleaching in off-focus planes as compared to single-photon confocal imaging (Brakenhoff et al., 1996; Mertz, 1998; Patterson and Piston, 2000). The constructive side of photobleaching is its ability to reveal valuable information about biological system dynamics such as diffusion (Davis and Bardeen, 2002; McNally and Smith, 2002; Delon et al., 2004). In this respect, MPE allows better control of photobleaching-based experiments than the single-photon case (Mazza et al., 2007). Overall, proper balance between SNR and photobleaching effects is necessary for qualitative imaging. Unfortunately, one drawback to MPE microscopy is that it provides both lower resolution and excitation efficiency with respect to the linear case. Today, it is worth noting the implementation of augmented optical resolution techniques directed toward multiphoton microscopy (Diaspro, 2010). The diffraction barrier, as summarized by Abbe's formula (Abbe, 1884) is still the main limiting factor, so it becomes imperative to adapt superresolution techniques to multiphoton imaging. A number of superresolution techniques have become apparent over the last decade. Hell et al. (1992) showed that by utilizing 4pi geometry within a nonlinear excitation scheme resolution could be improved (Hell et al., 2009). Since this approach is optically based, it is still limited by diffraction even though it provides superresolution. Theoretically, there is no limit to the ultimate superresolution attainable since the "actor" influencing the resolution performance is no longer dominated by optics. Approaching superresolution by using stimulated emission depletion (STED; Moneron and Hell, 2009), images can be formed at an unlimited resolution beyond the diffraction limit based on the transition properties of the systems used. Also, Betzig et al. (2006), Zhuang et al. (Rust et al., 2006), and Hess et al. (2006) have demonstrated that photophysical properties of the target fluorophore can be used for achieving superresolution. Other promising approaches are structured illumination (Gustafsson, 2005), aperture engineering (di Francia, 1952; Neil et al., 2000; Mondal and Diaspro, 2008; Mondal, 2010), and computational deconvolution-based techniques (van kempen et al., 1997; Mondal et al., 2007, 2008). Another technique of potential interest is surface plasmon-based nanoimaging, which relies on the behavior of light under the influence of plasmons allowing superlensing (Kawata et al., 2009). Most of these techniques have been demonstrated for both single- and mutiphoton fluorescence microscopy. Multiphoton molecular excitation as a single quantum event was first predicted and theorized by Maria-Goppert Mayer (1931). The first experimental demonstration of two-photon excitation (2PE) was reported by Kaiser and Garret (1961) on CaF_2: Eu^{2+}. Two-photon absorption has been demonstrated in a solution of NADH (Rounds et al., 1966) and in cells (Berns, 1976). In the mid-1970s Sheppard and Kompfner (1978) reported about the possibility of 2PE fluorescence imaging of biological samples. 2PE fluorescence microscopy is probably the most exciting development in the field of microscopy in the last two decades and, most important, after the application on living cells by Denk et al. (1990; Xu et al., 1996) in conjunction with imaging in thick 3D samples (So et al., 2000; Chirico et al., 2003; Helmchen and Denk, 2005).

Point Spread Function for Single- and Multiphoton Imaging

The point spread function (PSF) characterizes an imaging system and, more important, determines its resolving power (Bianco and Diaspro, 1989). In addition, knowledge of the PSF allows 3D computational optical sectioning (Diaspro et al., 1990; van Kempen et al., 1997). Here we show some aspects of the emerging multiphoton imaging system and make a comparison with state-of-art confocal and two-photon microscopy treated as 3D image formation systems (Diaspro, 2002). Following the remarkable work by Wolf and Richards (1959), the x, y, and z components of an electric field at the focus of the objective lens (of aperture angle, α) are given by the following:

$$\begin{cases} E_x = -iC(I_0 + I_2 \cos 2\varnothing) \\ E_y = -iCI_2 \sin 2\varnothing) \\ E_z = -2CI_2 \cos\varnothing \end{cases} \qquad (6.1)$$

wherein the diffraction integrals over the aperture angle α are

$$I_0(u,v) = \int_{\theta=0}^{\alpha} (1 + \cos\theta)\sin\theta J_0\left(\frac{v\sin\theta}{\sin\alpha}\right) \times \cos^{1/2}\theta\, e^{i\left(\frac{u\cos\theta}{\sin^2\alpha}\right)} d\theta$$

$$I_1(u,v) = \int_{\theta=0}^{\alpha} \sin^2 \theta J_1 \times \cos^{1/2}\theta e^{i\left(\frac{u\cos\theta}{\sin^2\alpha}\right)} d\theta$$

$$I_2(u,v) = \int_{\theta=0}^{\alpha} (1-\cos\theta)\sin\theta J_2\left(\frac{v\sin\theta}{\sin\alpha}\right) \times \cos^{1/2}\theta e^{i\left(\frac{u\cos\theta}{\sin^2\alpha}\right)} d\theta$$

where the parameter φ is the angle between the incident electric field and direction of observation; $u = \left(\frac{2\pi}{\lambda}\right)z\sin^2\alpha$ and $v = \left(\frac{2\pi}{\lambda}\right)(x^2+y^2)\sin\alpha$ are, respectively, the longitudinal and transverse coordinates.

Confocal microscopy is a smart combination of widefield microscopy and a pinhole. The pinhole prevents light from off-focus planes from reaching the detector, thereby exclusively allowing light to be detected emanating from the geometrical focus. This is a huge advantage because there are many benefits for the optical sectioning of biological specimens. Unfortunately, constraints related to limited resolution still remain.

For laser scanning confocal microscopy, the 3D PSF is given by the following (Wolf and Richards, 1959; Bertero and Boccacci, 1998):

$$h(x,y,z) = \left|h_{\lambda_{ill}}(x,y,z)\right|^2 \times \left(k(x,y) \otimes \left|h_{\lambda_{det}}(x,y,z)\right|^2\right), \tag{6.2}$$

wherein the excitation PSF is

$$h_{ill}(u,v,\varnothing) = \left|\overline{E}(u,v,\varnothing)\right|^2 = I_0^2 + I_2^2 + 2I_1I_2\cos(2\varnothing) + 4I_1^2\cos^2\varnothing$$

for linearly polarized light and $= I_0^2 + 2I_1^2 + I_2^2$ for randomly polarized light. The detection PSF, $h_{\lambda_{det}}(u,v,\varnothing) = \left|I_0\right|^2 + 2\left|I_1\right|^2 + \left|I_2\right|^2 . k(x,y)$, describes the geometry of the pinhole, $\vec{h}_{\lambda_{ill}}$, and the amplitude of the illumination light in the focal region while $\vec{h}_{\lambda_{det}}$ is the amplitude for the detection ($\vec{h}_{\lambda_{det}}$ is similar to $h_{\lambda_{ill}}$, but is calculated at a red-shifted emission wavelength λ_{det}).

The lateral (*XY*) and axial (*XZ* and *YZ*) views of the system PSF for confocal microscopy are shown in Figure 6.1A. The dimension (full-width at half-maxima, of the spot; FWHM) of the lateral and axial components of the system PSF are, respectively, r=λ/2*n* sin (α) and r=λ/*n* sin² (α), where λ, α, and *n* denote the wavelength, the aperture angle of the lens, and

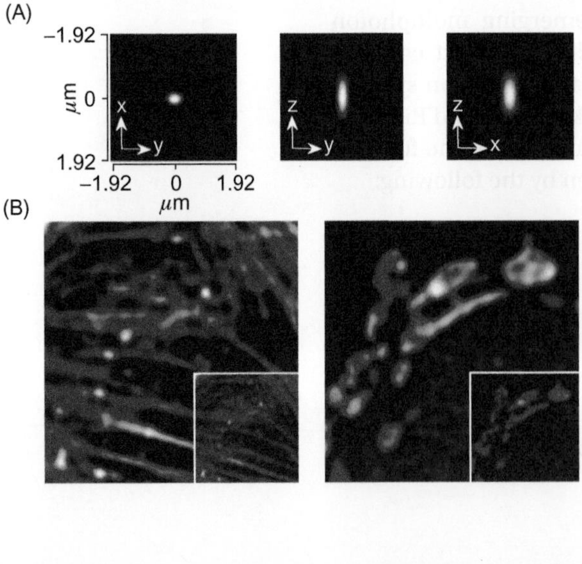

Figure 6.1 (A) Three-dimensional confocal PSF. (B) Reconstructed confocal image of F-actin and mitochondria. Insets show raw data.

(A)
(B)

the refractive index of objective immersion oil. The lateral and axial sampling is chosen as 30 and 90 nm. Raw (insets) and reconstructed confocal images of F-actin (tagged with BODIPY FL phalloidin) and mitochondria (tagged with MitoTracker Red CMXRos; Molecular Probes) are shown in Figure 6.1B. The excitation and emission wavelengths for F-actin imaging are 488 and 520 nm, respectively, whereas for mitochondria their values are 543 and 600 nm.

2PE, as a special, widely utilized case of MPE, benefits from the intrinsic optical-sectioning property, which is due to nonlinear intensity dependence. Thus, for 2PE, a pinhole is redundant, so the effective PSF is simply the illumination PSF multiplied with the emission PSF. The excitation PSF for randomly polarized light illumination is given by the following:

$$h^{2PE}_{\lambda_{ill}} = \left|\vec{E}(x,y,z)\right|^4 = \left[|I_0|^2 + 2|I_1|^2 + |I_2|^2\right]^2, \qquad (6.3)$$

and the detection PSF (corresponding to emitted randomly polarized fluorescent light) is simply $h^{2PE}_{\lambda_{det}}(u,v,\varnothing) = |I_0|^2 + 2|I_1|^2 + |I_2|^2$. For 2PE microscopy, an illumination wavelength of 750 nm was used. The corresponding emission wavelength is 675 nm. For two-photon images, an oil immersion [refractive index (RI) = 1.5] objective of numerical aperture (NA) 1.4 was used. A lateral sampling of 40 nm and axial sampling of 120 nm was used for obtaining 2PE images. In Figure 6.2, 2PE imaging from two different fluorophores and its comparison with single-photon excitation is shown. Figure 6.2E shows the 2PE image obtained with the rhizome of lily-of-the-valley (*Convallaria majalis*). Localization due to the 2PE process is clearly evident from Figure 6.2C and D, whereas the single-photon excitation process shows an X-type excitation profile (see Figure 6.2A and B), thereby

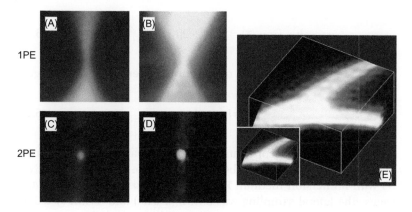

Figure 6.2 Excitation profile (along optical axis) of two different fluorophores for both single- (A, B) and two-photon (C, D) excitation. (E) Three-dimensional reconstructed 2PE image of *C. majalis.* Inset shows raw data.

unnecessarily exciting off-focus (top and bottom) layers resulting in increased photobleaching.

Super Resolution Techniques for Multiphoton Fluorescence Microscopy

Recently, several approaches have been implemented toward achieving optical superresolution (Diaspro, 2010). Superresolution is highly desirable for MPE microscopy, since MPE suffers from low resolution because of the requirement of longer excitation wavelengths. Summarized below are some of the high-resolution techniques for multiphoton fluorescence microscopy.

Two-photon 4pi Microscopy

4π-geometry was proposed (Cremer and Cremer, 1978), suggesting the maximum possible aperture angle for high-resolution imaging. This has been further developed and effectively used (Hell et al., 1994; Hell, 2007) for 4pi superresolution microscopy.

The excitation PSF for 4pi (A-type) microscopy is the following (Schrader et al., 1998):

$$h_{ill}(u,v,\varnothing) = |\bar{E}(u,v,\varnothing) + \bar{E}(-u,v,\varnothing)|^4$$

$$= \begin{cases} [Re\{I_0\}^2 + Re\{I_2\}^2 + 2Re\{I_1\}Re\{I_2\}\cos(2\varnothing) \\ +4Re\{I_1\}^2\cos^2\varnothing, \textit{for linear polarized light}; \\ [Re\{I_0\}^2 + 2Re\{I_1\}^2 + Re\{I_2\}^2]^2, \\ \qquad \textit{for circular polarized light.} \end{cases} \quad (6.4)$$

The detection PSF is simply as follows:

$$h_{\lambda_{det}}(u,v,\varnothing) = |I_0|^2 + 2|I_1|^2 + |I_2|^2. \quad (6.5)$$

The system PSF for 2PE-4pi microscopy is given by the following:

$$h = h_{\lambda_{ill}}^2 \times h_{\lambda_{det}}. \quad (6.6)$$

For 2PE-4pi microscopy, an illumination wavelength of 910 nm was used and emission was observed at 520 nm. An objective of NA 1.35 with immersion medium RI of 1.45 was used for obtaining 4pi images. The lateral sampling was 33 nm (x-axis) and 60 nm (y-axis),

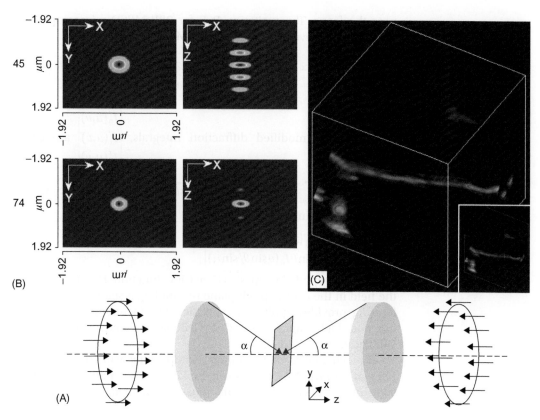

Figure 6.3 (A) Optical configuration for 2PE-4pi microscopy. (B) 2PE-4pi excitation PSF for aperture angle 45° and 740 nm. (C) Three-dimensional reconstructed 4pi-2PE of mitochondrial network. Inset shows raw 3D data.

respectively, whereas the axial sampling was 61 nm. Figure 6.3C shows the 4pi images of a mitochondrial network tagged with green fluorescent protein in stably transfected cell lines, that is, hepatoma (HepG2) and insulinoma (INS1E). The optical setup is shown schematically in Figure 6.3A. The excitation PSF at two different aperture angles is shown in Figure 6.3B, thereby demonstrating the effectiveness of high NA objectives. The main concern of 4pi microscopy is the cross talk introduced by refraction pattern side lobes. However, the effect of side lobes can be substantially minimized computationally (Schrader et al., 1998; Mondal et al., 2007, 2008).

Multiple Excitation Spot Optical Microscopy

Apodizing and aperture engineering techniques are well known for surpassing the resolution limit (di Francia, 1952; Neil et al., 2000; Mondal and Diaspro, 2008). Spatial filters are used just before the back aperture of the objective lens for the desired field pattern at and near the focus. The excitation PSF for 2PE-MESO microscopy (see Figure 6.4A) is given by the following (Mondal, 2009, 2010):

$$h_{ex}(u,v,\propto -\varnothing_1) = |E_1(u,v,\propto -\varnothing_1) + E_2(u,v,\propto -\varnothing_1)|^2$$
$$= |ReI_0'|^2 + 2|ReI_1'|^2 + |ReI_2'|^2 \tag{6.7}$$

where, the modified diffraction integrals, $\begin{bmatrix} I_0'(u,v) \\ I_1'(u,v) \\ I_2'(u,v) \end{bmatrix}$ are given

by $I_m'(u,v) = \int_{\theta=0}^{\alpha} B_m(\theta)H(\theta - \theta_1)\sqrt{\cos\theta}\, e^{i\frac{u\cos\theta}{\sin^2\alpha}} d\theta$ and the vector

$$B = \begin{bmatrix} (1+\cos\theta)\sin\theta J_0(v\sin\theta/\sin\alpha) \\ \sin^2\theta J_1(v\sin\theta/\sin\alpha) \\ (1-\cos\theta)\sin\theta J_0(v\sin\theta/\sin\alpha) \end{bmatrix}.$$

Such an apodizing mask (Heaviside function H (*)) truncates the field in the central pupil plane resembling a spatial filter. Spatial filtering provides application-specific PSFs including depth imaging and high lateral resolution (Dilipkumar and Mondal, 2011; Dilipkumar et al., 2011).

The schematics of the optical setup along with the excitation PSFs at various NAs are shown in Figure 6.4. Multiple excitation spots of nanometric dimensions must be obtained. This mimics the

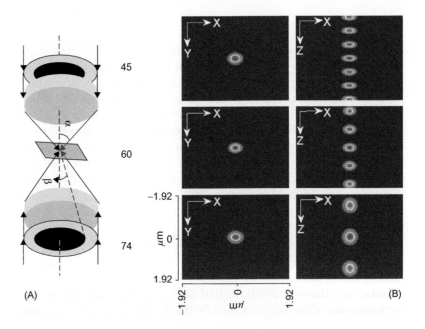

Figure 6.4 (A) Optical setup for 2PE-MESO PSF. (B) Corresponding PSF for various aperture angles.

possibility of simultaneous multilayer imaging provided an appropriate detection scheme is adopted. Confocal θ detection is an appropriate detection platform under these circumstances. An axial resolution of 210 nm (FWHM) is obtained at an aperture angle 60°, which is approximately four times better than the diffraction limit. This technique can be a boon to multispot fluorescence correlation spectroscopy and multiple optical trap systems.

Multiphoton Localization Techniques

Another approach used to solve the problem of limited spatial resolution in fluorescence microscopy is based on detection of single molecules followed by localization. In the past few years several localization-based techniques have been developed, such as PALM (Betzig et al., 2006), FPALM (Hess et al., 2006), STORM (Rust et al., 2006), PALMIRA (Egner et al., 2007), and GSDIM (Folling et al., 2008). The idea behind all of these techniques is the detection of single molecules and localization of their position and exploiting the capability of fluorophores to change their spectral properties under light illumination or to cycle between emissive and "dark" states. In localization-based techniques several processes, such as photoactivation, photoconversion, or transition to the triplet dark state, can be exploited to control the number of excited molecules to have a sparse subset of molecules "switched on" in the field of view. This "on-off" transition or spectral change can be light induced and can further be attributed to triplet state formation, photoisomerization, or electron transfer (Dedecker, 2008). Thanks to these properties a sparse subset of molecules can be driven to the "bright state" wherein the position of each single molecule is localized with high precision. The total localization precision in 2D cases can be obtained by summing the contribution due to photon-counting noise, that of pixilation noise and, finally, background noise as shown by the following (Thompson et al., 2002):

$$\sigma_{xy}^2 = \frac{s^2 + \dfrac{a^2}{12}}{N} + \frac{8.\pi.s^4.b^2}{a^2.N^2}, \qquad (6.8)$$

where N is the number of photons collected, s is the standard deviation of the Gaussian intensity PSF, a represents the effective pixel size, and b is the background noise.

This process can be repeated for all of the molecules and the final high-resolution image can be obtained mapping all of the localized positions. This superresolution technique can be implemented both on widefield and total internal reflection fluorescence microscope imaging systems. In widefield illumination schemes, both

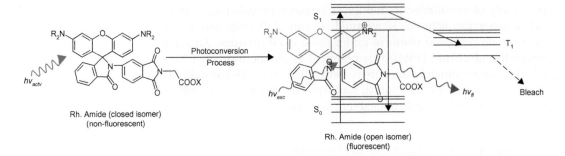

Figure 6.5 Photoconversion of rhodamine (Rh) amide followed by excitation of the fluorescent isomer.

photoconversion and photobleaching are introduced in the entire excitation volume and all of the molecules above and below the focal plane undergo photoconversion. Therefore, all of the molecules within the illumination volume are not available for further activation cycles. Within this scenario, inclined illumination and 2PE (Tokunaga et al., 2008) have been demonstrated to be useful tools for spatial confinement of the activation process, which has been used to increase the SNR in single-molecule detection techniques (Makio et al., 2008).

One of the basic aspects of single molecule localization is represented by the use of suitable fluorescent markers that undergo conformational changes when irradiated at the appropriate wavelength. In general, organic markers such as a derivative of Rhodamine 590 suitable for single- and two-photon-induced molecular switching, or phoactivatable proteins can be used (Schneider et al., 2005). The rhodamine amide exists in two convertible isomeric forms: a nonfluorescent and a fluorescent red color isomer (Belov et al., 2010). Upon irradiation with 375 nm the closed isomer gets converted to a fluorescent one. Next, laser illumination at 532 nm wavelength is used to excite the fluorescent conformational state until photobleaching occurs (the off state). The same effect can be obtained with two-photon activation at a wavelength of 750 nm followed by excitation at 532 nm (Folling et al., 2008). A schematic diagram of the switching mechanism is shown in Figure 6.5.

As far as single molecule superresolution is concerned, the fluorescent probes must have high quantum yields and the spectra of both fluorescent and nonfluorescent isomers must be well separated and thermally stable so that spontaneous conversion is minimal. This makes sure that only the light-controlled activation is possible. Moreover, the switching rate must be high and controllable, and it must be verified that the activation and inactivation rates are balanced such that only a fraction of the probe molecules are activated in time and available for readout. Additionally, the detected events must be sparsely distributed such that the distance between them is well above the diffraction limit of the system. Having all of these properties in a single fluorophore is rare; hence, localization techniques are limited to a selected number of probes. Folling et al. (2008) reported 5 μm diameter, amino-modified silica beads (see Figure 6.6) stained on the surface with an amino-reactive modification of the fluorescent marker PC-Rh590. Usually frames are acquired until a sufficient number of single events is collected to form a meaningful image. With a frame time of 10 ms, total image acquisition time is approximately a few minutes. The reported lateral resolution reaches down to 13 nm. This is approximately a 20-fold improvement over the classical diffraction limit. Such a capability has several applications in fundamental biophysics ranging from understanding protein migration in cells (Huang et al., 2010), and subcellular complex dynamics (Blosser et al., 2009; Planchon et al., 2011) as well as many more applications. Recently, new approaches for individual molecule localization techniques oriented to super-resolution imaging of thick samples have been developed to improve the imaging depth capability (Cella et al., 2011; York et al., 2011). Within this scenario, 2PE coupled with individual molecule localization techniques can provide a suitable tool to reduce scattering effects and extend the application of superresolution imaging to thick samples.

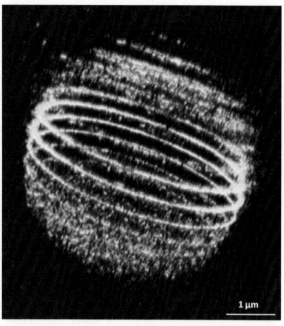

Figure 6.6 Three-dimensional reconstructed 2PA-PALMIRA image from 16 stacks of a 5 μm silica bead stained on the surface with PC-Rh590. 2PA-PALMIRA has been performed using 647 nm CW laser light (power up to 350 MW/cm^2). A localization accuracy of 15 nm has been obtained. (*From Folling et al., 2008.*)

Multiphoton STED Microscopy

Among superresolution techniques, STED is one of the prominent techniques. This method employs two diffraction limited time-spaced beams (Gaussian and doughnut) for achieving high resolution. The Gaussian beam excites the sample from singlet ground state S_0 to singlet excited state S_1. This is immediately (a time shorter than the lifetime of the molecule) followed by illumination with a red-shifted doughnut-shaped beam that causes depletion of molecules at the periphery of the Gaussian beam, thereby enabling fluorescence emission from the central region only. Depending on the molecule, particular excitation and depletion wavelengths are selected. Stimulated emission instantly returns an excited molecule to the ground state by forcing it to emit a photon identical to the one used to induce the transition, quenching the fluorescence.

The saturation of this transition provides the nonlinear response that allows the squeezing of the effective fluorescent spot beyond the diffraction limit. The resolution of the STED microscope is thus a function of the spatial distribution and magnitude of the intensity of the depleting light, with no theoretical limit to the ultimate achievable value (Hell and Wichmann, 1994). The resulting resolution obeys the formula $d_{sted} = \lambda / \left(2NA\sqrt{1 + I/I_s} \right)$, where I is the intensity at the maximum of the STED doughnut and I_s is the effective saturation intensity that can be defined as the intensity at which the probability of fluorescence emission is reduced by half. This technique has also been demonstrated in conjunction with 2PE (Ding et al., 2009; Moneron and Hell, 2009; Bianchini et al., 2011). Two-photon excitation has allowed 3D fluorescence imaging of thick biological specimens (e.g., brain, skin, muscle, etc.; Bianchini and Diaspro, 2012). The combination of 2PE and STED microscopy will provide the additional advantage of superresolution imaging deep within tissues or living animals. Here "deep" refers to the distance beyond the scatter length of photons, where performance of single-photon imaging is substantially degraded. In this method the excitation process is nonlinear while the fluorescence is still quenched by one-photon processes. The annular illumination patterns can be formed using oil or water immersion objectives despite the index of refraction mismatch between the immersion solution and the tissue (Ding et al., 2009). Figure 6.7 shows images of 40 nm yellow-green fluorescent beads (FluoSpheres carboxylate-modified microspheres; Invitrogen) acquired by confocal (Figure 6.7A), STED (Figure 6.7B), 2PE (Figure 6.7C), and 2PE-STED (Figure 6.7D) techniques. The images were acquired sequentially with a custom-adapted Leica TCS STED-CW microscope (Leica Microsystems). 2PE was performed with an ultrafast mode-locked Ti:sapphire laser (Chameleon Ultra II; Coherent) at 750 nm. One-photon excitation was performed at 488 nm and fluorescence acquired in the 500–550 nm spectral range, using the internal avalanche photodiode detector of the instrument. In the STED mode (Figure 6.7B and D), the doughnut-shaped 592 nm CW laser beam was superimposed at a typical power setting of 300 mW at the sample. The comparison of images of the nanoparticles clearly shows a substantial resolution enhancement when STED is added. A line profile across the typical image of a particle displays a FWHM of about 80 nm in the STED case. Considering the 40 nm diameter size of the beads, this implies an optical resolution of <80 nm in the focal plane. To increase the contrast and reduce the noise, a linear deconvolution could be carried out. Figure 6.7E and F show the deconvolved images of the squared region in Figure 6.7C and D. Gaussian models with an FWHM of 80 and 250 nm were used for the deconvolution of 2PE-STED and 2PE pictures, respectively. The superresolution capability of 2PE-STED is evident from the zoomed images, especially when compared to results acquired with traditional

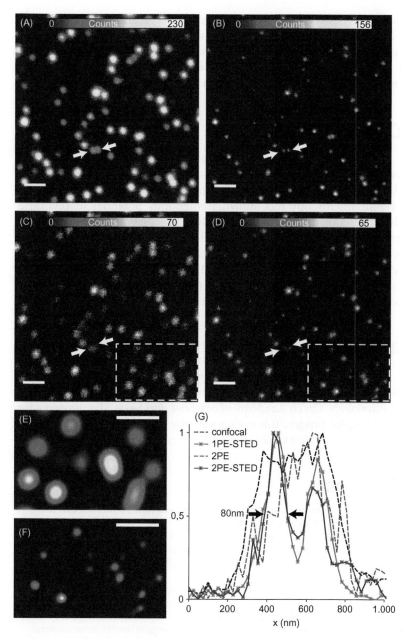

Figure 6.7 Comparison of (A) confocal, (B) STED, (C) 2PE, and (D) 2PE-STED images of 40 nm fluorescent microspheres (yellow-green fluorescent FluoSpheres carboxylate-modified microspheres; Invitrogen) mounted on a coverglass. (A–D) Raw pictures of confocal, STED, 2PE and 2PE-STED. (E) and (F) show linearly deconvolved enlargements of the squared areas marked in (C) and (D). Scale bars, 1 μm. Normalized line profiles taken between the arrows in (A–D) are plotted in (G) in black, red, green, and blue, respectively. Please see color plates at the back of the book.

diffraction-limited 2PEF microscopy. Several further improvements in STED 2PEF microscopy are still possible. First, the use of an annular mask in combination with a helical phase mask (Klar et al., 2000) would ideally improve the resolution in both radial and axial directions, as has been demonstrated for conventional STED microscopy (Willig et al., 2007). Second, the advent of new synthetic and genetically encoded fluorophores with lower I_s will allow higher resolution with less laser power (Belov et al., 2010). Lastly, an efficient use of the power of STED in combination with new methods that exploit lifetime information (Vicidomini et al., 2011) will continue to improve the resolution even deep in thick and complex samples.

Conclusions

In this chapter we have described the fundamental mechanism behind multiphoton microscopy and included recent advances in this field. With the advent of superresolution techniques, multiphoton microscopy has shown tremendous potential in revealing the mechanism of various biological processes. Of all the different techniques, the intrinsic optical sectioning capability gives MPE a step ahead with respect to other available imaging techniques. Multiphoton-based 4pi, STED, and localization techniques (PALM, GSDIM, FPALM, STORM, etc.) are the techniques of the future that will be used to better understand biological processes.

Acknowledgments

We acknowledge financial support from the Indian Institute of Science and BRNS (DAE). We acknowledge Lydie Hlavata at Institute of Physiology, Czech Academy of Sciences and Leica Microsystems for providing 4pi-images. This work is partially supported by IIT, IFOM, and Italian MIUR PRIN 2008JZ4MLB for P.P.M., A.D., F.C.Z., P.B., and E.R.

References

Abbe, E., 1884. Note on the proper definition of the amplifying power of a lens or a lens-system. J R Microsc Soc 4, 348–351.

Belov, V.N., Wurm, C.A., Boyarskiy, V.P., Jakobs, S., Hell, S.W., 2010. Rhodamines NN: a novel class of caged fluorescent dyes. Angew Chem Int Ed Engl 49, 3520–3523.

Berns, M.W., 1976. A possible two-photon effect in vitro using a focused laser beam. Biophys J 16, 973–977.

Bertero, M., Boccacci, P., 1998. Introduction To Inverse Problems in Imaging. IOP Publishing, Bristol, UK.

Betzig, E., Patterson, G.H., Sougrat, R., Lindwasser, O.W., Olenych, S., Bonifacino, J.S., et al., 2006. Imaging intracellular fluorescent proteins at nanometer resolution. Science 313, 1642–1645.

Bianchini, P., Diaspro, A., 2012. Fast scanning STED and two-photon fluorescence excitation microscopy with continuous wave beam. J Microsc 245, 225–228.

Bianchini, P., Harke, B., Galiani, S., Diaspro, A., 2011. Two-photon excitation STED-CW microscopy. Proc SPIE 7903, 1–5.

Bianco, B., Diaspro, A., 1989. Analysis of the three dimensional cell imaging obtained with optical microscopy techniques based on defocusing. Cell Biophys 15, 189–200.

Blosser, T., Yang, J., Stone, M., Narlikar, G., Zhuang, X., 2009. Dynamics of nucleosome remodeling by individual ACF complexes. Nature 462, 1022–1027.

Brakenhoff, G.J., Muller, M., Ghauharali, R.I., 1996. Analysis of efficiency of two-photon versus single-photon absorption of fluorescence generation in biological objects. J Microsc 183, 140–144.

Cella Zanacchi, F., Lavagnino, Z., Perrone Donnorso, M., Del Bue, A., Furia, L., Faretta, M., et al., 2011. Live-cell 3D superresolution imaging in thick biological samples. Nat Methods 8, 1047–1049.

Chirico, G., Cannone, F., Baldini, G., Diaspro, A., 2003. Two-photon thermal bleaching of single fluorescent molecules. Biophys J 84, 588–598.

Cremer, C., Cremer, T., 1978. Considerations on a laser scanning microscope with high resolution and depth of field. Microsc Acta 81, 31–44.

Davis, S.K., Bardeen, C.J., 2002. Using two-photon standing waves and patterned photobleaching to measure diffusion from nanometers to microns in biological systems. Rev Sci Instrum 73, 2128–2135.

Dedecker, P., 2008. Diffraction-unlimited optical microscopy. Mater Today 11, 12–21.

Delon, A., Usson, Y., Derouard, J., Biben, T., Souchier, C., 2004. Photobleaching, mobility and compartmentalisation: inferences in fluorescence correlation spectroscopy. J Fluoresc 14, 255–267.

Denk, W., Strickler, J.H., Webb, W.W., 1990. 2-photon laser scanning fluorescence microscopy. Science 248, 73–76.

di Francia, G.T., 1952. Super-gain antennas and optical resolving power. Nuovo Cimento 9, 426–435.

Diaspro, A., 2002. Confocal and Two-Photon Microscopy. Wiley-Liss, New York.

Diaspro, A., 2004. Confocal and multiphoton microscopy. In: Palumbo, G., Pratesi, R. (Eds.), Lasers and Current Optical Techniques in Biology. RsC-Royal Society of Chemistry, Cambridge, UK, pp. 429–478.

Diaspro, A., 2010. Nanoscopy and Multidimensional Optical Microscopy. CRC Press, Boca Raton, FL, pp. 1–448.

Diaspro, A., Robello, M., 2000. Two-photon excitation of fluorescence for three-dimensional optical imaging of biological structures. J Photochem Photobiol B 55, 1–8.

Diaspro, A., Sartore, M., Nicolini, C., 1990. Three-dimensional representation of biostructures imaged with an optical microscope: I. Digital optical sectioning. Image Vis Comput 8, 130–141.

Diaspro, A., Chirico, G., Collini, M., 2005. Two-photon fluorescence excitation in biological microscopy and related techniques. Q Rev Biophys 38, 97–166.

Dilipkumar, S., Mondal, P.P., 2011. High resolution multiple excitation spot optical microscopy. AIP Adv 1, 022128.

Dilipkumar, S., Diaspro, A., Mondal, P.P., 2011. Spatial filter based 3D resolution improvement and polarization properties of multiphoton multiple- excitation-spot-optical microscopy. Rev Sci Instrum 82, 063705.

Ding, J.B., Takasaki, K.T., Sabatini, B.L., 2009. Supraresolution imaging in brain slices using stimulated-emission depletion two-photon laser scanning microscopy. Neuron 63, 429–437.

Drobizhev, M., Makarov, N.S., Tillo, S.E., Hughes, T.E., Rebane, A., 2011. Two photon absorption properties of fluorescent proteins. Nat Methods 8, 393–399.

Egner, A., Geisler, C., von Middendor, C., Bock, H., Wenzel, D., Medda, R., et al., 2007. Fluorescence nanoscopy in whole cells by asynchronous localization of photoswitching emitters. Biophys J 93, 3285–3290.

Esposito, A., Federici, F., Usai, C., Cannone, F., Chirico, G., Collini, M., et al., 2004. Notes on theory and experimental conditions behind two-photon excitation microscopy. Microcs Res Tech 63, 1217.

Folling, J., Belov, V., Riedel, D., Schnle, A., Egner, A., Eggeling, C., et al., 2008a. Fluorescence nanoscopy with optical sectioning by two-photon induced molecular switching using continuous-wave lasers. Chemphyschem 9, 321–326.

Folling, J., Bossi, M., Bock, H., Medda, R., Wurm, C.A., Hein, B., et al., 2008b. Fluorescence nanoscopy by ground-state depletion and single-molecule return. Nat Methods 5, 943–945.

Goppert-Mayer, M., 1931. Uber elementarakte mit zwei quantensprungen. Ann Phys Lpz 9, 273–283.

Gustafsson, M.G.L., 2005. Nonlinear structured-illumination microscopy: wide-field fluorescence imaging with theoretically unlimited resolution. Proc Nat Acad Sci USA 102, 13081–13086.

Hell, S.W., 2007. Far-field optical nanoscopy. Science 316, 1153–1158.

Hell, S.W., Stelzer, E.H.K., 1992. Fundamental improvement of resolution with a 4Pi confocal fluorescence microscope using two photon excitation. Opt Comm 93, 77–282.

Hell, S.W., Wichmann, J., 1994. Breaking the diffraction resolution limit by stimulated emission: stimulated- emission-depletion fluorescence microscopy. Opt Lett 19, 780–782.

Hell, S.W., Schmidt, R., Egner, A., 2009. Diffraction-unlimited three-dimensional optical nanoscopy with opposing lenses. Nat Photonics 3, 381–387.

Hell, S.W., Stelzer, E.H.K., Lindek, S., Cremer, C., 1994. Confocal microscopy with an increased detection aperture: true-B 4Pi confocal microscopy. Opt Lett 19, 222–224.

Helmchen, F., Denk, W., 2005. Deep tissue two-photon microscopy. Nat Methods 2, 932–940.

Hess, S.T., Girirajan, T.P.K., Mason, M.D., 2006. Ultra-high resolution imaging by fluorescence photoactivation localization microscopy. Biophys J 91, 4258–4272.

Huang, B., Babcock, H., Zhuang, X., 2010. Breaking the diffraction barrier: Super-resolution imaging of cells. Cell 143, 1047–1058.

Kaiser, W., Garret, C.G.B., 1961. Two-photon excitation in $CaF2:Eu^{2+}$. Phys Rev Lett 7, 229–231.

Kawata, S., Inouye, Y., Verma, P., 2009. Plasmonics for near-field nano-imaging and superlensing. Nat Photonics 3, 388–394.

Klar, T.A., Jakobs, S., Dyba, M., Egner, A., Hell, S.W., 2000. Fluorescence microscopy with diffraction resolution barrier broken by stimulated emission. Proc Natl Acad Sci USA 97, 8206–8210.

Koester, H.J., Baur, D., Uhl, R., Hell, S.W., 1999. Ca^{2+} fluorescence imaging with pico- and femtosecond two-photon excitation: signal and photodamage. Biophys J 77, 2226–2236.

Konig, K., 2000. Multiphoton microscopy in life sciences. J Microsc 200, 83104.

Konig, K., Becker, T.W., Fischer, P., Riemann, I., Halbhuber, K.J., 1999. Pulse-length dependence of cellular response to intense near-infrared laser pulses in multi-photon microscopes. Opt Lett 24, 113–115.

Lippitz, M., Erker, W., Decker, H., van Holde, K.E., Basche, T., 2002. Two-photon excitation microscopy of tryptophan-containing proteins. Proc Nat Acad Sci USA 99, 2772–2777.

Mazza, D., Cella, F., Vicidomini, G., Krol, S., Diaspro, A., 2007. Role of three-dimensional bleach distribution in confocal and two-photon fluorescence recovery after photobleaching experiments. Appl Opt 46, 7401–7411.

McNally, J.G., Smith, C.L., 2002. Photobleaching by confocal microscopy. In: Diaspro, A. (Ed.), Confocal and Two-Photon Microscopy: Foundations, Applications, and Advances. Wiley-Liss, New York, pp. 525–538.

Mertz, J., 1998. Molecular photodynamics involved in multi-photon excitation fluorescence. Eur Phys J D3, 53–66.

Mertz, J., Xu, C., Webb, W.W., 1995. Single-molecule detection by two-photon-excited fluorescence. Opt Lett 20, 2532–2534.

Mondal, P.P., 2009. Multispot point spread function for multiphoton fluorescence microscopy. Rev Sci Instrum 80, 096104.

Mondal, P.P., 2010. Multiple excitation nano-spot generation and confocal detection for far-field microscopy. Nanoscale 2, 381–384.

Mondal, P.P., Diaspro, A., 2007. Reduction of higher-order photobleaching in two-photon excitation microscopy. Phys Rev E Stat Nonlin Soft Matter Phys 75, 061904.

Mondal, P.P., Diaspro, A., 2008. Lateral resolution improvement in two photon excitation microscopy by aperture engineering. Opt Comm 281, 1855–1859.

Mondal, P.P., Gilbert, R.J., So, P.T.C., 2010. Photobleaching reduced fluorescence correlation spectroscopy. Appl Phys Lett 97, 103704.

Mondal, P.P., Vicidomini, G., Diaspro, A., 2007. Markov random field aided bayesian approach for image reconstruction in confocal microscopy. J Appl Phys 102, 044701.

Mondal, P.P., Vicidomini, G., Diaspro, A., 2008. Image reconstruction for multi-photon fluorescence microscopy. Appl Phys Lett 92, 103902.

Moneron, G., Hell, S.W., 2009. Two-photon excitation STED microscopy. Opt Express 17, 14567.

Neil, M.A.A., Juskaitis, R., Wilson, T., Laczik, Z.J., 2000. Optimized pupil-plane filters for confocal microscope point-spread function engineering. Opt Lett 25, 245–247.

Patterson, G.H., Piston, D.W., 2000. Photo-bleaching in two-photon excitation microscopy. Biophys J 78, 2159–2162.

Planchon, T.A., Gao, L., Milkie, D.E., Davidson, M.W., Galbraith, J.A., Galbraith, C.G., et al., 2011. Rapid three-dimensional isotropic imaging of living cells using Bessel beam plane illumination 8, 417–423.

Rounds, D.E., Olson, R.S., Johnson, F.M., 1966. Two-photon absorption in reduced nicotinamideadenine denucleotide (NADH). Northeast Electronics Research and Engineering Meeting (NEREM) Rec. 8:158.

Rust, M.J., Bates, M., Zhuang, X., 2006.). Subdiffraction-limit imaging by stochastic optical reconstruction microscopy (STORM). Nat Methods 3, 793–795.

Schneider, M., Barozzi, S., Testa, I., Faretti, M., Diaspro, A., 2005. Two-photon activation and excitation properties of PA-GFP in the 720-920-nm region. Biophys J 89, 1346–1352.

Schrader, M., Hell, S.W., van der Voort, H.T.M., 1998. Three-dimensional super-resolution with a 4Pi-confocal microscope using image restoration. J Appl Phys 84, 4033.

Sheppard, C.J., Kompfner, R., 1978. Resonant scanning optical microscope. Appl Opt 17, 2879–2885.

So, P.T.C., Dong, C.Y., Masters, B.R., Berland, K.M., 2000. Two-photon excitation fluorescence microscopy. Annu Rev Biomed Eng 2, 399–429.

Thompson, R.E., Larson, D.R., Webb, W.W., 2002. Precise nanometer localization analysis for individual fluorescent probes. Biophys J 82, 2775–2783.

Tokunaga, M., Imamoto, N., Sakata Sogawa, K, 2008. Highly inclined thin illumination enables clear single-molecule imaging in cells. Nat Methods 5, 159–161.

van Kempen, G.M.P., van Vliet, L.J., Verveer, P.J., van der Voort, H.T.M., 1997. A quantitative comparison of image restoration methods for confocal microscopy. J Microsc 185, 354–365.

Vicidomini, G., Moneron, G., Han, K.Y., Westphal, V., Ta, H., Reuss, M., et al., 2011. Sharper low-power STED nanoscopy by time gating. Nat Methods 8, 571–573.

Willig, K.I., Harke, B., Medda, R., Hell, S.W., 2007. STED microscopy with continuous wave beams. Nat Methods 4, 915–918.

Wolf, E., Richards, B., 1959. Electromagnetic diffraction in optical systems. II. Structure of the image field in an aplanatic system. Proc Roy Soc A 253, 358–379.

Xu, C., Zipfel, W., Shear, J.B., Williams, R.M., Webb, W.W., 1996. Multiphoton fluorescence excitation: new spectral windows for biological nonlinear microscopy. Proc Natl Acad Sci USA 93, 10763–10768.

York, A.G., Ghitani, A., Vaziri, A., Davidson, M.W., Shroff, H., 2011. Confined activation and subdiffractive localization enables whole-cell PALM with genetically expressed probes. Nat Methods 8, 327–333.

Appendix

Single-photon excitation was well known for centuries, but the excitation using two photons was first conceptualized by Maria Goppert-Mayer (1931). The occurrence of such a rare event was predicted by quantum mechanics but not realized until the advent of pulsed lasers. The first experimental realization of such an event along with its application in biological imaging was from the group of Watt Webb (Denk et al., 1990), followed by several developments (Diaspro et al., 2005).

Initially, the system is in the ground state ($|\varnothing_i\rangle$). An incoming radiation field perturbs the perturbation-free, time-independent Hamiltonian. The resulting Hamiltonian consists of time-dependent perturbation and perturbation-free, time-independent Hamiltonian $H'=H_0+V$. The resulting state can be decomposed in terms of stationary state basis set as the following:

$$|\psi,t\rangle_I = \sum_n C_n(t)|\varnothing_n\rangle. \qquad (6.A.1.1)$$

where $C_n=\langle\varnothing_n|U_I(t)|\varnothing_i\rangle$ is the time-evolution operator (Esposito et al., 2004), $V(t)=A_1 e^{-i\omega t}$, and $A_1 = \dfrac{e}{mc}\bar{p}.\hat{q}A_0 e^{i\vec{k}.\vec{r}}$.

Transition Probability for Single-Photon Excitation

Single-photon excitation is associated with the first non-zero term of the expansion. $C_n^{(1)}$ is the probability that the system makes a transition from $|\varnothing_i\rangle$ to the excited state $|\varnothing_n\rangle$ by the absorption of a single photon. The transition probability is given by the following:

$$p^{(1)}(|\varnothing_i\rangle \rightarrow |\varnothing_n\rangle) = \left|C_n^{(1)}\right|^2 = \frac{2\pi}{\hbar}|\langle\varnothing_n|V|\varnothing_i\rangle|^2\delta(E_{ni}-\hbar\omega) \qquad (6.A.1.2)$$

where $E_{ni}=(E_n-E_i)$.

Single-photon cross-section probability can be rewritten in terms of photon number $\left(\dfrac{1}{\hbar\omega}\right)$ and molecular cross-section as shown by the following:

$$\sigma_s = \frac{2\pi e^2}{m^2 c^2 \omega}\left|\left\langle\varnothing_n|\hat{\epsilon}.\bar{p}e^{i\frac{\omega}{c}\widehat{n.\vec{r}}}|\varnothing_i\right\rangle\right|^2 \delta(E_{ni}-\hbar\omega) \qquad (6.A.1.3)$$

$$= \frac{4\pi^2 e^2}{c} \frac{\omega_{ni}^2}{\omega} |\langle \varnothing_n | x | \varnothing_i \rangle|^2 \, \delta(E_{ni} - \hbar\omega) \qquad (6.A.1.4)$$

And is the single-photon cross-section.

Transition Probability for MPE

MPE is associated with the Nth non-zero term of the expansion (Eq. 6.A.1.1). The excitation is via multiple intermediate virtual states. The transition probability is given by the following:

$$p^{(N)}(|\varnothing_i\rangle \to |\varnothing_n\rangle) = |C_n^{(N)}|^2.$$

Again, we can write this in terms of photon number $\left(\dfrac{1}{\hbar\omega}\right)$ and molecular cross-section, that is,

$$p^{(N)}(|\varnothing_i\rangle \to |\varnothing_n\rangle) = \left(\frac{1}{\hbar\omega}\right)^N \sigma_N. \qquad (6.A.1.5)$$

The molecular cross-section is given by the following:

$$\sigma_{MPE}(|\varnothing_i\rangle \to |\varnothing_n\rangle) \propto$$
$$\sum_{|\varnothing_{N-1}\rangle |\varnothing_{N-1}\rangle \ldots |\varnothing_1\rangle} |\langle \varnothing_n | x | \varnothing_{N-1}\rangle \| \langle \varnothing_{N-1} | x | \varnothing_{N-2}\rangle | \ldots |\langle \varnothing_1 | x | \varnothing_i \rangle||^2$$
$$\delta \left(\omega_{n\,N-1} + \omega_{N-1\,N-2} + \ldots + \omega_{1i - N\hbar\frac{\omega}{N}} \right). \qquad (6.A.1.6)$$

From an experimental viewpoint, 2PE is the most utilized non-linear process. The molecular cross-section for 2PE is given by the following:

$$\sigma_{2PE}(|\varnothing_i\rangle \to |\varnothing_n\rangle) = \frac{8\pi^3 e^4}{\hbar^2 c^2} \frac{\omega_{nm}^2 \omega_{mi}^2}{\omega^2} \times$$
$$\left| \sum_{||\varnothing_m\rangle} \frac{\langle \varnothing_n | x | \varnothing_m \rangle \| \langle \varnothing_m | x | \varnothing_i \rangle \|}{(\omega_{ni} - \omega)^2} \right|^2 \qquad (6.A.1.7)$$
$$\delta \left(\omega_{nm} + \omega_{mi} - 2\hbar\frac{\omega}{2} \right)$$

And is the two-photon cross-section. Additionally, it should be remembered that for 2PE we have used light of wavelength $2\lambda = \omega/2$.

Simultaneously a $\leq 10^{-16}$ s interaction of two or more photons with the molecule results in an MPE process. Lifetimes of the virtual intermediate states determine the time scale for photon coincidence. Approximate cross-section for n-photon absorption is given by the following (Xu et al., 1996):

$$\sigma_n = A^n \Delta \tau^{n-1} \qquad (6.A.1.8)$$

So, the approximate cross-sections for single- and two-photon absorption are $\sigma_1 \approx A = 10^{-17}$ (s/photon) and $\sigma_2 \approx A^2 \Delta \tau = 10^{-49}$ (s/photon), respectively (the dipole transition length is $0.1\,\text{Å}$ for $A = 10^{-17}$ and the lifetime of virtual states is around 10^{-14} s). An extremely small molecular cross-section makes it difficult to realize multiphoton processes. Typical single- and two-photon cross-sections for a fluorescent probe such as DAPI are $1.3 \times 10^{-16}\,\text{cm}^2$ at 345 nm and $1.6 \times 10^{-49}\,\text{cm}^2$ at 700 nm. Collections of two-photon cross-sections can be found in Xu et al. (1996) and Esposito et al. (2004).

THE CELL AT MOLECULAR RESOLUTION: PRINCIPLES AND APPLICATIONS OF CRYO-ELECTRON TOMOGRAPHY

Rubén Fernández-Busnadiego[1], and Vladan Lucic[2]

[1]*Yale University School of Medicine, New Haven, Connecticut, USA,*
[2]*Max-Planck-Institute of Biochemistry, Martinsried, Germany*

CHAPTER OUTLINE
Introduction: Cellular Landscapes at Molecular Resolution   141
The Cryo-ET Method   143
    The Cryo Preparation   143
    Imaging in EM   145
    Tomography   148
    Resolution, Noise, and Radiation Damage   149
Detection, Identification, and Hybrid Methods   151
    Visual Identification and Cryo-ET of Intact Cells   151
    Correlative LM and Cryo-ET   156
    Labeling Strategies for Correlative and Direct Identification
        Approaches   161
    Linking Structure and Physiology: Functional Manipulations in Cryo-ET   163
    Uncovering Information by Image Processing   169
Conclusions   175
Acknowledgments   178
References   178

Introduction: Cellular Landscapes at Molecular Resolution

Electron microscopy (EM) has been instrumental for the development of cell biology, as it offered the first visualizations of intracellular architecture and organelles (Porter et al., 1945; Palade, 1955). By

Cellular Imaging Techniques for Neuroscience and Beyond.
DOI: http://dx.doi.org/10.1016/B978-0-12-385872-6.00007-6

revealing the mechanisms of fundamental processes such as synaptic transmission, EM has also played a critical role in neuroscience. Pioneering EM work provided definitive proof of the neuron doctrine by showing the discontinuity between presynaptic and postsynaptic terminals (Robertson, 1953; Sjostrand, 1953; Estable et al., 1954). Synaptic vesicles, first visualized by EM (De Robertis and Bennett, 1954; Palade and Palay, 1954), were hypothesized to mediate the observed quantal nature of neurotransmitter release (Del Castillo and Katz, 1954) as later EM studies demonstrated (Heuser et al., 1979). Today, however, cell biology and neuroscience have shifted their focus to light microscopy (LM) techniques that allow the dynamic study of specific molecules by fluorescent tags and enable real-time imaging in living cells. Furthermore, as reviewed in this volume, novel LM approaches that go beyond the diffraction barrier now reach resolutions of tens of nanometers. In the face of these developments, EM does retain the following key advantages: (1) single nanometer resolution, one order of magnitude higher than the most advanced LM techniques and (2) comprehensive ultrastructural information, visualizing not only a few chosen structures but their entire cellular context, that is, revealing both cellular morphology and, simultaneously, the localization of the molecules of interest.

But to what degree of detail can EM describe cellular architecture? EM techniques based on sample dehydration and heavy-metal staining offer a reliable description of, essentially, the membranous compartments of the cell and its major macromolecules. Cryo-electron tomography (cryo-ET) goes one step further by providing 3D visualization of molecular complexes in their native, fully hydrated environment. Whereas higher resolution approaches such as x-ray and EM crystallography, NMR, or single particle cryo-EM can be pursued when working with large numbers of copies of a (purified) given molecule, cryo-ET allows visualization of nonrepetitive pleiomorphic structures such as cells or organelles at a resolution better than 5 nm. Therefore, it is ideally suited to reveal cellular organization at molecular resolution.

Technical details that are important for understanding images obtained by cryo-ET, as well as the potential and the limitations of this technique, are covered in the following section. These include cryo-preservation (vitrification), image formation, ET, and the specific issues that arise when vitrified samples are imaged in EM (Figure 7.1). The high sensitivity of vitrified specimens to radiation damage effectively limits the electron exposure, resulting in a poor signal-to-noise ratio (SNR). The unprecedented fidelity of cryo-ET visualization of cellular landscapes can also turn into a double-edged sword for the microscopist, as the amount of data in cryo-tomograms and its complexity is often overwhelming. In fact, the identification of the objects of interest remains challenging for cryo-EM as a whole. The section Detection, Identification, and Hybrid Methods will focus on strategies to overcome this problem: the most prominent intracellular structures (e.g., actin filaments) can be directly identified by visual inspection (see Visual Identification and Cryo-ET of Intact Cells), but even in these cases the search procedure can be tedious. Correlative LM and cryo-ET (see Correlative LM and Cryo-ET) can ease this problem by recording in LM the location of the features of interest labeled with fluorescent dyes, and then pointing cryo-ET directly to those positions. Current labeling techniques applicable to cryo-ET, including the development of clonable electron-dense tags, are discussed in the section Labeling Strategies for Correlative and Direct Identification Approaches. Pharmacological and genetic manipulations (see Linking Structure and Physiology: Functional Manipulations in Cryo-ET) can be used to identify the structures of interest and understand their physiological roles by disrupting them or inducing morphological changes. Last, computational approaches (see Uncovering Information by Image Processing) are crucial toward the ultimate goal: extracting the maximum amount of biologically relevant information from cryo-tomograms. Instead of giving a comprehensive review of cryo-ET applications, priority will be given here to those that brought significant methodological advances, and also to applications investigating eukaryotic and in particular neuronal cells.

The Cryo-ET Method

The Cryo Preparation

Liquid water is obviously the ideal environment for biological samples, but its considerable rate of evaporation in a vacuum makes it is incompatible with imaging in the electron microscope column. Despite best efforts, the introduction of an EM hydration stage (Parsons, 1974; Parsons et al., 1974) did not overcome the problem in the long run and was superseded by methods that, one way or another, dehydrate the sample. Today, these methods typically include chemical fixation to stabilize the sample followed by dehydration, embedding in a resin, and contrasting with heavy metals.

Cryo-preservation (vitrification) takes a different path. Instead of being removed, liquid water is solidified, which provides physical fixation of the sample. Furthermore, in this preparation biological samples

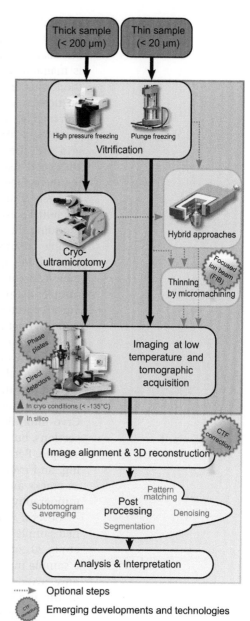

Figure 7.1 Workflow of cryo-electron microscopy, with selected emerging methods under active development. (*Reproduced from* Vanhecke et al., 2011, *with permission.*).

are embedded in a crystal-free state of solid water (vitreous water or amorphous ice). That is, not only the samples are frozen-hydrated but also vitrified; that is, the natural distribution of water molecules is maintained, as the rapid cooling (within microseconds) prevents the formation of hexagonal ice crystals which have deleterious consequences to sample ultrastructure (Dubochet et al., 1988). The vitrification method is dictated by the sample type and its thickness (Figure 7.1). Samples of 20 μm or thinner can be vitrified by plunge freezing: an EM grid with the sample is plunged into a cryogen (a liquid at $< -160°C$ with good heat-transfer characteristics; e.g., ethane or propane) allowing the achievement of cooling speeds in the order of 105 °C/s. Liquid nitrogen is unsuitable for vitrification, because it boils quickly upon the initial contact with a sample, producing a layer of nitrogen gas surrounding the sample that drastically impedes heat propagation and sample cooling. High pressure freezing (HPF) combines rapid freezing with the application of high pressure (2000 bar) which lowers the melting point of water and acts as a physical cryoprotectant. Biological samples of up to 200 μm thickness can be vitrified using this technique, but cryoprotectants are often needed (Studer et al., 2008). Vitrified samples ought not to be warmed above −135°C, since at this temperature water undergoes a phase transition resulting in cubic ice crystal formation, which damages the ultrastructure. Hence, the sample must be consistently kept at liquid nitrogen temperature.

Samples thicker than 0.5–1 μm are too thick for imaging in EM and require a thinning step under cryo-conditions (i.e., below −135°C). Cryo-ultramicrotomy (cryo-sectioning) produces thin (50–500 nm) sections, but causes artifacts, notably anisotropic compression of the sample. Performing cryo-ultramicrotomy demands skills and training, and even then the yield is mostly modest. Other artifacts such as cracks and crevasses can locally render the sample unsuited for tomography and lower the yield of good samples.

Several new cryo-micromachining methods to thin down vitrified samples are under active development (Rigort et al., 2010; Rigort et al., 2012a). Cryo-planing, an approach where the upper layers of a sample mounted on an EM grid are removed by cryo-ultramicrotomy, avoids typical cryo-sectioning artifacts. The remaining sample on the grid shows no compression, but only features close to the grid can be imaged. Focused ion beam (FIB) milling is more precise than ultramicrotomy-based methods. Here, material at selected areas of the sample is ablated with a focused beam of gallium ions at a precision of tens of nanometers. Currently, there are no indications that the localized release of a high amount of energy by the ion beam causes significant heating or other damage.

Because water, the main constituent of biological material, is maintained in hydrated vitreous samples, they are considered to represent the native structure more faithfully than dehydrated samples.

These cryo-preserved biological samples are generally characterized by smooth and continuous membranes and non-extracted cytoplasm (see for example Figures 7.3A,D and 7.6C,D). Artifacts such as aggregation and structural rearrangements arising from chemical fixation and dehydration are avoided (Dubochet and Sartori Blanc, 2001). The biological material is observed directly, unlike in the heavy-metal-stained samples where uneven stain distribution and a mixture of positive and negative contrast interfere with data interpretation. In order to overcome the difficulties of sectioning or thinning vitrified samples (as discussed in the previous paragraph), high pressure frozen samples may be dehydrated by freeze substitution and embedded in plastic resin. While samples prepared using this method are dehydrated, the initial chemical fixation is avoided. This method is already established in other branches of cell biology, and its increased use in neuroscience is a promising development.

Currently, cryo-preservation ensures the best preservation of biological material. However, the sensitivity of vitrified material to electron irradiation makes low-dose imaging conditions compulsory. Working with vitrified samples is challenging and requires state-of-the art instrumentation and a skilled operator in order to reap the advantages.

Imaging in EM

Electron microscopes used for cryo-ET are typically equipped with a field emission gun yielding a highly coherent electron beam, and are operated at intermediate electron accelerating voltages (100–300 kV). Although electrons accelerated at higher voltages penetrate deeper and allow imaging of thicker samples, their detection tends to be poorer. Consequently, 300 kV is typically used for imaging intact cells, while cellular fractions and cryo-thinned samples may be imaged at lower voltages.

Magnetic lenses of an electron microscope can focus an electron beam like optical lenses do in a light microscope. The imaging system of an electron microscope comprises condenser lenses to adjust the sample illumination and an objective lens system that provides the initial magnification, as well as projection lenses that provide further magnification. Stigmator lenses correct the aberrations of the main lenses that arise because lenses cannot be made perfectly rotationally symmetrical. In addition, deflector lenses are used to tilt or shift the beam.

The interaction between electrons and specimen results in elastic and inelastic scattering of the electrons. Kinetic energy is conserved by definition in elastic scattering. It occurs when the electrons interact with the cell nuclei; they essentially do not lose energy during this interaction. Elastically scattered electrons are well localized, contain high-resolution information, and are useful for imaging. Inelastically scattered electrons damage the specimen by depositing part of their energy in the sample, thus causing structural damage. This limits the electron exposure that can be applied before structural changes within the specimen occur. In addition, the low angular spread and generally poor localization and coherence of inelastic scattering events add noise and contaminate the high-resolution information.

In biological samples, at 100–300 kV acceleration voltage, inelastic scattering is about three times more likely to occur than elastic scattering. Consequently, multiply scattered electrons tend to undergo at least one inelastic scattering. While sample damage can only be avoided by using a sufficiently low electron exposure, the noise caused by inelastic scattering can be reduced by removing inelastically scattered electrons with an energy filter that operates in the zero energy-loss mode. This filter contains a system of lenses that work like a magnetic prism, separating electrons by their energy, with a slit that allows the passage of the electrons at a specified energy range. It can be integrated into the column (Ω-filter) or placed as a postcolumn filter immediately before the detector.

The image contrast is in general due to the difference between the density (atomic numbers, to be precise) of the biological material and the embedding medium (vitreous water). Because unstained biological material essentially consists of atoms of low atomic number, the contrast and, consequently, the SNR in cryo-ET are generally lower than with stained samples. Amplitude contrast, which is due to electrons scattering at angles that are too high to be collected and detected, is the main contrast-forming mechanism for stained samples, but it plays only a minor role for unstained biological specimens due to the lower scattering angles of lighter elements.

Phase contrast is the main contrast mechanism for unstained samples. It is essentially an interference effect, similarly to phase contrast in LM. However, it is induced in EM by negative defocusing (underfocusing), that is, imaging at an offset from the optical image plane. This form of imaging is described mathematically by the microscope contrast transfer function (CTF), a (damped) sinusoidal function of the spatial frequencies (Figure 7.2). The CTF attenuates low frequencies as well as (high) frequencies close to its first zero, and generates negative contrast at some higher frequencies (Figure 7.2, top). This effectively limits the resolution in tomograms to the first zero of the CTF. Phase plates promise to improve contrast without sacrificing resolution and are currently under development. Aberration correctors that compensate for spherical and chromatic aberration were recently introduced and might also prove beneficial in the future, particularly for higher resolution data.

The automated feedback loops used in tomographic acquisition work by immediately accessing recorded images; therefore they require a digital camera. Electron detectors currently used in cryo-ET are based on a charge-coupled device (CCD). The electrons are

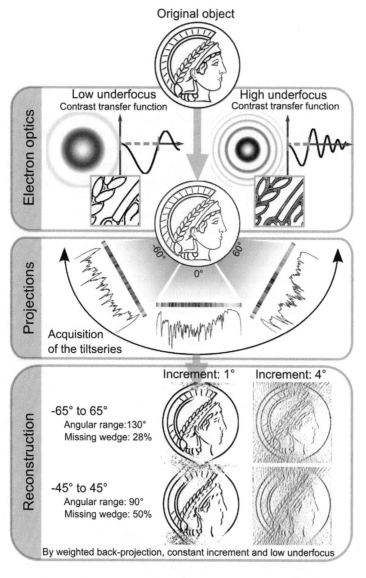

Figure 7.2 Principle of electron tomography. Top, a 2D object with details illustrating the influence of the CTF (shown in both line-graph and 2D kernel forms) at low and high underfocus. Middle, one-dimensional projections of the object at different angles shown as projection images and line graphs. Bottom, tomographic reconstructions of tilt series with different tilt range and angular increments obtained by weighted back-projection. (*Reproduced from Vanhecke et al., 2011, with permission.*)

converted to photons by a scintillator array, followed by photon detection by a CCD. Higher thickness of the scintillator layer improves the sensitivity, which is highly desirable for thicker samples, but leads to poorer signal localization. In contrast to CCD-based cameras, direct detector devices (DDD) detect high-energy electrons directly, avoiding the distortions induced by the generation and transfer of photons. Faster readout and better transfer properties promise to make DDDs superior to steadily improving CCD cameras, especially at intermediate voltages, albeit at a cost of a limited lifetime (McMullan et al., 2009; Milazzo et al., 2011).

Tomography

In tomographic acquisition, a set of images is recorded at different spatial orientations of the sample. This is made possible by a computerized specimen stage that allows precise translation in the x-, y-, and z-axis as well as tilting of the specimen. Due to a large depth of focus, each of these images is a projection through the sample. These images are then computationally merged into a 3D image, a tomogram (Figure 7.2, bottom). This is in contrast with confocal LM where acquired images represent z-slices of a sample. In ET, z-slices can be computationally extracted only after a tomogram is reconstructed.

In ET, the most common data acquisition scheme is the single-axis tilt series, where a specimen is rotated (tilted) around a fixed axis. The tilt range is usually limited to ±70° due to physical restrictions of the instrument, and may be further limited to about ±60° due to other factors like sample thickness, resulting in an incomplete set of projections (Figure 7.2, middle). Because each projection image is represented in the Fourier space as a central slice perpendicular to the direction of the beam, single-tilt tomographic series have a wedge-shaped region in the Fourier space where structural information is missing. This missing wedge causes distortions in tomograms such as elongation in the direction of the electron beam. Approaches like dual-axis tilt acquisition, where the sample is tilted around two orthogonal axes, or a conical tilt scheme where the sample is rotated around the optical axis at a constant tilt angle, reduce the amount of missing information, but they are rarely used in cryo-ET because of technical difficulties.

The two most commonly used angular increment schemes for single-axis tilt are the constant increment and the Saxton scheme (Saxton et al., 1984). The Saxton scheme uses a decreasing tilt increment for higher tilt angles, and although it is theoretically advantageous because it reduces the anisotropicity of the resolution caused by the apparent increase of thickness of a realistic (slab-like) sample geometry, the constant increment appears to perform better in the presence of noise. The maximum attainable isotropic resolution d of a cylindrical sample with diameter D imaged at a finite angular increment $\Delta\alpha$ is determined from Crowther's criterion as $d = D\, \Delta\alpha$ (Crowther et al., 1970). To compensate for the increased apparent thickness at higher tilt angles, the electron exposure is typically increased with increased tilt angle.

As we will discuss in the following sections, biological samples are sensitive to electron beam exposure, so the total electron exposure of the sample has to be controlled. The Hegerl-Hoppe dose fractionation theorem provides the underlying theoretical principle of 3D imaging. It states that by spreading the total electron dose among projections of a tomographic series, the information in the third dimension can be gained without loss of resolution in the x-y plane

(Hegerl and Hoppe, 1976). At the practical level, the development of automated schemes allows tilt series acquisition in low-dose mode by correcting positioning and focusing errors that are due to mechanical imperfections of the stage (Koster et al., 1997). This is accomplished with the help of additional images that are typically recorded at a distance to the feature of interest, causing no or only minimal additional beam exposure of the feature of interest. Currently, acquisition schemes using prediction procedures and/or stage movement calibration allow faster data acquisition and require less operator input (Ziese et al., 2002; Mastronarde, 2005; Zheng et al., 2007).

To correct for translations in the image plane and possibly for errors due to imprecise tilting and imperfect tilt axis orientation in respect to the optical axis, the projections of a tilt series need to be aligned. The most straightforward way to align the image stack is by selecting fiducial markers, such as high-contrast particles (typically gold), which are easily recognizable even in very noisy images. Alternatively, fiducial-less alignment procedures rely on the correlation of intrinsic features of projections and thus have a limited applicability for noisy cryo-ET images.

The reconstruction process transforms the aligned tilt series into a 3D image or tomogram. Currently, the most commonly used 3D reconstruction procedure is the weighted back-projection. It essentially re-projects the images to the 3D space and corrects for the uneven radial distribution of information in the Fourier space that this re-projection causes (Hoppe et al., 1986; Radermacher, 1992). While this approach attempts to solve the reconstruction problem exactly, the system that needs to be solved is both underdetermined and overconstrained. Iterative reconstruction algorithms such as algebraic reconstruction technique and simultaneous iterative reconstruction technique attempt to minimize the difference between tilt series projections and re-projections of the 3D image (Gordon et al., 1970). Such algorithms can incorporate additional constraints (such as maximum entropy) and may further improve tomogram quality. Reconstructions with the aforementioned algorithms are done in real space, as Fourier space approaches are still computationally expensive and are used only for small volumes.

Resolution, Noise, and Radiation Damage

Having discussed the technical aspects of imaging-vitrified samples in EM, we are now ready to assess their contribution to the resolution. Thanks to the very short wavelength of electrons, electron microscopes reached nanometer resolution decades ago, whereas today the resolution of the instrument is better than 1Å. Several technical parameters limit the nominal resolution of tomograms, and the most significant is the pixel size at the specimen level and the position of the first zero of the CTF (1/(3.2–5.5 nm) for typical underfocus values in the range of 5 to 15 μm at 300 kV). In addition, the limited angular range and finite angular increment further limit the resolution and render it anisotropic. The missing Fourier information along the direction of the electron beam causes elongation and blurring in that direction in the 3D reconstruction of the object. Some of the abovementioned limitations can be alleviated by decreasing the pixel size or the angular increment while keeping the total electron exposure unchanged. This would, however, decrease the electron exposure in individual projections and increase the relative contribution of the CCD noise, resulting in decreased contrast and SNR and negatively affecting the accuracy of the alignment and reconstruction.

Arguably, the typically low SNR of cryo-tomograms presents the strongest limitation on the usable resolution. Biological material is only slightly denser than its aqueous environment, resulting in generally poor contrast in projection images. The poor sampling of electron scattering and limitations of CCD camera performance are the main contributors to noise. Oversampling (binning) is advisable to compensate for the inadequate transfer function of CCDs and to increase SNR, but unfortunately it also decreases the nominal resolution limit.

The electron radiation sensitivity of biological samples is a fundamental constraint on the maximal electron illumination of the sample. Inelastic electron scattering can break covalent bonds and cause ionization. High-resolution information is reduced even at low electron exposure (10–20 $e^-/Å^2$). Increasing the exposure leads to a progressive loss of resolution and can cause bubbling of the sample at exposures higher than 150–200 $e^-/Å^2$. This bubbling is likely caused by the accumulation of hydrogen ions, formed by breaking organic carbon-hydrogen bonds within the sample, which at some point exerts a strong force on the sample (Leapman and Sun, 1995). While imaging the sample at lower temperature slows down this process, decreasing the temperature from the liquid nitrogen to the liquid helium temperature (4 K) does not seem to be advisable for cryo-ET, because contrast is decreased at liquid helium temperature (Comolli and Downing, 2005; Glaeser, 2008).

The actual determination of the parameters for a tomographic acquisition usually starts with the estimate of the total electron exposure. As mentioned in the previous paragraph, the SNR receives two contributions with increased electron exposure: one that increases the signal and the other that increases the noise because of the radiation damage of the sample, yielding an optimal exposure at which the SNR is maximized (Hayward and Glaeser, 1979). Given that during imaging the higher resolution information is lost earlier (at lower exposures) than the lower resolution information, the target resolution of a tomogram determines the optimal electron exposure. While the precise dependence of the optimal exposure on resolution has been determined for catalase crystals (Baker and Rubinstein, 2010), the optimal exposure in cryo-ET appears to be sample dependent and is most often determined by heuristic methods. The other acquisition parameters are determined with the idea that the resolution should not be decreased further.

Another general approach for increasing the resolution in cryo-tomograms is to reduce sample thickness, as the number of multiply scattered electrons increases with sample thickness. These electrons are not used for imaging but they increase sample damage. Consequently, the resolution can be increased by imaging cellular fractions and reconstituted systems, albeit at the price of lower biological fidelity. On the other hand, thin sections of vitrified cells obtained by cryo-ultramicrotomy or cryo-micromachining provide higher resolution, but are still very difficult to produce and handle. Therefore, combining cryo-ET data of intact cells and appropriate thin samples may be the solution to couple higher resolution information with *in situ* imaging of molecular complexes.

Detection, Identification, and Hybrid Methods

Visual Identification and Cryo-ET of Intact Cells

The most straightforward way to identify structures visualized in cryo-tomograms, as well as in any other type of EM images, is visual recognition. However, its application is limited to larger structures such as organelles, membrane-bound cellular compartments, and cytoskeleton. While cellular cryo-ET allows *in situ* observation of cellular complexes at the molecular level, its applications to intact eukaryotic cells are particularly challenging due to high sample thickness and are often limited to visual detection. Still, in many cases the resolution is sufficient to gain relevant insights into the architecture and function of cellular and/or subcellular structures.

Plasmodium

Malaria, an infectious disease widespread in tropical and subtropical regions, represents one of the major challenges of global health today. It is caused by a protozoan parasite, *Plasmodium*, which has thus become an important research target. Because *Plasmodium* is less than 1 μm wide, most of its parts are amenable to cryo-ET even in the intact form. In a recent study, several cryo-tomograms of different parts of *Plasmodium* were recorded, and the location of the major organelles and microtubules was determined (Kudryashev et al., 2010c). This work also showed that these organelles and microtubules are linked to the inner membrane complex via numerous small filaments (Figure 7.3A–C). In addition to being at the technical forefront of experiments applying cryo-ET in intact eukaryotic cells, this work on *Plasmodium* exemplifies the approach where the primary motivation for studying a given cell type is its relevance as a whole, as opposed to the importance of subcellular structures like actin or microtubules that are shared by many cell types.

Red blood cells (erythrocytes) are the natural host of *Plasmodium*. Recently, the periphery of intact human erythrocytes infected by *Plasmodium* was been studied by cryo-ET (Cyrklaff et al., 2011).

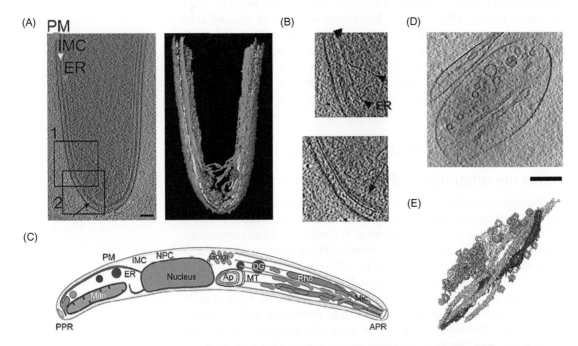

Figure 7.3 Tomography of whole eukaryotic cells. (A) Tomographic slice and a surface rendering of *Plasmodium* posterior end showing plasma membrane (PM; blue), inner membrane complex (IMC; yellow), endoplasmic reticulum (ER; green), cytoplasmic vesicles (red), and the proximal polar ring (black arrow). Boxes indicate regions enlarged in (B). Scale bar, 100 nm. (B) Enlarged boxes from (A). Top, ER membranes (black arrowheads). Bottom, continuity between the PPR (black arrow) and the subpellicular network. (C) Overall architecture of a *Plasmodium* sporozoite. Ap, apicoplast; APR, apical polar ring (at the front end); DG, dense granule; ER, endoplasmic reticulum; IMC, inner membrane complex; Mic, micronemes; Mito, mitochondrion; MT, microtubule; Rho, rhoptries; PPR, proximal polar ring (at the posterior end), PM, plasma membrane. (D) Tomographic slice showing a presynaptic terminal from a dissociated hippocampal culture (DIV 12). Scale bar, 200 nm. (E) Manual segmentation of the presynaptic terminal shown in (D). Green, microtubules; blue, smooth ER; orange, synaptic vesicles; yellow, larger vesicles. (*A–C reproduced from* Kudryashev et al., 2010c, *with permission.*) Please see color plates at the back of the book.

The authors found that *Plasmodium*-infected erythrocytes developed an actin-based cytoskeletal structure connecting erythrocyte membrane with the Maurer's cleft, an elaborate set of membranes generated by the parasite in the host cell cytoplasm. Interestingly, this cytoskeletal structure was aberrant in erythrocytes suffering from hemoglobinopathies that protect patients from malaria's severe symptoms.

Neurons

Primary neurons are a challenging target under extensive investigation in their own right, given their fundamental role in many physiological and pathophysiological mechanisms. Images of cryo-sectioned organotypic hippocampal slices provided the first glimpse at vitrified neuronal synapses (Zuber et al., 2005). Smooth membranes, dense but well-resolved material in the synaptic cleft, and well-distributed cytoplasmic material could be observed. However, cryo-sectioning artifacts, notably compression, were also obvious. In dissociated neuronal cultures, the vitrified layer at neuronal cell bodies and their immediate surroundings are too thick to be examined without thinning procedures. However, arguably the most important functional parts of neurons, axons and dendrites are often thin enough to be suitable samples for cellular tomography (Garvalov et al., 2006; Lucic et al., 2007) allowing direct cryo-ET investigation of the architecture of neuronal complexes *in situ* without prior thinning.

Few processes grow from cultured neuronal cells during the early developmental stages, but these processes become predominant after 7–10 days *in vitro* (DIV). Microtubules are found in all dendrites and axons from only 1 in the thinnest axons to over 10 in processes with larger diameter. Cryo-tomograms of neuronal processes revealed that microtubules contain lumenal densities organized in discrete globular particles (Garvalov et al., 2006). In both dendrites and axons, tomograms showed tubules and sacks of the smooth endoplasmic reticulum (ER) winding around microtubules (Figure 7.3D and E). Mitochondria and the elements of the secretory system, from early to late endosomes such as multivesicular bodies, were also observed (Fernández-Busnadiego et al., 2011).

Dissociated cultures of mature neurons are one of the most important preparations for the investigation of synaptic organization. For cryo-ET, cultures have to be grown at low density, and even then they often develop a dense network of processes that upon vitrification results in a large part of the grid surface being covered with thick ice. Searching for features of interest and recording tilt series are rendered considerably more difficult, and this, consequently, decreases the experimental yield. However, in regions of thin ice, also present in the preparation, it is possible to observe characteristic presynaptic features like clusters of synaptic vesicles and tubulovesicular membranous compartments (Figure 7.3D and E). Moreover, elements of synaptic molecular complexes such as distinct filaments within the postsynaptic density and structures protruding from synaptic vesicle membranes were visualized in their natural cellular environment (Lucic et al., 2007). Nevertheless, the sensitive nature of the cultured neurons complicates their handling for cryo preparations. A related and promising cell line is the neuroblastoma/glioma hybrid cell line NG108-15. These cells change their morphology depending on the presence of growth medium and develop long neurites, taking on a neuron-like morphology, which allows imaging in cryo-ET (Sartori et al., 2007). In any case, cryo-ET of intact neurons is progressing and is approaching the level where it can be applied to verify structural observations obtained in synaptosomes (discussed in the section "Linking Structure and Physiology: Functional Manipulations in Cryo-ET") and put them in a more general cellular context, avoiding the technical difficulties associated with vitrified slice cultures.

Cytoskeleton

While most types of eukaryotic cells are too thick to be imaged *in toto*, some cells have peripheral regions that are considerably thinner and thus suitable for imaging in cryo-ET. This applies in general to motile and adherent cells, where actin dynamics in those thin regions at least partially underlies motility and surface adherence. It does not come as a surprise that the visualization of actin at the peripheral part of *Dictyostelium*, a slime mold cell, marked the beginning of cellular cryo-tomography (Medalia et al., 2002). Further investigations in the same system focused on filopodia (Medalia et al., 2007), slender cytoplasmic extensions involved in sensing the environment and responding to environmental cues. The authors showed that actin filaments were not continuous from the filopodia tip down to the base, and suggested that elongation and bending of filopodia occurs through extensive rearrangement of actin filaments throughout the filopodia. Filopodia were tentatively assigned to extending and retracting states on the basis of the configuration of the actin network.

In some cases, the picture of actin obtained by cryo-ET appears to differ from what findings based on other methods suggested. For example, *Plasmodium* cells seem to contain many fewer actin filaments than previously expected (Kudryashev et al., 2010b). This discrepancy might be due to the regulation of filament formation turnover or to the difficulty of unambiguously discerning short actin filaments in the dense environment of a thick cell. A recent study of lamellipodia (thin sheet-like protrusions that lead cell movement) investigated four types of migratory cells using cryo-ET as well as other EM preparations (Urban et al., 2010). Although a large number of actin filaments in lamellipodia were visualized, only a few branches were detected. These findings suggested that most of the apparent filament junctions arise from overlapping filaments that could be resolved in 3D. This conclusion contradicted the currently accepted dendritic nucleation model of the actin organization in lamellipodia, which states that lamellipodia protrusions are driven by branches that constantly nucleate on pre-existing filaments. While it appears that actin branching is not as extensive as previously thought, the contribution of short branches to lamellipodia extension is not completely clear at this point (Small et al., 2011; Yang and Svitkina, 2011).

Microtubules and microtubule-associated proteins have been the focus of great interest in the cryo-EM community. Until now, however, purified or reconstituted systems predominantly have been studied. As microtubules are often localized in a more central area of the cell body than actin filaments, the investigation of microtubules by cryo-ET of intact cells is limited to cells with sufficiently flat regions. In a comprehensive study, the authors compared several cell types and observed that small globular particles of about 7 nm in diameter abound in the lumen of microtubules in neuronal processes and P19 multipotent cells, but not in Ptk2 and HeLa cells (Garvalov et al., 2006). The existence of these lumenal densities was also demonstrated in vitreous sections of CHO cells (Bouchet-Marquis et al., 2007). The luminal density of *Plasmodium* microtubules is different from that of neurons, while some microtubules also appear thicker on the lumenal side (Cyrklaff et al., 2007). This additional density might be due to proteins that possibly play a role in microtubule stabilization.

Other Cell Types

Fibroblasts are interesting for cryo-ET because their thickness gradually decreases from several micrometers at the level of the nucleus to only 50 nm at the cell edges. They can spread up to 100 μm, resulting in major parts of the cell being thinner than 500 nm (Koning et al., 2008). Because fibroblasts are highly motile, their cytoskeleton was carefully investigated by cryo-ET. Investigators also detected microtubule lumenal particles and quantified different types of microtubule ends (Koning et al., 2008). In addition, actin and intermediate filaments were imaged, as well as various organelles, confirming the power of cellular tomography in detecting detailed subcellular structures. Fibroblasts were also used to study the architecture of focal adhesion complexes (Patla et al., 2010). It was shown that these complexes contain small, torus-shaped particles attached to actin filaments. Immunolabeling of cells having the dorsal membrane sheared off identified vinculin as a component of these particles, in agreement with the data previously obtained from super-resolution fluorescence microscopy. Well-developed molecular tools exist for several relatively flat cell lines commonly used in cell biology (such as Ptk2 and Cos cells), making them good candidates for the combination of genetic approaches and cryo-ET.

Correlative LM and Cryo-ET

Correlative LM and cryo-ET is a hybrid approach that allows imaging of the same feature over multiple length scales, combining LM with the ability to obtain high-resolution information in EM. Depending on the fluorescently labeled protein, LM can provide additional information regarding the functional state, subcellular localization, or molecular environment of the molecules visualized in cryo-ET and, consequently, help their identification. Approaches that combine LM with other EM preservation methods have already demonstrated their relevance for the integration of structure and function and are becoming important tools in cell biology (reviewed in Koster and Klumperman, 2003).

The large array of fluorescent probes currently available together with the widespread use of fluorescence microscopy has provided new insights into the identity, spatial distribution, and dynamics of molecules involved in many cellular processes. However, due to current technical limitations such as insufficient resolution and the fact that only a few different fluorescently labeled molecules can be imaged simultaneously, many questions regarding the molecular and supramolecular architecture of the labeled objects remain unanswered. Cryo-ET is one of the more promising techniques at hand to solve some of these complex problems regarding cellular architecture and function, especially when complemented with light microscopy.

In principle, the procedure applied in correlative LM and cryo-ET seems straightforward. First, a feature of interest is identified on an LM image, and then it is traced through a series of EM images of increasing magnification until it is successfully resolved at the desired level of detail. However, new challenges and constraints arise when this seemingly straightforward approach is applied to samples that are vitrified in a thin layer of ice. As discussed earlier, vitrified samples are sensitive to electron irradiation and thus easily damaged, and, consequently, imaging in cryo-ET has to be done using the lowest possible electron exposures. The electron exposure spent locating a feature of interest needs to be minimized to allow for sufficient electron exposure to be applied to record a tomographic series.

Some types of samples allow a straightforward approach. For example, most organelles are large enough to be clearly identifiable, and a correlation can be established on visual cues only, based on a limited number of low-dose images. It was also shown that cultured neurons in their early developmental stages are sufficiently sparse to allow for a visual correlation of cryo-EM and LM in a direct manner (Garvalov et al., 2006). In a different study, fenestrated areas of cultured endothelial cells were thin enough to be visualized by cryo-ET, but also sufficiently large to provide straightforward orientation within the field of vision, thus tomograms could be correlated to LM images without resorting to additional information (Braet et al., 2007).

In more complicated cellular landscapes, distinct morphological features are not readily identifiable at low EM magnifications, while searching for a feature of interest at high magnifications is infeasible due to the small field of view. These applications require a purely computational, coordinate transformation-based approach to navigate the search in cryo-EM to the position of an LM feature of interest, such as a fluorescent spot. In this case, the LM feature needs to be correlated to the coordinates of the EM stage so that a tomogram can be recorded at that position without the need of images at lower EM magnifications. The coordinate-transformation–based method requires higher reliability and precision of the EM stage than the visual cues-assisted correlation. In addition, this approach is also useful when technical issues preclude the effective use of the low magnifications in EM.

The method of choice for establishing a coordinate transform between the pixel-based coordinate system of the LM images and the physical coordinates of the EM goniometric stage is the orientation by visually detectable landmarks in the light and electron microscope, for example, grid bars. Once in place, the coordinate transformation can be used to retrieve the exact EM stage positions of features of interest identified in LM. This approach was used in mature neuronal cultures to establish a correlation between synaptic vesicles labeled by FM fluorescent dye and visualized in cryo-ET (Lucic et al., 2007). The correlation error determined in the test examples was between 1 and 3 μm.

In some of the previous examples, LM images were recorded in live cultures prior to freezing, requiring adherent cells to establish a proper correlation. Thanks to the recent development of cryo-LM holders it is now possible to image vitrified samples in LM. These cryo-holders are specifically designed to ensure that the EM grid is kept under cryo conditions (<−150°C) in the holder to keep the sample vitreous during LM imaging (Sartori et al., 2007; Schwartz et al., 2007; van Driel et al., 2009; Rigort et al., 2010) (Figure 7.4A and B). Consequently, the sample is in the same state and features of interest are kept in the same position during both LM and EM imaging, which is a prerequisite for establishing a proper correlation. This is

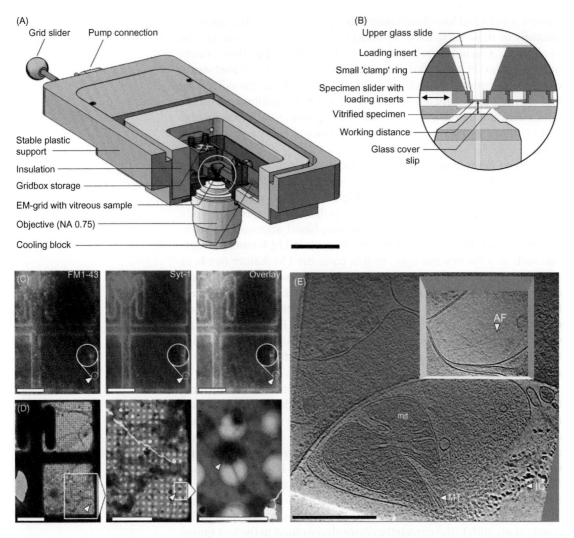

Figure 7.4 Correlative LM and cryo-ET. (A) Cut-away perspective view of the cryo-LM holder with a 63× long distance objective (NA 0.75), depicting housing and insulation requirements. Scale bar, 40 mm. (B). Magnified view of the central part (white circle, A), pointing at the imaging position and specimen slider with four loading positions for vitrified EM grids. (C) Cryo-fluorescent labeling of endocytic markers in a cultured hippocampal neuron (9 DIV). Left, FM 1–43; center, anti-synaptotagmin 1 antibody; right, overlay. Round insets show the spot of interest (white arrowhead) at higher magnification and with increased contrast. Scale bar, 50 µm. (D) Cryo-EM images, at increasing magnifications, at the position (indicated by the red cross) that was correlated to the fluorescence spot indicated in (C). Scale bars: left, 50 µm; center, 30 µm; right, 5 µm. (E) Slice from a cryo-electron tomogram of the neuronal varicosity indicated by an arrowhead in (D). Mitochondrion (mit), microtubule (MT). Indented inset: actin filaments (AF) in a different z-slice. Scale bar, 250 nm. (*A and B reproduced from* Rigort et al., 2010, *with permission. C–E reproduced from* Fernández-Busnadiego et al., 2011, *with permission.*) Please see color plates at the back of the book.

important not only for highly motile cells, but also for imaging rapidly moving cellular processes, such as cytoskeleton-based transport. The rate of cooling inside cryo-LM holders is automatically controlled and does not require manual handling. While all models allow fluorescence imaging, it is often useful to record transmission light images, for example, to get an estimate of the ice thickness. The design started by Sartori et al. (2007) and continued by Rigort et al. (2010) allowed transillumination from the beginning of the development, while recent modifications of the other two models also include this capability (Briegel et al., 2010). Presently, LM cryo-holders are limited to air objectives with a working distance of about 2 mm or higher, effectively limiting the resolution that can be achieved. The current generation of cryo-holders for LM has been improved with respect to stability and contamination rate, allowing continuous imaging over several hours, which is necessary for comprehensive cryo-LM imaging.

All cryo-LM holders described so far are separate devices that can be installed on standard light microscopes. All parts of the light microscope remain at ambient temperature and only the sample is kept in the holder under cryo conditions. Consequently, there is a considerable temperature difference between the objective lens and the sample that needs to be maintained at the shortest possible distance. One possibility to further decrease this distance and increase the resolution is to place the objective lens (or a part of it) at liquid nitrogen temperature. We believe the earliest implementation of this concept has been published in Makita et al. (1980), but the immersion of the objective in the liquid nitrogen causes persistent bubbling of liquid nitrogen, resulting in suboptimal imaging conditions (van Driel et al., 2009). A light microscope was essentially built inside a chamber kept under cryo-conditions as a further development of this concept, with the goal to use it for x-ray tomography and cryo-LM (Le Gros et al., 2009). Both the sample and the objective lens were immersed in liquid propane, which has a similar refraction coefficient to oil, thus making it possible to use short working distance oil immersion objectives. Although the risk of damage to the equipment at cryo-temperatures was an initial concern, the authors noted that objective lenses are robust toward immersion in cryogen and that no damage of the optical system was detected even after a number of alternating cycles between room and cryogenic temperature. In another development, a fully integrated correlative LM and cryo-EM system was constructed by mounting a specially designed laser scanning fluorescence microscope module on one of the side ports of an electron microscope. For imaging in LM, the grid has to be tilted by 90° (Agronskaia et al., 2008). These developments make the full power of the diffraction-limited LM available for the correlative approach. On the other hand, experimental setups that do not require custom-made microscopes or significant hardware adjustments tend to be more flexible and applicable to a wider range of biological problems. Fluorophore properties are not expected to be different under cryogenic conditions, because the energy levels of the molecules are determined by the laws of quantum mechanics and, consequently, do not depend on temperature. However, it appears that fluorophore bleaching is reduced in vitrified samples (Schwartz et al., 2007; Gruska et al. 2008). This is not entirely unexpected, given that proximity of, or collision between, oxygen and a fluorophore, which are the major causes of quenching and photobleaching in the liquid state (Cantor and Schimmel, 1980; Lichtman and Conchello, 2005), are greatly reduced under cryo conditions. It might be beneficial that cryo-fluorescence imaging precedes acquisition of a cryo-tomogram, because EM imaging may reduce fluorescence (Lucic et al., 2007; van Driel et al., 2008).

Grid damage and deformation are unavoidable consequences of the many transfer steps and frequent grid handling that is necessary in the correlative microscopy methods. Unfortunately, the gold grids that are typically used for cell cultures are particularly soft, which translates into a reduced yield and lower precision of the correlation. This problem was greatly alleviated by the introduction of the AutoGrid (FEI) support ring that drastically improves frozen grid handling and preservation.

Rapid freezing tends to be unreliable for thicker samples, because it yields ice of irregular thickness and does not always provide optimal preservation. Since this problem has no straightforward solution, it is advisable to image a large portion or even a whole grid in LM to increase the chances that the same spot is imaged in both LM and EM. As grids are usually slightly bent, z-stacks have to be recorded at each position to obtain in-focus images, significantly increasing the time needed for imaging and handling of the images. A recent study showing that a significant percentage of mammalian kidney cells had plasma membranes compromised during the blotting procedure underscores the importance of a careful morphological assessment of vitrified cells in both LM and EM (Lepper et al., 2010). The high speed of the freezing procedure prevents long-range propagation of damage, and therefore only damage to regions of the cell in the vicinity of the spot of interest is relevant and needs to be avoided.

In an application of the correlative procedure on dissociated neuronal cultures, a synaptic bouton fluorescently labeled with the endocytotic marker FM1-43 and an antibody against the extracellular domain of synaptotagmin, a transmembrane presynaptic protein, was located and recorded in cryo-ET (Figure 7.4C–E;

Fernández-Busnadiego et al., 2011). While the endocytic marker provided a stronger signal, the antibody labeling was a more specific indicator for sites of synaptic vesicle exo-endocytosis.

These demonstrations provide an important "proof of concept" for the entirely software-based correlative LM and cryo-ET approach, which can currently provide information about the environment or the functional state of features visualized in cryo-ET. Improving cryo-LM imaging to allow single molecule imaging or the application of LM superresolution methods would significantly increase the potential of the correlative approach and might allow the identification of individual fluorescent molecules in cryo-tomograms.

Labeling Strategies for Correlative and Direct Identification Approaches

Labeling methods determine to a large extent the applicability and the type of information that can be obtained from correlative LM and cryo-ET. The fluorescent labeling methods have to be noninvasive and the labeling has to be done before freezing. Immunolabeling techniques suitable for living cells are mostly restricted to proteins with extracellular epitopes, because the immunolabeling of intracellular epitopes requires methods that induce significant structural changes, such as membrane permeabilization or pore forming (Humbel et al., 1998). Of particular interest is labeling with quantum dots, which possess strong fluorescence and are visible in cryo-tomograms. However, steric hindrance may prohibit their access to regions with high protein density, such as cellular junctions.

Nevertheless, immunolabeling methods combined with techniques where a fluorescent or an electron-dense label is conjugated to a protein of interest may have a wider application to isolated cellular components. For example, nuclear pore complexes and the positions of gold-labeled proteins involved in nuclear import were visualized in cryo-tomograms of isolated nuclei, and superimposing the tomograms revealed the spatial distribution of the labeled protein (Beck et al., 2007). The combination of cryo-tomography with computer simulations of the biochemical reactions underlying nuclear import allowed the authors to obtain a dynamic picture of cargo traversing the nuclear pore, showing how a systems biology approach can provide insights into dynamic cellular processes.

Many genetically encoded or nontoxic fluorescent indicators can be introduced into living cells to identify the subcellular localization of a molecule of interest. Similarly, a clonable label for EM would represent a direct way of localizing supramolecular structures within their cellular context. While such an approach would at first glance be analogous to the use of green fluorescent protein, the dynamic picture present in live cell imaging would still be missing. Small metal-binding proteins such as metallothionein can be fused to cellular proteins and act as nucleation sites for heavy atoms that need to be added to the surrounding medium and taken up by cells (Mercogliano and DeRosier, 2007; Nishino et al., 2007). However, this approach is not completely straightforward, because it is difficult to distinguish between nonspecific uptake/precipitation and bona fide labeling. It also needs to be considered that metallothioneins occur naturally in many prokaryotes and eukaryotes, where their expression is induced in response to the presence of some toxic metals. Relatively high concentrations of metal salts in the growth medium might be cytotoxic or have other undesired or unpredictable physiological effects. Despite these potential problems, small gold clusters were detected upon expression of metallothionein fusion proteins in *Escherichia coli*. The cellular distribution of these clusters varied with the expression level and was clearly different from the pattern observed in the absence of metallothionein fusion protein expression, arguing for the specific detection of the expressed protein (Diestra et al., 2009). A related approach focused on the small iron-binding protein ferritin. Expression of ferritin fusion proteins in *E. coli* grown in iron-rich medium allowed the detection of small iron particles in cryo-tomograms (Wang et al., 2011). It needs to be verified that the iron-loaded, ferritin fusion protein has the correct localization and that the procedure causes minimal cellular perturbations, but the low toxicity of iron is the clear advantage of this method.

The elemental composition of a sample can be mapped directly by electron energy loss spectroscopy (EELS). This method relies on the use of energy filters to detect the energy loss of an incident electron upon inelastic scattering from specific elements. Considering that EELS requires a relatively high electron irradiation, the detection of phosphate- and carbon-rich bodies present in vitrified whole bacteria was not entirely expected (Comolli et al., 2006). Even though this method can bring only limited biological information, it might be quite useful in combination with other related techniques. One must also bear in mind that with the labeling and detection strategies described in this section a very small subset of the proteome can be visualized simultaneously, as opposed to a computational pattern recognition-based approach with a theoretically comprehensive database of unique templates. On the other hand, information on the 3D subcellular distribution of one particular protein may be valuable if morphology is well preserved.

Different approaches are available and many more could potentially be applied to improve the labeling and identification of cellular molecular complexes in cryo-ET. While some of them are likely to be significantly improved in the future, it seems unlikely that any of these methods will reach universal applicability anytime soon. Therefore, researchers will likely continue to adapt the most promising methods within the available spectrum to their needs.

Linking Structure and Physiology: Functional Manipulations in Cryo-ET

The use of pharmacological and genetic manipulations seems an obvious approach to discern specific subcellular components and macromolecular complexes. However, finding statistically significant differences between treatments often demands a fair number of tomograms recorded and analyzed for each condition.

Structures of Distinct Shape

In some cases, manipulations affect structures that can be easily detected and identified in cryo-tomograms. For example, the outer membrane of wild-type corynebacteria and mycobacteria was revealed in two studies using vitrified cells (Figure 7.5A; Hoffmann et al., 2008; Zuber et al., 2008), while in earlier work on dehydrated samples the same structure could not be detected. Furthermore, mycolic acid-deficient mutants lacked the outer membrane, demonstrating the importance of these lipids in outer membrane formation. These data allowed the authors to propose membrane organization models that differed substantially from previous work.

The morphology of the cytoplasmic membrane, the outer membrane, and the associated complexes were investigated in a comparative approach involving three related species of *Borrelia*, the bacteria that causes Lyme disease. Partial removal of the outer membrane of *Borrelia* by an antibody directed against an outer surface protein led to the conclusion that the peptidoglycan layer localized between the two membranes is strongly linked to the cytoplasmic membrane (Kudryashev et al., 2009).

Cytoskeletal filaments are thin and do not have a particularly strong contrast in cryo-tomograms of intact bacterial cells. However, due to their characteristic shape these filaments can be routinely detected by visual inspection of consecutive 2D tomographic slices. Four different categories of filament bundles that differ in their dimensions and cellular localization were observed in *Caulobacter crescentus*, a thin bacterium that can be imaged *in toto* in cryo-ET. A systematic knock-out approach identified homologs of all three eukaryotic cytoskeletal families (actin, tubulin, and intermediate filaments), as well as other proteins among the constituents of bacterial filaments (Ingerson-Mahar et al., 2010).

The large range of genetic approaches available in *E. coli* can be used to complement cryo-ET, but these cells are often too thick to be

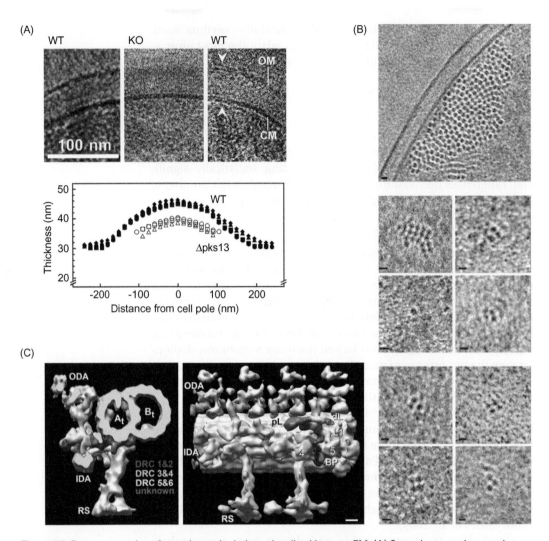

Figure 7.5 Recent examples of genetic manipulations visualized by cryo-EM. (A) Cryo-electron micrographs of vitreous sections from *Corynebacterium glutamicum*. Wild-type cells imaged at high (top left) and low (top right) defocus. The bilayer structure of the cytoplasmic membrane (CM) and of the outer membrane (OM) is resolved (arrowheads). Top middle, cryo-section of the mycolic acid-lacking mutant *C. glutamicum* Δpks13 at low defocus. Bottom, thickness of the cell walls determined from several cells as measured from the surface of the CM to the outer surface of the cell wall (filled symbols, wild type cells; open symbols, mutant cells). (*Reproduced from* Hoffmann et al., 2008, *with permission.*) (B) Cryo-electron micrographs of vitreous sections of cells overexpressing ParM (top) or expressing high (middle) or low (bottom) ParM copy plasmids showing end-on views of ParM filaments. (*Reproduced from* Salje et al., 2009, *with permission.*) (C) Architecture of the dynein regulatory complex (DRC) in cross-section (left) and as overview (right) visualized by subtomogram averaging. ODA; outer dynein arms; IDA, inner dynein arms; RS, radial spokes; At, A-tubule; Bt, B-tubule; BP, base plate protrusion; dL, distal lobe; L1, linker protrusion 1; pL, proximal lobe. The locations of DRC subunits (see color legend, left) were obtained from mutants lacking specific DRC regions. Scale bars: A, 100 nm; B and C, 10 nm. (*Reproduced from* Heuser et al., 2009, *with permission.*) Please see color plates at the back of the book.

imaged *in toto*. The importance of filaments composed of the actin homolog ParM for DNA replication in *E. coli* was shown by imaging cryo-sectioned cells where the amount of ParM was controlled using plasmids with different expression levels (Figure 7.5B; Salje et al., 2009). A relatively elaborate genetic approach that included wild-type, loss-of-function mutants and different expression levels of two chemosensory system proteins (CheA and CheW) was used to describe for the first time the supramolecular organization of chemoreceptors in *E. coli* (Zhang et al., 2004, 2007). Periodic structures located at cell poles were directly visualized in cryo-tomograms, but it was necessary to use the genetic manipulations and immuno-EM to identify chemotaxis receptors as the main constituents of this structure. Also, several studies have contributed to the molecular dissection of the components of the eukaryotic flagellum using deletion or ligand binding strategies (Figure 7.5C; see Uncovering Information by Image Processing) (Heuser et al., 2009; Movassagh et al., 2010).

Presynaptic Cytomatrix

The general strategy of the studies mentioned in the previous section was to identify structures based on changes of their copy number upon genetic manipulations. Here, we will address in some detail the analysis of presynaptic architecture employing pharmacological manipulations to bring synapses to a variety of physiological conditions, and monitoring the corresponding structural changes to build functional models (Fernández-Busnadiego et al., 2010).

The presynaptic terminal hosts a complex machinery responsible for the transduction of electric signals into neurotransmitter release, and understanding the molecular architecture of this machinery is critical to elucidate its mechanisms of action. Early EM studies revealed that, within the presynaptic terminal, synaptic vesicles were embedded in a dense filamentous network known as the presynaptic cytomatrix (Figure 7.6A; Landis et al., 1988; Hirokawa et al., 1989). More recent work using electron tomography provided additional ultrastructural detail (Figure 7.6B; Siksou et al., 2007) and highlighted the influence of EM sample preparation on the resulting observations (Siksou et al., 2011).

Vitrified presynaptic terminals were investigated by cryo-ET using two complementary preparations: rat cerebrocortical synaptosomes and vitreous sections from rat hippocampal organotypic slices (Fernández-Busnadiego et al., 2010). Synaptosomes are a cell fraction of isolated synapses and a well-established model for neurotransmitter release (Whittaker, 1984; Nicholls and Sihra, 1986; Harrison et al., 1988). Synaptosomes are advantageous for cryo-ET because their moderate thickness (usually 300–500 nm) allows for vitrification by plunge-freezing, resulting in high-quality tomographic data. On the other hand, organotypic slices provide a close approximation to

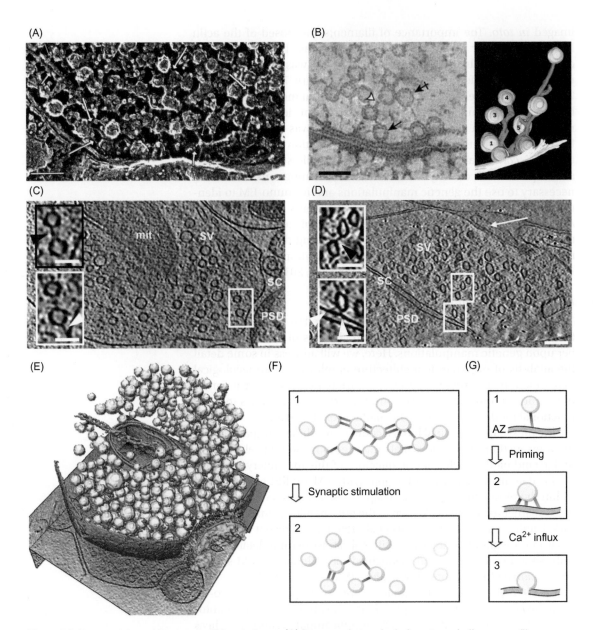

Figure 7.6 Presynaptic architecture by EM techniques. (A) Presynaptic terminal of a rat cerebellar mossy fiber upon quick-freezing, deep-etching, and platinum replication, showing abundant presynaptic filaments (arrows). (*Reproduced from* Hirokawa et al., 1989, *with permission*.) (B) Rat hippocampal presynaptic terminal upon high-pressure freezing, freeze substitution, and heavy-metal staining, imaged by ET. Filaments linking vesicles to each other and to the AZ can be seen in the tomographic slice (left) and 3D rendering (right). (*Reproduced from* Siksou et al., 2007, *with permission*.) (C–G). Presynaptic morphology visualized in tomograms of vitrified synapses visualized by cryo-ET. (*From* Fernández-Busnadiego et al., 2010). SV, synaptic vesicle; mit, mitochondrion; SC, synaptic cleft; PSD, postsynaptic density. (C) Synaptosome. Upper inset. connector linking two vesicles (black arrowhead).

nervous tissue but require high-pressure freezing and laborious cryo-ultramicrotomy procedures.

In both preparations, most synaptic vesicles were linked to each other by short (<40 nm) filaments termed connectors, and most vesicles in the proximity of the active zone (AZ) were linked to it by similar strands (tethers; Figure 7.6C and D). Longer filaments were scarce, indicating that connectors and tethers were the most prominent structural features of the presynaptic cytomatrix in vitrified specimens.

To investigate the function and the identity of connectors and tethers, presynaptic architecture was monitored under the following pharmacological treatments: hypertonic sucrose, causing specific release of the readily releasable pool (RRP), that is, the vesicles that can be immediately exocytosed upon Ca^{2+} influx (Rosenmund and Stevens, 1996) and membrane depolarization by KCl, causing large-scale exocytosis and mobilizing the reserve pool as well (Ashton and Ushkaryov, 2005; Rizzoli and Betz, 2005). Modulatory treatments included tetanus toxin, which prevents synaptic vesicle fusion by cleaving the vesicular SNARE protein synaptobrevin/VAMP (Schiavo et al., 2000), as well as the phosphatase inhibitor okadaic acid, which alters the phosphorylation state of synapsin (Jovanovic et al., 2001), and increases vesicle mobility (Betz and Henkel, 1994). Organotypic slices were used as control for possible artifacts introduced by synaptosome extraction.

Objective and comprehensive detection of connectors and tethers, a prerequisite for the analysis of these structures, was achieved by a customized automated segmentation procedure (Figure 7.6E; described in the section "Uncovering Information by Image Processing"). In addition, the amount and complexity of data required a software approach for the analysis of morphological characteristics, gray-scale values, and localization of connectors and tethers.

The results of this analysis indicate that synaptic vesicle clustering is mainly mediated by short intervesicular connectors. Upon stimulation, a large number of connectors are removed in a phosphorylation-dependent process, thereby increasing vesicle mobility (Figure 7.6F). Even though the molecular identity of the connectors remains

---

Figure 7.6 (Continued) Lower inset, tether linking a vesicle to the AZ (white arrowhead; same vesicles as in the upper inset at another z-slice). (D) Organotypic slice. Synaptic vesicles are compressed along the cutting direction (white arrow). Upper inset, connector linking two vesicles (black arrowhead). Lower inset, tethers linking a vesicle to the AZ (white arrowheads). (E) Three-dimensional rendering of a synaptosome and corresponding tomographic orthoslices. Synaptic vesicles (yellow), connectors (red), tethers (blue), mitochondria (light blue), microtubule (dark green), plasma membrane (purple), cleft complexes (green), PSD (orange). Proposed models for connector (F) and tether (G) function in synaptic vesicle mobilization and fusion. The number and topology of connectors and tethers shown in (F) and (G) are based on average values. Shaded vesicles in (F) (bottom) represent the decrease in vesicle number upon prolonged stimulation. Scale bars: A–D: 100 nm in main panels, 50 nm in insets. Please see color plates at the back of the book.

unknown, the data suggest that connectors are formed by more than one molecular species and that these are differentially distributed within the presynaptic terminal. While synapsin may be implicated in connector formation away from the AZ (Hirokawa et al., 1989), other molecules, such as synaptobrevin, may be involved in the formation of connectors in the vicinity of the AZ.

Currently, the mechanisms of vesicle association and fusion at the AZ remain controversial because of the disparity of the results obtained with different EM sample preparation techniques. In heavy-metal–stained samples, vesicles making direct membrane contact with the AZ are commonly referred to as "docked," and a subset of them is thought to form the RRP. In contrast, vesicles making direct membrane contact with the AZ were extremely scarce in vitrified synaptosomes and hippocampal slices, whereas a large majority of vesicles in the vicinity of the AZ were linked to it by tethers. Most of these tethers were shorter than 10 nm, suggesting that they might have been obscured by heavy-metal staining in previous studies. Therefore, it appears that direct membrane contact between vesicle and AZ occurs only transiently during vesicle fusion and that tethering is the main mechanism of vesicle association with the AZ. Furthermore, even though densities were observed on the AZ of vitrified synapses, they did not form a regular arrangement and their positions were not correlated with those of tethered vesicles, arguing against the presynaptic web model.

Two distinct populations of synaptic vesicles were differentiated based on the effects of hypertonic sucrose and tetanus toxin on the length and number of their tethers to the AZ. Long (>5 nm) tethers were not affected by these treatments, indicating that they do not participate directly in vesicle fusion. Under both conditions, the number of vesicles with multiple short (<5 nm) tethers was significantly reduced, suggesting that these vesicles constitute the RRP and that the formation of short tethers requires SNARE complex assembly. The model of vesicle tethering and fusion emerging from these observations is summarized in Figure 7.6G. In the first step, long tethers, which do not require full SNARE complex assembly for their formation, mediate the initial association of the vesicle to the AZ. Once the vesicle is held in the proximity of the AZ, shorter SNARE complex-dependent tethers may form. The formation of a sufficient number of these short tethers constitutes the priming step by which vesicles become readily releasable. Synaptic stimulation causes a Ca^{2+} influx that may trigger direct membrane contact of primed vesicles and the AZ, immediately followed by fusion. This model could clarify the discrepancy between vesicle docking and availability for release; whereas in dehydrated samples, all vesicles in the vicinity of the AZ seem to be docked, these actually correspond to a superposition of different tethering states in vitrified specimens, and therefore to vesicles with different states of assembly of the fusion machinery. Recently, a genetic approach was adopted to obtain clues on the molecular identity of tethers. The family of RIM proteins plays a key role in AZ organization and synaptic vesicle priming for exocytosis (Deng et al., 2011; Kaeser et al., 2011). Cryo-ET was employed to study presynaptic terminals in mice lacking RIM1α, the most abundant RIM isoform (Fernández-Busnadiego et al., 2012). Interestingly, two morphologically distinct groups of terminals were detected, one having only minor alterations compared to wild-type animals (in terms of vesicle distribution and cytomatrix architecture), while the other one showed major disturbances, suggesting a different molecular composition of these synapses. Furthermore, the formation and maturation of tethers were impaired in the absence of RIM1α. These observations also serve as an illustration of the higher degree of structural complexity of genetic phenotypes in mammalian systems in comparison to prokaryotes.

Uncovering Information by Image Processing

The biological complexity and low SNR often turn the detection of meaningful information in cryo-tomograms into a challenging task. Computational approaches for improving the resolution and detecting and identifying structures of interest in cryo-tomograms are outlined in this section.

CTF Correction

We mentioned earlier that due to the oscillatory nature of the CTF, contrast may be inverted and information may be suppressed at higher frequencies. CTF correction is a procedure that aims to restore the distorted high-resolution information in cryo-tomograms. The CTF is determined and corrected on each individual projection comprising a tilt series before the tomogram is reconstructed (Fernández et al., 2006). However, due to the defocus gradient present at (tilted) projections and the low SNR, this process is often challenging. So far, the need for CTF correction in tomography has been somewhat underappreciated, but it might become mandatory in order to improve the resolution further.

Denoising

Denoising algorithms aim to reduce the noise while preserving the features of interest, thus improving the SNR. Image denoising is an important step that usually precedes the interpretation of tomograms. In Gaussian filtering, images are essentially slightly smoother or blurred by spreading the gray-scale (image) values at every pixel to the nearby pixels by diffusion. This removes some noise, but the higher frequencies of the features of interest are also removed. Nonlinear anisotropic diffusion overcomes this limitation by preventing diffusion over the edges present in a tomogram. These edges are detected based on the local image gradient values, and are expected to delineate the structures imaged (Fernandez and Li, 2003; Frangakis and Hegerl, 2001). This procedure preserves and enhances the structures of interest while reducing the noise in the surrounding area. Nonlinear anisotropic diffusion is a local method, meaning that the value at a given pixel is determined by the values of pixels in a small region centered at that specific pixel. In nonlocal means filtering on the contrary, a structure receives contributions from similar structures found in the whole tomogram, or in a large part thereof. The higher the similarity between the structures, the stronger is the contribution. The idea here is that repeating structures essentially get averaged, which decreases the noise in these structures. This method is therefore particularly suited to filamentous structures such as actin fibers (Rigort et al., 2012b). Depending on the specific method used, characteristic properties of the structure of interest, such as size, shape, or gray-scale gradient, may need to be specified as input parameters.

Segmentation

In image segmentation, regions of an image corresponding to features of interest (segments) are detected and separated from the background. This allows easier computational manipulations of segments, such as visualization or selective image processing. In manual segmentation, structures of interest are visually detected and traced by hand, marking their contours on successive slices in the volume, sometimes with the help of image processing tools that allow completion of partially segmented forms. For example, lipid membranes, which tend to be easily identifiable due to their high contrast and smooth appearance, are most often segmented manually in cryo-tomograms. This is a time-consuming procedure requiring a lot of focus on details and precise sensorimotor coordination of the computer operator. It thus represents one of the bottlenecks preventing larger scale tomographic analysis of cells. Due to the high noise, computational segmentation is a difficult task in cryo-tomograms, even for membranes.

Several segmentation algorithms have been proposed in the last decade, but their application remained sparse. Two recent membrane segmentation approaches appear promising. The first of these was developed for the segmentation of bacterial membranes. This segmentation procedure proceeds slice by slice, starting with a manually segmented one. In each slice the membrane is inferred based on a template representing the membrane and on the expected membrane shape, which is assumed to be ellipsoidal. The template and the expected shape are determined in each slice from the already segmented slices, as well as from the previously segmented tomograms (Moussavi et al., 2010). The second approach works directly in 3D. It uses local properties to determine membrane-like pixels, which are then analyzed using global information to determine which pixels actually constitute the membranes (Martínez-Sánchez et al., 2011).

Visual detection of finer, nonuniform structures in cryo-tomograms is difficult and error prone. To detect such structures in an objective, reproducible, and comprehensive way it is necessary to develop an automated segmentation method. Such a procedure, based on thresholding and connectivity, was developed for segmentation of synaptic adhesion complexes (Lucic et al., 2005); it is generally applicable to structures that contact one or more already segmented features (e.g., membranes). This method was further developed to include some aspects of the watershed transform (Soille, 2003), as this strongly reduced the influence of variable background values. It was also used for segmentation of thin filaments that link synaptic vesicles to each other and to the plasma membrane (Fernández-Busnadiego et al., 2010). After detection of the structures of interest, their features, such as gray-scale levels and morphological characteristics, as well as location and relationship to other structures, were analyzed for each tomogram. This allowed the statistical analysis of differences between features for different sample conditions.

Recently, an automated actin segmentation procedure was presented (Rigort et al., 2012). The algorithm searches for occurrences of cylindrical structures with a diameter equal to that of actin filaments by template matching (described in more detail later in this section). A tracing procedure is then used to segment complete actin filaments and remove falsely detected filaments.

Subtomogram Averaging

Subtomogram averaging is a technique used to improve the resolution of selected features in tomography. Subtomograms containing a particle of interest (a protein or a protein complex) are aligned and averaged. In this manner one can obtain a higher resolution structure of the particle while minimizing the noise and the effects caused by the missing wedge.

This procedure is conceptually similar to the EM single particle approach, where a 3D structure is obtained by aligning and averaging 2D images of a particle of interest. However, due to a larger number of degrees of freedom during the alignment step (three translations and three rotations in 3D, vs. two translations and one rotation in 2D), subtomographic averaging tends to be more influenced by the initial model used, and more likely to converge to a local minimum. On the positive side, unlike other methods aimed at higher resolution structures such as x-ray crystallography and single particle EM, subtomographic averaging does not require extensive purification (or crystallization). Furthermore, particles can be imaged *in situ*, and even large complexes can be investigated.

The nuclear pore complex is an example of a system where subtomographic averaging was successfully applied on a very large protein assembly that is difficult to investigate using other structural biology methods (Beck et al., 2007). The average nuclear pore subunit structure was obtained by averaging the individual subunits of the complex from cryo-tomograms of intact, transport-competent *Dictyostelium* nuclei. Copies of the average structure were then rejoined to obtain the structure of the whole complex, which revealed new structural features, as well as two different conformational states.

Isolated eukaryotic flagella have proven to be an extraordinary system allowing the molecular dissection of its components combining genetic, pharmacological, and computational tools. These flagella consist of relatively thin (~260 nm) but long arrays with longitudinal periodicity, which are ideal targets for subtomogram averaging. A number of studies have used this approach to unravel the structural basis of force generation by dynein (Nicastro et al., 2006), focusing on dynein arm architecture (Bui et al., 2008) or subunit organization of the dynein regulatory complex (Figure 7.5C; Heuser et al., 2009). Dynein conformational changes upon nucleotide binding were studied by subtomogram averaging followed by 3D classification of apo and ligand-bound particles (Movassagh et al., 2010).

Cryo-tomograms obtained from vitreous sections tend to have a relatively high SNR, making them a promising system for subtomogram averaging. For example, Al-Amoudi and colleagues (2007) showed that cadherin molecules form a densely packed periodic structure at the desmosomal junction, as opposed to the disordered structure seen using freeze-substituted material. In addition, it was shown that cadherins are tightly packed via alternating *cis*- and *trans*-interactions. The somewhat unpredictable degree of cryo-sectioning-induced compression may be a problem for averaging structures that have different orientations or come from different tomograms. While compression clearly deforms lipid membranes, it is not clear at this point if or to what extent macromolecular complexes are affected. Ribosomes, for example, do not seem to be compressed (Pierson et al., 2011).

Subtomogram averaging can also be applied to molecular complexes imaged *in situ*, that is, in intact cells. For example, over-expression of serine chemoreceptors in a strain of *E. coli* lacking many chemotaxis-related proteins ensured a high copy number and low background that benefited the averaging of these receptors (Khursigara et al., 2008). Averaging the receptors imaged from four different states of the chemotactic system yielded two distinct receptor conformations. Different occurrences of the two conformations in the four states led the authors to suggest that these conformations correspond to the active and inactive kinase states.

Even more demanding was the determination of the structure of the *Borrelia* flagellar motor *in situ* (Figure 7.7D). A high number of tomograms was needed, because only a few copies of the motor are present in each tomogram. Two complementing studies yielded similar structures with a high level of structural detail, which allowed a precise localization of some molecules within the complex (Liu et al., 2009; Kudryashev et al., 2010a). The structures also showed some differences, possibly due to procedures that differed to some extent, particularly in the treatment of symmetries and the missing wedge. It also became evident that two distinct types of motors exist, but their exact roles are not clear at this point.

Template Matching

Another method to identify structures in a tomogram is template matching, where the volume is searched for occurrences of a template of choice (e.g., a macromolecular complex) by cross-correlation. Templates are based on higher resolution structures typically obtained from x-ray crystallography or single particle EM. During the search a template is both translated (over the entire tomogram or a region of interest) and rotated. The template has to be adjusted to account for the imaging conditions (CTF, detector performance), as well as for the orientation of the missing wedge. While template matching is a powerful method that could allow identification of a large number of molecules visualized both *in vitro* and *in situ*, it is currently limited by the availability of higher resolution structures that are large enough (roughly 10 nm or larger) to be used as templates. In an early example, ribosomes were successfully mapped in tomograms of the thin bacterium *Spiroplasma melliferum* (Ortiz et al., 2006).

Template matching also provides the orientation of the structure of interest. This information was used to show that stalled ribosomes isolated from *E. coli* have preferred relative orientation and distance in respect to their nearest neighbors (Figure 7.7A; Brandt et al., 2009). In a subsequent publication these results were confirmed by the visualization of ribosomes engaged in active translation (polysomes) in protrusions of intact glial cells (Figure 7.7B; Brandt et al., 2010). This particular arrangement was lost upon treatment with puromycin, an antibiotic known to stop translation and dissociate polysomes. Furthermore, structural differences obtained by subtomographic

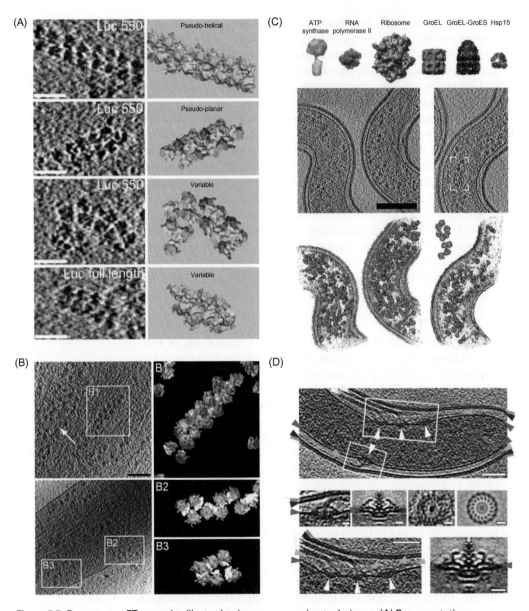

Figure 7.7 Recent cryo-ET examples illustrating image processing techniques. (A) Representative topologies of polyribosomes from *E. coli* lysates. Left, tomographic slices of polyribosomes; right, corresponding localization of ribosomes obtained by template matching (large ribosomal subunit, blue; small subunit, yellow; red cones point to the peptide exit tunnel; dark gray ribosomes were not unambiguously assigned to the polysomes). The three top cases represent polyribosomes stalled with Luc 550, and the bottom case represents a nonstalled full-length Luc. Scale bars, 50 nm. (*Reproduced from* Brandt et al., 2009, *with permission.*) (B) Same as in (A) except that ribosomes were imaged in intact human cells. (*Reproduced from* Brandt et al., 2010, *with permission.*) (C) Template matching in intact *Leptospira interrogans* cells: top, macromolecule template library; middle, representative tomographic subvolumes and a group of ribosomes

averaging of active and stalled ribosomes allowed the identification of ribosome-associated complexes.

Thanks to their high RNA content, ribosomes possess strong contrast in cryo-tomograms, which allows their easy visual detection. Proteins generally have lower contrast, so validation of the hits produced by template matching becomes an important task. This is related to the fact that the cross-correlation function at the heart of the pattern matching approach returns a score (typically between 0 and 1) at each position of the analyzed tomogram, and that the conversion of the score to one of the two possible outcomes (hit or miss) is not always straightforward. To solve this problem Beck and coworkers (2009) (Figure 7.7C) adapted a proteomics approach to derive a more elaborate scoring function for statistically reliable determination of template matching hits. This scoring function was optimized on test data (*in silico*) using different templates (including basic geometrical shapes) and template matching on the background. The results for several proteins ranging from 200 to 2000 kDa were validated by quantitative mass spectroscopy. An important result of this study is that the performance of the electron detector places the strongest constraint on the applicability of template matching.

Conclusions

Cryo-ET occupies a unique position among imaging and structural biology methods. Because vitrification provides faithful preservation of biological material, molecular complexes are imaged in a fully hydrated, close-to-physiological state, at the resolution in the range of single nanometers. Thus, cryo-ET has been used to reveal the cellular organization at molecular resolution and to identify molecular complexes in intact cells directly from their structural signature.

Figure 7.7 (Continued) resembling the pseudoplanar relative orientation can be seen in the bracketed region; bottom, corresponding localization of macromolecules. Scale bars: 200 nm. (*Reproduced from* Beck et al., 2009, *with permission.*) (D) Single and averaged *Borrelia garinii* flagellar motors. Top, cryo-electron tomogram slice showing four motors (white arrowheads). Colored arrowheads point to the surface layer (light blue), outer membrane sheath (dark blue), peptidoglycan (green), cytoplasmic membrane (magenta), and periplasmic flagellar filaments (yellow). Middle, side, and top views of single motors and the corresponding 16-fold rotationally averaged structures. Bottom, peptidoglycan binding to the motor. Enlargement of the three motors from the top left and average structure from the middle rotated by 11° around the vertical axis (right). Scale bars: 100 nm (top and bottom left) and 20 nm (middle and bottom right). (*Reproduced from* Kudryashev et al., 2010a. *with permission.*) Please see color plates at the back of the book.

Sample preparation, including both biological and EM aspects, is arguably the most important step in cryo-ET. Although its progress in this respect is slow and often uneventful, it is persistent. Recent progress in focused ion beam milling deserves particular mentioning, as it promises to alleviate the limitations imposed by specimen thickness and increase the range of specimens amenable to cryo-ET.

Cryo-tomograms of intact cells are essentially images of their entire proteome at molecular resolution, enabling the mapping of the spatial relationships of the macromolecules within the cell. Sample thickness determines to a large extent the usability and the resolution that can be obtained. While only some cells have regions that are thin enough to be imaged in intact form, many other cells can be vitrified and then thinned using one of the existing and emerging micromachining methods. As fractionated and reconstituted systems can be quite thin, higher resolution information obtained from these systems complements well the *in situ* data from cryo-ET of intact cells.

Perhaps paradoxically, the universal detection or the ability to visualize the complete molecular content of a system under study, which is one of the big advantages of cryo-ET, significantly complicates the identification of molecular components in complex cellular landscapes. It is then no surprise that very diverse identification strategies have been designed and that new ones are in development.

Identification by visual inspection of cryo-tomograms is suited for molecular complexes of moderate size, characteristic shape, or strong contrast, such as organelles, cytoskeletal filaments, or ribosomes. Labeling by electron-dense labels is a relatively straightforward molecular identification method. While current labeling methods are limited to extracellular epitopes and isolated components, recent developments of genetically encoded, electron-dense labels raise hope that these labels could reach broader applicability in the future. Nevertheless, direct labeling methods are expected to remain limited to the number of labels that could be introduced simultaneously.

Fluorescent labeling is used in correlative LM and cryo-ET, where the location of fluorescent puncta can be determined with sufficient accuracy to enable subsequent tomography. Alternatively, the correlative approach can provide information about subcellular localization, or the functional state of features imaged in cryo-ET. Several recent hardware developments enable LM imaging of vitrified samples, ensuring that the features of interest are in the same state during imaging in LM and in EM. Future adaptations of cryo-LM for single molecule and superresolution LM imaging modes may reach the necessary precision for the direct detection of single fluorescently labeled molecules in cryo-tomograms.

Specific perturbations of cellular systems add a functional dimension to cryo-ET and can be used for the identification of molecular components of the structures visualized. In some cases, the target of genetic manipulations can be detected in tomograms in a straightforward manner, leading to their molecular identification, as was the case for bacterial filaments. Pharmacological manipulations of presynaptic terminals affected many molecules, but changed the functional state of the release machinery in a defined manner leading to a structural model of neurotransmitter release. Further work is needed to improve this model to the point where specific molecules can be assigned to the structures detected in cryo-tomograms. The development of image processing tools for cryo-tomograms plays a key role in cryo-ET data interpretation. Different tools, many of which were adapted from single particle reconstruction and other branches of image processing, are at the disposition of a researcher, who needs to choose the most appropriate ones for a problem at hand. CTF-correction aims to correct for the nonlinear imaging properties of the microscope. Denoising enhances structure(s) of interest while removing the higher frequency information. Automated image segmentation is necessary for the objective and comprehensive detection of structures of interest, in particular where a large number of structures need to be analyzed further. For example, manual dendritic spine quantification in two-photon microscopy was identified as a major source of discrepancies between laboratories (Holtmaat and Svoboda, 2009).

As we said earlier, high-resolution information is present in cryo-tomograms, but it is buried in noise. As noise is probabilistic in origin, statistical methods are needed to uncover the information. Subtomographic averaging of sufficiently abundant larger molecules or complexes found in cryo-tomograms is a classical example of this approach: averaging increases the structural information and decreases random noise, leading to higher resolution of the averaged structure. Template matching aims to identify structures of interest in cryo-tomograms based on pre-existing higher resolution structures obtained from other methods. A statistical approach is often needed to reliably discriminate against false positives and false negatives. This method has the potential to identify a significant number of molecular complexes in an intact cell based on their structural signature, but to reach its full power probably requires higher resolution tomograms than what is achievable today.

Long-awaited developments such as phase plates or aberration correctors are expected to bring substantial improvements in tomogram resolution (Figure 7.1). Development of direct electron detectors, as well as improvements of CCD-based detectors, are of particular importance, because electron detectors might be the strongest resolution-limiting factor today.

Therefore, we expect that the sample preparation and hardware improvements, together with better detection and identification methods and hopefully the wider use of cryo-ET, will increase the current stream of discoveries by cryo-ET and bring a fresh perspective to cell and neurobiology.

Acknowledgments

We thank Gabriela J. Greif for critical reading of the manuscript.

References

Agronskaia, A.V., Valentijn, J.A., van Driel, L.F., Schneijdenberg, C.T., Humbel, B.M., van Bergen en Henegouwen, P.M., et al., 2008. Integrated fluorescence and transmission electron microscopy. J Struct Biol 164, 183–189.

Al-Amoudi, A., Díez, D.C., Betts, M.J., Frangakis, A.S., 2007. The molecular architecture of cadherins in native epidermal desmosomes. Nature 450, 832–837.

Ashton, A.C., Ushkaryov, Y.A., 2005. Properties of synaptic vesicle pools in mature central nerve terminals. J Biol Chem 280, 37278–37288.

Baker, L.A., Rubinstein, J.L., 2010. Radiation damage in electron cryomicroscopy. Methods Enzymol 481, 371–388.

Beck, M., Lucic, V., Forster, F., Baumeister, W., Medalia, O., 2007. Snapshots of nuclear pore complexes in action captured by cryo-electron tomography. Nature 449, 611–615.

Beck, M., Malmstrom, J.A., Lange, V., Schmidt, A., Deutsch, E.W., Aebersold, R., 2009. Visual proteomics of the human pathogen *Leptospira interrogans*. Nat Methods 6, 817–823.

Betz, W.J., Henkel, A.W., 1994. Okadaic acid disrupts clusters of synaptic vesicles in frog motor nerve terminals. J Cell Biol 124, 843–854.

Bouchet-Marquis, C., Zuber, B., Glynn, A.M., Eltsov, M., Grabenbauer, M., Goldie, K.N., et al., 2007. Visualization of cell microtubules in their native state. Biol Cell 99, 45–53.

Braet, F., Wisse, E., Bomans, P., Frederik, P., Geerts, W., Koster, A., et al., 2007. Contribution of high-resolution correlative imaging techniques in the study of the liver sieve in three-dimensions. Microsc Res Tech 70, 230–242.

Brandt, F., Carlson, L.A., Hartl, F.U., Baumeister, W., Grunewald, K., 2010. The three-dimensional organization of polyribosomes in intact human cells. Mol Cell 39, 560–569.

Brandt, F., Etchells, S.A., Ortiz, J.O., Elcock, A.H., Hartl, F.U., Baumeister, W., 2009. The native 3D organization of bacterial polysomes. Cell 136, 261–271.

Briegel, A., Chen, S., Koster, A.J., Plitzko, J.M., Schwartz, C.L., Jensen, G.J., 2010. Correlated light and electron cryo-microscopy. Methods Enzymol 481, 317–341.

Bui, K.H., Sakakibara, H., Movassagh, T., Oiwa, K., Ishikawa, T., 2008. Molecular architecture of inner dynein arms in situ in *Chlamydomonas reinhardtii* flagella. J Cell Biol 183, 923–932.

Cantor, C., Schimmel, P., 1980. Biophysical chemistry, vol. 2. W.H. Freeman & Company, New York, p. 447.

Comolli, L.R., Downing, K.H., 2005. Dose tolerance at helium and nitrogen temperatures for whole cell electron tomography. J Struct Biol 152, 149–156.

Comolli, L.R., Kundmann, M., Downing, K.H., 2006. Characterization of intact subcellular bodies in whole bacteria by cryo-electron tomography and spectroscopic imaging. J Microsc 223, 40–52.

Crowther, R.A., DeRosier, D.J., Klug, A., 1970. The reconstruction of a three-dimensional structure from projections and its application to electron microscopy. Proc R Soc Lond A 317, 319–340.

Cyrklaff, M., Kudryashev, M., Leis, A., Leonard, K., Baumeister, W., Menard, R., et al., 2007. Cryoelectron tomography reveals periodic material at the inner side of subpellicular microtubules in apicomplexan parasites. J Exp Med 204, 1281–1287.

Cyrklaff, M., Sanchez, C.P., Kilian, N., Bisseye, C., Simpore, J., Frischknecht, F., et al., 2011. Hemoglobins S and C interfere with actin remodeling in *Plasmodium falciparum*-infected erythrocytes. Science 334, 1283–1286.

De Robertis, E.D., Bennett, H.S., 1954. A submicroscopic vesicular component of Schwann cells and nerve satellite cells. Exp Cell Res 6, 543–545.

Del Castillo, J., Katz, B., 1954. Quantal components of the end-plate potential. J Physiol 124, 560–573.

Deng, L., Kaeser, P.S., Xu, W., Südhof, T.C., 2011. Rim proteins activate vesicle priming by reversing autoinhibitory homodimerization of Munc13. Neuron 69, 317–331.

Diestra, E., Fontana, J., Guichard, P., Marco, S., Risco, C., 2009. Visualization of proteins in intact cells with a clonable tag for electron microscopy. J Struct Biol 165, 157–168.

Dubochet, J., Sartori Blanc, N., 2001. The cell in absence of aggregation artifacts. Micron 32, 91–99.

Dubochet, J., Adrian, M., Chang, J.J., Homo, J.C., Lepault, J., McDowall, A.W., et al., 1988. Cryo-electron microscopy of vitrified specimens. Q Rev Biophys 21, 129–228.

Estable, C., Reissig, M., De Robertis, E., 1954. Microscopic and submicroscopic structure of the synapsis in the ventral ganglion of the acoustic nerve. Exp Cell Res 6, 255–262.

Fernandez, J.J., Li, S., 2003. An improved algorithm for anisotropic nonlinear diffusion for denoising cryo-tomograms. J Struct Biol 144, 152–161.

Fernández, J.J., Li, S., Crowther, R.A., 2006. CTF determination and correction in electron cryotomography. Ultramicroscopy 106, 587–596.

Fernández-Busnadiego, R., Zuber, B., Maurer, U.E., Cyrklaff, M., Baumeister, W., Lucic, V., 2010. Quantitative analysis of the native presynaptic cytomatrix by cryoelectron tomography. J Cell Biol 188, 145–156.

Fernández-Busnadiego, R., Asano, S., Sakata, E., Zürner, M., Schoch, S., Baumeister, W. et al., 2012. Structural analysis of synaptic vesicle priming by RIM1α (submitted).

Fernández-Busnadiego, R., Schrod, N., Kochovski, Z., Asano, S., Vanhecke, D., Baumeister, W., et al., 2011. Insights into the molecular organization of the neuron by cryo-electron tomography. J Electron Microsc (Tokyo) 60 (Suppl. 1), S137–S148.

Frangakis, A.S., Hegerl, R., 2001. Noise reduction in electron tomographic reconstructions using nonlinear anisotropic diffusion. J Struct Biol 135, 239–250.

Garvalov, B.K., Zuber, B., Bouchet-Marquis, C., Kudryashev, M., Gruska, M., Beck, M., et al., 2006. Luminal particles within cellular microtubules. J Cell Biol 174, 759–765.

Glaeser, R.M., 2008. Retrospective: radiation damage and its associated "information limitations". J Struct Biol 163, 271–276.

Gordon, R., Bender, R., Herman, G.T., 1970. Algebraic reconstruction techniques (ART) for three-dimensional electron microscopy and x-ray photography. J Theor Biol 29, 471–481.

Gruska, M., Medalia, O., Baumeister, W., Leis, A., 2008. Electron tomography of vitreous sections from cultured mammalian cells. J Struct Biol 161, 384–392.

Harrison, S.M., Jarvie, P.E., Dunkley, P.R., 1988. A rapid Percoll gradient procedure for isolation of synaptosomes directly from an S1 fraction: viability of subcellular fractions. Brain Res 441, 72–80.

Hayward, S.B., Glaeser, R.M., 1979. Radiation damage of purple membrane at low temperature. Ultramicroscopy 4, 201–210.

Hegerl, R., Hoppe, W., 1976. Influence of electron noise on three-dimensional image reconstruction. Z Naturforschung 31a, 1717–1721.

Heuser, T., Raytchev, M., Krell, J., Porter, M.E., Nicastro, D., 2009. The dynein regulatory complex is the nexin link and a major regulatory node in cilia and flagella. J Cell Biol 187, 921–933.

Heuser, J.E., Reese, T.S., Dennis, M.J., Jan, Y., Jan, L., Evans, L., 1979. Synaptic vesicle exocytosis captured by quick freezing and correlated with quantal transmitter release. J Cell Biol 81, 275–300.

Hirokawa, N., Sobue, K., Kanda, K., Harada, A., Yorifuji, H., 1989. The cytoskeletal architecture of the presynaptic terminal and molecular structure of synapsin 1. J Cell Biol 108, 111–126.

Hoffmann, C., Leis, A., Niederweis, M., Plitzko, J.M., Engelhardt, H., 2008. Disclosure of the mycobacterial outer membrane: cryo-electron tomography and vitreous sections reveal the lipid bilayer structure. Proc Natl Acad Sci USA 105, 3963–3967.

Holtmaat, A., Svoboda, K., 2009. Experience-dependent structural synaptic plasticity in the mammalian brain. Nat Rev Neurosci 10, 647–658.

Hoppe, W., Schramm, H.J., Sturm, M., Hunsmann, N., Gassmann, J., 1986. Three-dimensional electron microscopy of individual biological objects. 1. methods. Z Naturforsch A31, 645–655.

Humbel, B.M., de Jong, M.D., Muller, W.H., Verkleij, A.J., 1998. Pre-embedding immunolabeling for electron microscopy: an evaluation of permeabilization methods and markers. Microsc Res Tech 42, 43–58.

Ingerson-Mahar, M., Briegel, A., Werner, J.N., Jensen, G.J., Gitai, Z., 2010. The metabolic enzyme CTP synthase forms cytoskeletal filaments. Nat Cell Biol 12, 739–746.

Jovanovic, J.N., Sihra, T.S., Nairn, A.C., Hemmings Jr., H., Greengard, P., Czernik, A.J., 2001. Opposing changes in phosphorylation of specific sites in synapsin I during Ca^{2+}-dependent glutamate release in isolated nerve terminals. J Neurosci 21, 7944–7953.

Kaeser, P.S., Deng, L., Wang, Y., Dulubova, I., Liu, X., Rizo, J., et al., 2011. Rim proteins tether Ca^{2+} channels to presynaptic active zones via a direct PDZ-domain interaction. Cell 144, 282–295.

Khursigara, C.M., Wu, X., Zhang, P., Lefman, J., Subramaniam, S., 2008. Role of HAMP domains in chemotaxis signaling by bacterial chemoreceptors. Proc Natl Acad Sci USA 105, 16555–16560.

Koning, R.I., Zovko, S., Barcena, M., Oostergetel, G.T., Koerten, H.K., Galjart, N., et al., 2008. Cryo electron tomography of vitrified fibroblasts: microtubule plus ends *in situ*. J Struct Biol 161, 459–468.

Koster, A.J., Klumperman, J., 2003. Electron microscopy in cell biology: integrating structure and function. Nat Rev Mol Cell Biol 4 (Suppl.), S6–S10.

Koster, A.J., Grimm, R., Typke, D., Hegerl, R., Stoschek, A., Walz, J., et al., 1997. Perspectives of molecular and cellular electron microscopy. J Struct Biol 120, 276–308.

Kudryashev, M., Cyrklaff, M., Baumeister, W., Simon, M.M., Wallich, R., Frischknecht, F., 2009. Comparative cryo-electron tomography of pathogenic Lyme disease spirochetes. Mol Microbiol 71, 1415–1434.

Kudryashev, M., Cyrklaff, M., Wallich, R., Baumeister, W., Frischknecht, F., 2010a. Distinct in situ structures of the *Borrelia* flagellar motor. J Struct Biol 169, 54–61.

Kudryashev, M., Lepper, S., Baumeister, W., Cyrklaff, M., Frischknecht, F., 2010b. Geometric constrains for detecting short actin filaments by cryogenic electron tomography. PMC Biophys 3, 6.

Kudryashev, M., Lepper, S., Stanway, R., Bohn, S., Baumeister, W., Cyrklaff, M., et al., 2010c. Positioning of large organelles by a membrane-associated cytoskeleton in *Plasmodium* sporozoites. Cell Microbiol 12, 362–371.

Landis, D., Hall, A., Weinstein, L., Reese, T., 1988. The organization of cytoplasm at the presynaptic active zone of a central nervous-system synapse. Neuron 1, 201–209.

Le Gros, M.A., McDermott, G., Uchida, M., Knoechel, C.G., Larabell, C.A., 2009. High-aperture cryogenic light microscopy. J Microsc 235, 1–8.

Leapman, R.D., Sun, S., 1995. Cryo-electron energy loss spectroscopy: observations on vitrified hydrated specimens and radiation damage. Ultramicroscopy 59, 71–79.

Lepper, S., Merkel, M., Sartori, A., Cyrklaff, M., Frischknecht, F., 2010. Rapid quantification of the effects of blotting for correlation of light and cryo-light microscopy images. J Microsc 238, 21–26.

Lichtman, J.W., Conchello, J.A., 2005. Fluorescence microscopy. Nat Methods 2, 910–919.

Liu, J., Lin, T., Botkin, D.J., McCrum, E., Winkler, H., Norris, S.J., 2009. Intact flagellar motor of *Borrelia burgdorferi* revealed by cryo-electron tomography: evidence for stator ring curvature and rotor/c-ring assembly flexion. J Bacteriol 191, 5026–5036.

Lucic, V., Yang, T., Schweikert, G., Forster, F., Baumeister, W., 2005. Morphological characterization of molecular complexes present in the synaptic cleft. Structure 13, 423–434.

Lucic, V., Kossel, A.H., Yang, T., Bonhoeffer, T., Baumeister, W., Sartori, A., 2007. Multiscale imaging of neurons grown in culture: from light microscopy to cryo-electron tomography. J Struct Biol 160, 146–156.

Makita, T., Hatsuoka, M., Watanabe, J., Sasaki, K., Kiwaki, S., 1980. Fluorescence microscopy of hydrated frozen-sections under liquid-nitrogen. *CryoLetters* 1, 438–444.

Martinez-Sanchez, A., Garcia, I., Fernandez, J.-J., 2011. A differential structure approach to membrane segmentation in electron tomography. J Struct Biol 175, 372–383.

Mastronarde, D.N., 2005. Automated electron microscope tomography using robust prediction of specimen movements. J Struct Biol 152, 36–51.

McMullan, G., Chen, S., Henderson, R., Faruqi, A.R., 2009. Detective quantum efficiency of electron area detectors in electron microscopy. Ultramicroscopy 109, 1126–1143.

Medalia, O., Beck, M., Ecke, M., Weber, I., Neujahr, R., Baumeister, W., et al., 2007. Organization of actin networks in intact filopodia. Curr Biol 17, 79–84.

Medalia, O., Weber, I., Frangakis, A.S., Nicastro, D., Gerisch, G., Baumeister, W., 2002. Macromolecular architecture in eukaryotic cells visualized by cryoelectron tomography. Science 298, 1209–1213.

Mercogliano, C.P., DeRosier, D.J., 2007. Concatenated metallothionein as a clonable gold label for electron microscopy. J Struct Biol 160, 70–82.

Milazzo, A.C., Cheng, A., Moeller, A., Lyumkis, D., Jacovetty, E., Polukas, J., et al., 2011. Initial evaluation of a direct detection device detector for single particle cryo-electron microscopy. J Struct Biol 176, 404–408.

Moussavi, F., Heitz, G., Amat, F., Comolli, L.R., Koller, D., Horowitz, M., 2010. 3D segmentation of cell boundaries from whole cell cryogenic electron tomography volumes. J Struct Biol 170, 134–145.

Movassagh, T., Bui, K.H., Sakakibara, H., Oiwa, K., Ishikawa, T., 2010. Nucleotide-induced global conformational changes of flagellar dynein arms revealed by in situ analysis. Nat Struct Mol Biol 17, 761–767.

Nicastro, D., Schwartz, C., Pierson, J., Gaudette, R., Porter, M.E., McIntosh, J.R., 2006. The molecular architecture of axonemes revealed by cryoelectron tomography. Science 313, 944–948.

Nicholls, D.G., Sihra, T.S., 1986. Synaptosomes possess an exocytotic pool of glutamate. Nature 321, 772–773.

Nishino, Y., Yasunaga, T., Miyazawa, A., 2007. A genetically encoded metallothionein tag enabling efficient protein detection by electron microscopy. *J Electron Microsc (Tokyo)* 56, 93–101.

Ortiz, J.O., Forster, F., Kurner, J., Linaroudis, A.A., Baumeister, W., 2006. Mapping 70S ribosomes in intact cells by cryoelectron tomography and pattern recognition. J Struct Biol 156, 334–341.

Palade, G., Palay, S.L., 1954. Electron microscope observations of interneuronal and neuromuscular synapses. Anat Rec 118, 335–336.

Palade, G.E., 1955. A small particulate component of the cytoplasm. J Biophys Biochem Cytol 1, 59–68.

Parsons, D.F., 1974. Structure of wet specimens in electron microscopy. improved environmental chambers make it possible to examine wet specimens easily. Science 186, 407–414.

Parsons, D.F., Matricardi, V.R., Moretz, R.C., Turner, J.N., 1974. Electron microscopy and diffraction of wet unstained and unfixed biological objects. Adv Biol Med Phys 15, 161–270.

Patla, I., Volberg, T., Elad, N., Hirschfeld-Warneken, V., Grashoff, C., Fassler, R., et al., 2010. Dissecting the molecular architecture of integrin adhesion sites by cryo-electron tomography. Nat Cell Biol 12, 909–915.

Pierson, J., Ziese, U., Sani, M., Peters, P.J., 2011. Exploring vitreous cryo-section-induced compression at the macromolecular level using electron cryo-tomography; 80s yeast ribosomes appear unaffected. J Struct Biol 173, 345–349.

Porter, K.R., Claude, A., Fullam, E.F., 1945. A study of tissue culture cells by electron microscopy: methods and preliminary observations. J Exp Med 81, 233–246.

Radermacher, M., 1992. Weighted back-projection methods. In: Frank, J. (Ed.), Electron Tomography. Plenum Press, New York, pp. 91–115.

Rigort, A., Bäuerlein, F.J., Leis, A., Gruska, M., Hoffmann, C., Laugks, T., et al., 2010. Micromachining tools and correlative approaches for cellular cryo-electron tomography. J Struct Biol 172, 169–179.

Rigort, A., Bäuerlein, F.J.B., Villa, E., Eibauer, M., Laugks, T., Baumeister, W., et al., 2012a. Focused ion beam micromachining of eukaryotic cells for cryoelectron tomography. Proc Natl Acad Sci USA 109, 4449–4454.

Rigort, A., Günther, D., Hegerl, R., Baum, D., Weber, B., Prohaska, S., et al., 2012b. Automated segmentation of electron tomograms for a quantitative description of actin filament networks. J Struct Biol 177, 135–144.

Rizzoli, S.O., Betz, W.J., 2005. Synaptic vesicle pools. Nat Rev Neurosci 6, 57–69.

Robertson, J.D., 1953. Ultrastructure of two invertebrate synapses. Proc Soc Exp Biol Med 82, 219–223.

Rosenmund, C., Stevens, C.F., 1996. Definition of the readily releasable pool of vesicles at hippocampal synapses. Neuron 16, 1197–1207.

Salje, J., Zuber, B., Lowe, J., 2009. Electron cryomicroscopy of E. coli reveals filament bundles involved in plasmid DNA segregation. Science 323, 509–512.

Sartori, A., Gatz, R., Beck, F., Rigort, A., Baumeister, W., Plitzko, J., 2007. Correlative microscopy: bridging the gap between light- and cryo-electron microscopy. J Struct Biol 160, 135–145.

Saxton, W.O., Baumeister, W., Hahn, M., 1984. Three-dimensional reconstruction of imperfect two-dimensional crystals. Ultramicroscopy 13, 57–70.

Schiavo, G., Matteoli, M., Montecucco, C., 2000. Neurotoxins affecting neuroexocytosis. Physiol Rev 80, 717–766.

Schwartz, C.L., Sarbash, V.I., Ataullakhanov, F.I., McIntosh, J.R., Nicastro, D., 2007. Cryo-fluorescence microscopy facilitates correlations between light and cryo-electron microscopy and reduces the rate of photobleaching. J Microsc 227, 98–109.

Siksou, L., Triller, A., Marty, S., 2011. Ultrastructural organization of presynaptic terminals. Curr Opin Neurobiol 21, 261–268.

Siksou, L., Rostaing, P., Lechaire, J.-P., Boudier, T., Ohtsuka, T., Fejtov, A., et al., 2007. Three-dimensional architecture of presynaptic terminal cytomatrix. J Neurosci 27, 6868–6877.

Sjostrand, F.S., 1953. The ultrastructure of the innersegments of the retinal rods of the guinea pig eye as revealed by electron microscopy. J Cell Physiol 42, 45–70.

Small, J.V., Winkler, C., Vinzenz, M., Schmeiser, C., 2011. Reply: Visualizing branched actin filaments in lamellipodia by electron tomography. Nat Cell Biol 13, 1013–1014.

Soille, P., 2003. Morphological Image Analysis. Springer-Verlag, New York.

Studer, D., Humbel, B.M., Chiquet, M., 2008. Electron microscopy of high pressure frozen samples: bridging the gap between cellular ultrastructure and atomic resolution. Histochem Cell Biol 130, 877–889.

Urban, E., Jacob, S., Nemethova, M., Resch, G.P., Small, J.V., 2010. Electron tomography reveals unbranched networks of actin filaments in lamellipodia. Nat Cell Biol 12, 429–435.

van Driel, L.F., Knoops, K., Koster, A.J., Valentijn, J.A., 2008. Fluorescent labeling of resin-embedded sections for correlative electron microscopy using tomography-based contrast enhancement. J Struct Biol 161, 372–383.

van Driel, L.F., Valentijn, J.A., Valentijn, K.M., Koning, R.I., Koster, A.J., 2009. Tools for correlative cryo-fluorescence microscopy and cryo-electron tomography applied to whole mitochondria in human endothelial cells. Eur J Cell Biol 88, 669–684.

Vanhecke, D., Asano, S., Kochovski, Z., Fernández-Busnadiego, R., Schrod, N., Baumeister, W., et al., 2011. Cryo-electron tomography: methodology, developments and biological applications. J Microsc 242, 221–227.

Wang, Q., Mercogliano, C.P., Lowe, J., 2011. A ferritin-based label for cellular electron cryotomography. Structure 19, 147–154.

Whittaker, V.P., 1984. The synaptosome In: Lajtha, A. (Ed.), Handbook of Neurochemistry, Structural Elements of the Nervous System, vol. 7. Plenum Press, New York and London, pp. 1–39.

Yang, C., Svitkina, T., 2011. Visualizing branched actin filaments in lamellipodia by electron tomography. Nat Cell Biol 13, 1012–1013.

Zhang, P., Bos, E., Heymann, J., Gnaegi, H., Kessel, M., Peters, P.J., et al., 2004. Direct visualization of receptor arrays in frozen-hydrated sections and plunge-frozen specimens of *E. coli* engineered to overproduce the chemotaxis receptor TSR. J Microsc 216, 76–83.

Zhang, P., Khursigara, C.M., Hartnell, L.M., Subramaniam, S., 2007. Direct visualization of *Escherichia coli* chemotaxis receptor arrays using cryo-electron microscopy. Proc Natl Acad Sci USA 104, 3777–3781.

Zheng, S.Q., Keszthelyi, B., Branlund, E., Lyle, J.M., Braunfeld, M.B., Sedat, J.W., et al., 2007. UCSF tomography: an integrated software suite for real-time electron microscopic tomographic data collection, alignment, and reconstruction. J Struct Biol 157, 138–147.

Ziese, U., Janssen, A.H., Murk, J.-L., Geerts, W.J.C., Van der Krift, T., Verkleij, A.J., et al., 2002. Automated high-throughput electron tomography by pre-calibration of image shifts. J Microsc 205, 187–200.

Zuber, B., Nikonenko, I., Klauser, P., Muller, D., Dubochet, J., 2005. The mammalian central nervous synaptic cleft contains a high density of periodically organized complexes. Proc Natl Acad Sci USA 102, 19192–19197.

Zuber, B., Chami, M., Houssin, C., Dubochet, J., Griffiths, G., Daffé, M., 2008. Direct visualization of the outer membrane of mycobacteria and corynebacteria in their native state. J Bacteriol 190, 5672–5680.

8

CELLULAR-LEVEL OPTICAL BIOPSY USING FULL-FIELD OPTICAL COHERENCE MICROSCOPY

Arnaud Dubois

Laboratoire Charles Fabry, UMR 8501, Institut d'Optique, CNRS, Univ Paris Sud 11, Palaiseau Cedex France

CHAPTER OUTLINE
Introduction   185
The FF-OCM Technique   187
Detection Sensitivity   189
Spatial Resolution   191
Sample Motion Artifacts   192
FF-OCM for High-Resolution "Optical Biopsy"   193
Conclusion   195
Acknowledgment   196
References   196

Introduction

The diagnosis and study of many diseases including cancer depend upon biopsy and histopathological analysis of tissue at the cellular level using an optical microscope. A large variety of stains and fluorescence dyes can be employed to enhance the contrast of the microscopy image to differentiate individual components within the tissue. However, the penetration depth of visible light is very limited; only thin and transparent samples can be visualized with conventional optical microscopy. The samples have to be cut into slices of a few micrometers thickness after freezing or embedding in paraffin, which is time-consuming and may introduce structural deformation or damage. Moreover, some biological components can deteriorate or even be removed during tissue processing.

Cellular Imaging Techniques for Neuroscience and Beyond.
DOI: http://dx.doi.org/10.1016/B978-0-12-385872-6.00008-8

The introduction of confocal microscopy was a revolution in microscopy. By rejecting light coming from out-of-focus regions, optical sections can be produced without requiring physical sectioning. By labeling with fluorophores, specimens can be visualized at high resolution in three dimensions. The imaging penetration depth of confocal microscopy is, however, limited to ~100 μm in scattering tissues, and the use of fluorophores may affect the viability of cells.

With great advances in the technology of ultrashort pulsed lasers, nonlinear optical effects such as harmonic generation or two-photon absorption have been applied successfully in the field of microscopy. The advantage of two-photon excitation fluorescence microscopy over linear (one-photon) fluorescence microscopy includes efficient background rejection, reduced photobleaching and photodamage (König, 2000), and deeper imaging penetration because of a longer excitation wavelength. However, the penetration is limited to several hundreds of microns in highly scattering tissues (Helmchen and Denk, 2005).

Optical coherence tomography (OCT) is another technique that can image deeper than confocal microscopy and even nonlinear microscopy. OCT is based on low-coherence interferometry to measure the amplitude of light backscattered by the imaged sample (Huang et al., 1991; Fujimoto et al., 1995; Fercher, 1996; Tearney et al., 1996). The most significant impact of OCT is in ophthalmology for *in situ* examination of the pathologic changes of the retina (Swanson et al., 1993; Wojkowski et al., 2002; Hitzenberger et al., 2003; Nassif et al., 2004) and measurement of the dimensions of the anterior chamber of the eye (Izatt et al., 1994b; Simpson and Fonn, 2008). OCT has also been applied successfully to imaging of various highly scattering tissues (Fujimoto, 2003). Ultrahigh axial resolution of ~1 μm can be achieved by using laser-based ultrabroad bandwidth light sources (Drexler et al., 1999; Povazay et al., 2002; Wang et al., 2003). However, OCT suffers from a limited transverse resolution, because relatively low numerical aperture (NA) lenses have to be used to preserve a sufficient depth of field for cross-sectional imaging. Several solutions have been implemented to improve the transverse resolution of OCT, including adjustment of focus while the depth is scanned (Drexler et al., 1999; Lexer et al., 1999), illumination of the sample with a Bessel beam (Ding et al., 2002; Leitgeb, 2006), or image postprocessing based on the inverse scattering theory (Ralson et al., 2005). The most efficient approach for achieving fine enough transverse resolution for cellular-level imaging in scattering tissues is to acquire *en face* images rather than cross-sectional images. With this configuration, there is no depth of field limitation.

Optical coherence microscopy (OCM) is a version of OCT that produces *en face* tomographic images. High NA optics can be used in OCM to achieve higher transverse spatial resolution. Two general approaches for OCM have been reported to date. The first approach is based on the combination of confocal microscopy with low-coherence interferometry (Hamilton and Sheppard, 1982; Gu and Sheppard, 1993). This association was revisited later using modern technology, leading to the scanning OCM technique (Izatt et al., 1994a; Podoleanu et al., 1998). Broadband coherence gating was shown to significantly enhance the imaging depth of conventional confocal microscopy (Izatt et al., 1994a; Kempe and Rudolph, 1996). Scanning OCM was applied successfully for cellular-level resolution imaging deep below the surface of various human tissues such as skin (Aguirre et al., 2003), oral mucosa (Clark et al., 2004), and colonic mucosa (Izatt et al., 1996; Kempe and Rudolph, 1996; Aguirre et al., 2010). The second approach to OCM involves full-field illumination and detection. Also called full-field optical coherence tomography (FF-OCT), full-field optical coherence microscopy (FF-OCM) is an alternative technique to scanning OCM, based on white-light interference microscopy (Dubois et al., 2002, 2004b; Laude et al., 2002; Vabre et al., 2002). FF-OCM produces tomographic images in the *en-face* orientation by arithmetic combination of interferometric images acquired with an area camera and by illuminating the whole field to be imaged with low-coherence light (temporally and spatially). The major advantage of FF-OCM is its ultrahigh imaging resolution in both transverse and axial directions by using a simple and robust experimental arrangement (Dubois et al., 2004a; Oh et al., 2006a).

This chapter provides an overview of the principle of FF-OCM. The system characteristics, including the detection sensitivity and spatial resolution, are reported in detail. Several of the challenges, advantages, and drawbacks of this technique are discussed and compared with scanning OCM and confocal microscopy. Images of normal and diseased human breast tissues are presented to illustrate the potential of FF-OCM for high-resolution anatomopathological examinations without sample processing.

The FF-OCM Technique

The experimental arrangement of conventional FF-OCM is based on a white-light interference microscope in the Linnik configuration (Gale et al., 1996), as represented schematically in Figure 8.1. It consists of a Michelson interferometer with a microscope objective placed in each arm. A fiber halogen light source provides uniform illumination of the microscope objective fields with spatially incoherent broadband light. A low reflectivity mirror (~20%) is placed in the

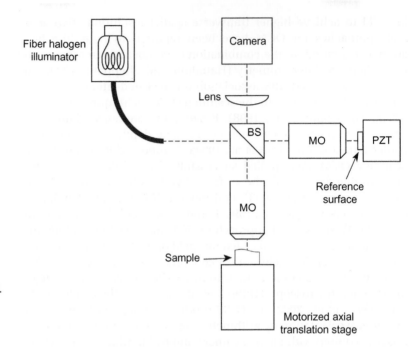

Figure 8.1 Schematic of the experimental setup of FF-OCM. MO, microscope objective; BS, beam-splitter cube; PZT, piezoelectric actuator.

reference arm of the interferometer in the object plane of the microscope objective. The interferometric images delivered by the Linnik microscope are acquired by a silicon-based area camera. Pixel binning can be applied to increase the dynamic range of the detection, and hence the detection sensitivity. This pixel binning is performed numerically after each image acquisition. Water-immersion microscope objectives (10×, NA 0.3) are employed to minimize dispersion mismatch in the interferometer arms as the imaging depth in the sample increases, and to reduce light reflection from the sample surface. A motorized translation stage is used to translate the sample in the axial direction to adjust the imaging depth. Two other motorized translation stages are used to displace the sample in orthogonal transverse directions for selecting the area to image and for creating high-resolution images of enlarged fields by image stitching.

The tomographic signal in FF-OCM is obtained by phase-shifting interferometry, a well-established method for phase measurements that can also be applied for measuring the amplitude of the interferometric signal (Kino and Chim, 1990; Caber, 1993; Larkin, 1996). This method consists of introducing a time modulation of the relative phase between the reference and sample waves and to acquire several interferometric images. In the latest FF-OCM system, *en face* tomographic images are produced by arithmetic combination of four interferometric images (Dubois et al., 2004b). The phase-shift is generated by the displacement of the reference mirror attached to a piezoelectric actuator. Each tomographic image corresponds to an image of the reflecting or backscattering structures of the sample located in a slice, perpendicular to the optical axis, which has a width equal to the coherence length. The tomographic images can be produced at a maximum rate equal to a quarter of the camera frame rate. However, in practice it is necessary to accumulate several interferometric images to improve the detection sensitivity, as will be shown in the following section. The tomographic images are then produced at a frequency on the order of 1 Hz.

Detection Sensitivity

The smallest signal that can be detected in FF-OCM, defined as the detection sensitivity, is a key parameter that affects the imaging contrast and the penetration depth. Several possible sources of noise need to be taken into account for the determination of the detection sensitivity. In addition to the shot noise induced by the fundamental photon noise, other kinds of noise originating from the camera may not be negligible, such as the readout noise and the dark noise (Lude et al., 2002; Oh et al., 2006b). Low-noise silicon-based cameras are almost shot-noise limited. In that case, assuming the well of the camera pixels to be full upon maximum illumination, it has been shown (Dubois et al., 2002; Laude et al., 2002) that the minimal detectable reflectivity can be expressed as the following:

$$R_{\text{min}} \sim \frac{(R_{\text{inc}} + R_{\text{ref}})^2}{2N\xi_{\text{sat}}R_{\text{ref}}}. \tag{8.1}$$

The parameter ξ_{sat} denotes the full-well capacity of the camera pixels. The sensitivity can be improved by increasing the full-well capacity; however, increasing the full-well capacity, while maintaining the image acquisition time requires the availability of a more powerful light source, which represents a practical limitation. Moreover, the maximal permissible exposure level of biological samples is limited. In FF-OCM, all of the light coming from the sample and collected by the microscope objective is detected by the camera. Due to the ultrashort coherence length of the light source, a large part of this light does not interfere. This proportion of light is represented in Equation (8.1) by the equivalent reflectivity R_{inc}. A camera with a high full-well capacity is required and image accumulation (parameter N) has to be applied to be able to discriminate the small interferometric signal from the whole detected signal. Since accumulating more images increases the acquisition time, this way of improving the detection sensitivity is limited if the sample is likely to move during the acquisition. *In vivo* imaging, in particular, may be incompatible with the accumulation of a large number of images. To minimize the parameter R_{inc}, all of the optical components are anti-reflection coated. When a biological sample is imaged, the reflection on the sample surface is minimized by index matching, which is achieved by using water-immersion objectives. The reference mirror reflectivity R_{ref} also has an influence on the detection sensitivity. By calculating the derivative of Equation (8.1) with respect to R_{ref}, one can easily establish that the optimal value of R_{ref} is reached when R_{ref} is equal to R_{inc}. In practice, the detection sensitivity of FF-OCM ranges between 75 and 90 dB (Dubois et al., 2004b).

Scanning OCM combines two physical mechanisms to achieve optical sectioning: confocal gating and coherence gating. Thanks to the confocal gate, light coming from out-of-focus regions in the sample is not received by the mono-detector. In FF-OCM, there is no confocal gate to eliminate this light that does not contribute to interference. This unwanted light is detected by the camera, which leads to a reduction of the useful dynamic of the camera. The effect of the coherence gate in scanning OCM then provides a very efficient rejection of the rest of unwanted light and improves the sectioning ability. Figure 8.2 compares typical theoretical axial point spread functions (PSF) resulting from confocal and coherence gates. At a depth of several tens of micrometers from the focus, the coherence gate rejects scattered light with orders of magnitude better efficiency than the confocal gate alone. The combined effects of the two gates are generally stronger than either gate individually, achieving greater image contrast and image penetration in scattering tissue. The confocal axial response alone, requiring high NA to be efficient, is affected by aberrations when focusing into tissue, which limits the performance of confocal microscopy for deep imaging. The efficiency of the

Figure 8.2 Theoretical axial responses resulting from coherence or confocal gates, in linear (A) and logarithmic (B) scales.

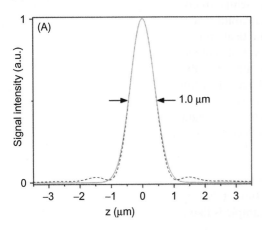

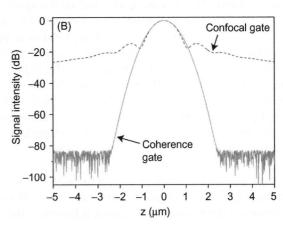

coherence gate may also degrade with depth due to dispersion mismatch in the interferometer arms. The consequences of this phenomenon, however, are generally less severe and can be corrected more easily.

Spatial Resolution

The axial resolution of conventional OCT is determined by the coherence function, i.e., the Fourier transform of the effective source power spectrum (Fercher, 1996). Since conventional OCT produces cross-sectional images (B scans), a large depth of focus is generally required, which imposes the use of low NA optics. FF-OCM acquires images in the *en face* orientation like a conventional microscope, which permits the use of high NA microscope objectives. In that case, the axial response is determined not only by the coherence function but also by the depth of focus, exactly like a confocal interference microscope (Chang and Kino, 1998). Assuming a Gaussian-shaped effective power spectrum, the theoretical axial resolution of FF-OCM without dispersion mismatch can be written as the following (Dubois et al., 2002):

$$\Delta z = \left[\frac{NA^2}{n_{im}\lambda} + \frac{n_{im}\pi}{2\ln 2}\left(\frac{\Delta\lambda}{\lambda^2}\right) \right]^{-1},$$
(8.2)

where n_{im} is the refractive index of the objective immersion medium; λ is the center wavelength; and $\Delta\lambda$ is the effective optical spectrum width (full-width at half-maximum; FWHM) given by the product of the light source, the spectral transmission of the optical components including the sample, and the immersion medium themselves; and the spectral response of the camera in use. The setup described here is characterized by a relatively low NA (0.3) and a broadband light source (halogen lamp); the axial resolution is then essentially imposed by the temporal coherence. The theoretical axial resolution was calculated to be equal to 0.75 μm ($\lambda = 750$ nm, $\Delta\lambda = 250$ nm). We have measured the interferogram response of our FF-OCM system. The coherence function that comprises the interference fringes looks Gaussian. In comparison with broadband sources used in ultrahigh-resolution OCT and OCM, such as ultrashort femtosecond lasers or supercontinuum fiber lasers (Drexler et al., 1999; Povazay et al., 2002; Wang et al., 2003; Wiesauer et al., 2005), a thermal light source provides a smoother spectrum without spikes or emission lines that could cause side lobes in the coherence function and create artifacts in the images. The experimental FWHM of the coherence function was found to be 0.8 μm. The coherence function broadens when dispersion mismatch occurs. This phenomenon leads to a degradation of the axial resolution as the imaging depth in the sample increases (Dubois et al., 2004b). For example, at a depth of 250 μm in human dermis, the axial resolution was estimated to be degraded by a factor of ~1.4.

The major advantage of FF-OCM is its ultrahigh axial resolution using a simple halogen lamp. The best axial resolution achieved in OCM, using state-of-the-art broad-bandwidth laser technology, is inferior by a factor of two (Wiesauer, 2005). Achieving ~1 μm axial resolution with confocal microscopy at the same wavelength requires objectives with an NA higher than 0.85. In practice, imaging inside scattering tissues is not possible with such microscope objectives because of the strong effects of optical aberrations that deteriorate both the axial and transverse resolutions.

The transverse resolution of an optical system can be defined as the width of the transverse PSF. Without any optical aberration, the PSF is described by the well-known Airy function whose FWHM is approximately the following:

$$\Delta r = \frac{\lambda}{2NA}. \tag{8.3}$$

The previous equation gives a theoretical diffraction-limited resolution of 1.3 μm, when microscope objectives with an NA of 0.3 are employed. We have measured the FWHM of the experimental transverse PSF of our FF-OCM to be 1.4 μm. FF-OCM and scanning OCM can offer the same transverse resolution, depending on the choice of the NA. An isotropic resolution on the order of 1 μm can be achieved with FF-OCM by using relatively low NA compared to a confocal microscope, which allows cellular-level imaging in scattering tissues.

Sample Motion Artifacts

In optical interferometry, path length modifications of only a few tens of nanometers generate significant phase changes that may cause a blurring of the measured interference signal if the detector is not fast enough. In FF-OCM, sample motions along the axial direction must not exceed a quarter of the center wavelength typically during the total image acquisition time to avoid interferometric signal washout, which represents a maximal permissible axial speed of 1–10 μm/s (Sacchet et al., 2010). The relatively long acquisition time of FF-OCM is currently a severe limitation for efficient real-time *in vivo* applications. For example, the speed of blood flow is a few hundreds of micrometers per second. Physiological motions such as the cardiac motion may even reach 100 mm/s. For these kinds of applications, a shorter acquisition time is definitely required. High-speed

FF-OCM systems have been proposed (Grieve et al., 2005; Moneron et al., 2005; Watanabe and Sato, 2008; Hrebesh et al., 2009), making *in vivo* imaging possible. However, the detection sensitivity of these systems was quite low. Real progress would be to use a camera where the ratio ξ_{sat}/T is higher (T=total image acquisition time), provided the brightness of the light source can be sufficiently increased to fill the pixel wells and the sample can tolerate the light exposure. Although the time to produce a single *en face* tomographic image is similar in FF-OCM and scanning OCM, this latest technique is less sensitive to axial motion since the acquisition time per pixel is considerably shorter. Because of this aspect, parallel acquisition of the image constitutes a drawback compared to point-by-point acquisition.

Transverse motion of the sample has other consequences in FF-OCM. The tomographic signal in FF-OCM is generated by the variation of the interferometric signal amplitude induced by the phase-shift introduced by the reference surface oscillation between interferometric images acquired sequentially in time. However, the intensity of light upon each pixel of the cameras may also vary if the backscattering structures of the sample move transversally. This variation of intensity generates a spurious signal, which superimposes onto the tomographic signal and may cause misinterpretations of the tomographic images. Axial motion of the sample has much less influence on this spurious signal, since it only causes a defocusing, which does not significantly change the intensity of light received by each pixel. We can consider that the motion of the scatterers within the sample become visible when the resulting intensity variation is greater than the minimal detectable reflectivity R_{min} given by Equation (8.1). For example, for a particle with equivalent reflectivity of 10^{-5} and an accumulation of five pairs of images, transverse motion at a speed lower than ~100 μm/s is not visible. The effect of transverse motion is clearly less severe than the effect of axial motion. Artifacts resulting from transverse motion can even be completely removed if the frames are acquired simultaneously and not sequentially in time. For that purpose, instantaneous phase-shifting interferometry was successfully applied in FF-OCM, at the price of an increase in system complexity, calibration, and cost (Akiba et al., 2003; Moneron et al., 2005; Sato et al., 2007; Sacchet et al., 2010). In scanning OCM, transverse motion does not generate a spurious signal as in FF-OCM, but may lead to a distortion of the image.

FF-OCM for High-Resolution "Optical Biopsy"

Although they are part of the standard method for disease assessment at the tissue and cellular levels, biopsy and histology have several drawbacks. FF-OCM suffers from high false negative rates due to sampling errors. The sample preparation is time-consuming, which may be awkward for preoperation examinations. Moreover, some tissues cannot be easily cut without hazard to the patient. Optical techniques that provide images *in situ* and in real time, without the need for tissue excision and processing, that have a spatial resolution close to that of histopathology are therefore of great interest. The high-resolution imaging capability of FF-OCM in the biomedical field has been extensively reported. Cellular-level resolution images of various ophthalmic tissues (Grieve et al, 2004), embryos (Dubois, et al., 2004b; Perea-Gomez et al., 2004; Oh et al., 2006a,b), human thyroid (Oh et al., 2006b), human esophagus (Dubois et al., 2004b), and plants (Dubois et al., 2004a; Boccara et al., 2007) have already been published.

In this chapter, we present images of human breast tissue obtained with an FF-OCM system developed by LLTech. All of the samples considered here are biopsy tissues. They were imaged *ex vivo*, with a drop of physiological serum inserted between the microscope objective and the samples. The microscope objectives, with an NA of 0.3, had a working distance of 3.3 mm. All of the tomographic images are in the *en face* orientation and are displayed in logarithmic scale. Figure 8.3 shows an FF-OCM image and the corresponding stained histological section. The downward arrow indicates the presence of a lobule and the upward arrow the presence of a milk duct with necrosis. Figure 8.4 compares normal and abnormal tissues. Image 8.4a shows the fibrous structure of healthy tissue with normal adipocytes. Image 8.4b shows nodular invasive carcinoma. Carcinomatous cells intermingled with trabecula of the

Figure 8.3 FF-OCM image (A) and corresponding histology (B) of human breast tissue *ex vivo*. The width of the images is 400 μm. (*Courtesy of LLTech.*)

(A)

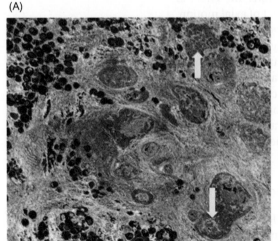

(B)

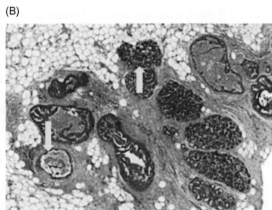

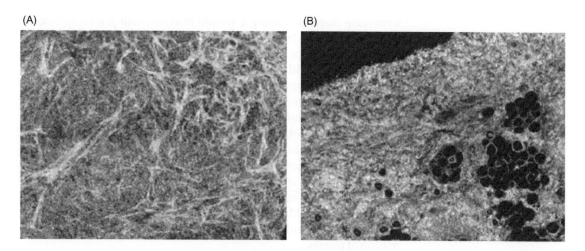

stroma reaction are visible. These images demonstrate the ability of FF-OCM to reveal architectural changes in tissue structures associated with cancer development. Figure 8.5 shows several features of human breast tissue such as duct with calcification (A), normal duct (B), and fat cells (C), illustrating the ultrahigh spatial resolution of FF-OCM.

Conclusion

FF-OCM is an alternative technique to conventional scanning OCM that provides ultrahigh resolution images in both axial and

Figure 8.4 FF-OCM images of human breast tissue showing healthy fibrous tissue with normal adipocytes (A) and diseased tissue with nodular invasive carcinoma (B). (*Courtesy of LLTech.*)

Figure 8.5 Several biological features of human breast tissues visualized with FF-OCM. Duct with calcification (A), normal duct (B), and fat cells (C). (*Courtesy of LLTech.*)

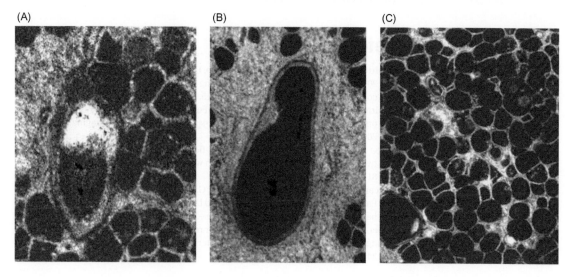

transverse directions using a simple tungsten-halogen lamp instead of a sophisticated laser-based source. Moreover, the parallel detection using a camera makes this technique attractive for its simplicity and robustness. With a spatial resolution approaching that of microscopy, this technology has the potential to replace conventional methods used in anatomopathology. FF-OCM, however, is more sensitive to motion than scanning OCM. *In vivo* imaging is still difficult with FF-OCM because of the appearance of artifacts resulting from sample motion. With the constant advances in the technologies of cameras and light sources, one can reasonably think that the acquisition speed of FF-OCM will be considerably improved in the near future. FF-OCM will then constitute a powerful tool for ultrahigh resolution *in vivo* imaging without any contrast agent, making *in situ* histopathology possible without the need for excision and histological processing of tissues.

Acknowledgment

This work would not have been possible without the invaluable contribution of postdoctoral associates and PhD students, including Delphine Sacchet, Gael Moneron, Kate Grieve, Julien Moreau, Elvire Guiot, Wilfrid Schwartz, and Laurent Vabre. I acknowledge the fruitful collaborations with multidisciplinary research teams. I am particularly grateful to Bertrand Leconte de Poly and Fabrice Harms at the company LLTech and to Claude Boccara for our impassioned discussions. This research has been supported by the Centre National de la Recherche Scientifique (CNRS) and the French Research Ministry (Ministère de la Recherche).

References

Aguirre, A.D., Chen, Y., Bryan, B., Mashimo, H., Huang, Q., Connolly, J.L., et al., 2010. Cellular resolution ex vivo imaging of gastrointestinal tissues with optical coherence microscopy. J Biomed Opt 15, 016025.

Aguirre, A.D., Hsiung, P., Ko, T.H., Hartl, I., Fujimoto, J.G., 2003. High-resolution optical coherence microscopy for high-speed, *in vivo* cellular imaging. Opt Lett 28, 2064–2066.

Akiba, M., Chan, K.P., Tanno, N., 2003. Full-field optical coherence tomography by two-dimensional heterodyne detection with a pair of CCD cameras. Opt Lett 28, 816–818.

Boccara, M., Schwartz, W., Guiot, E., Vidal, G., De Paepe, R., Dubois, A., et al., 2007. Early chloroplastic alterations analysed by optical coherence tomography during harpin-induced hypersensitive response. Plant J 50, 338–346.

Caber, P.J., 1993. Interferometric profiler for rough surfaces. Appl Opt 32, 3438–3441.

Chang, F.C., Kino, G.S., 1998. 325-nm interference microscope. Appl Opt 37, 3471–3479.

Clark, A.L., Gillenwater, A., Alizadeh-Naderi, R., El-Naggar, A.K., Richards-Kortum, R., 2004. Detection and diagnosis of oral neoplasia with an optical coherence microscope. J Biomed Opt 9, 1271.

Ding, Z., Ren, H., Zhao, Y., Nelson, J.S., Chen, Z., 2002. High-resolution optical coherence tomography over a large depth range with an axicon lens. Opt Lett 27, 243–245.

Drexler, W., Morgner, U., Kärtner, F.X., Pitris, C., Boppart, S.A., Li, X.D., et al., 1999. In-vivo ultrahigh-resolution optical coherence tomography. Opt Lett 24, 1221–1223.

Dubois, A., Moneron, G., Grieve, K., Boccara, A.C., 2004a. Three-dimensional cellular-level imaging using full-field optical coherence tomography. Phys Med Biol 49, 1227–1234.

Dubois, A., Grieve, K., Moneron, G., Lecaque, R., Vabre, L., Boccara, A.C., 2004b. Ultrahigh-resolution full-field optical coherence tomography. Appl Opt 43, 2874–2882.

Dubois, A., Vabre, L., Boccara, A.C., Beaurepaire, E., 2002. High-resolution full-field optical coherence tomography with a Linnik microscope. Appl Opt 41, 805–812.

Fercher, A.F., 1996. Optical coherence tomography. J Biomed Opt 1, 157–173.

Fujimoto, J.G., 2003. Optical coherence tomography for ultrahigh resolution in vivo imaging. Nat Biotechnol 21, 1361–1367.

Fujimoto, J.G., Brezinski, M.E., Tearney, G.J., Boppart, S.A., Bouma, B.E., Hee, M.R., et al., 1995. Optical biopsy and imaging using optical coherence tomography. Nat Med 1, 970–972.

Gale, D.M., Pether, M.I., Dainty, J.C., 1996. Linnik microscope imaging of integrated circuit structures. Appl Opt 35, 131–148.

Grieve, K., Dubois, A., Simonutti, M., Pâques, M., Sahel, J., Le Gargasson, J.F., et al., 2005. In vivo anterior segment imaging in the rat eye with high speed white light full-field optical coherence tomography. Opt Express 13, 6286–6295.

Grieve, K., Paques, M., Dubois, A., Sahel, J., Boccara, A.C., Le Gargasson, J.F., 2004. Ocular tissue imaging using ultrahigh-resolution full-field optical coherence tomography. Invest Ophthalmol Vis Sci 45, 4126–4131.

Gu, M., Sheppard, C.J.R., 1993. Fiber-optical confocal scanning interference microscopy. Opt Commun 100, 79–86.

Hamilton, D.K., Sheppard, C.J.R., 1982. A confocal interference microscope. Optica Acta 29, 1573–1577.

Helmchen, F., Denk, W., 2005. Deep tissue two-photon microscopy. Nat Methods 2, 932–940.

Hitzenberger, C.K., Trost, P., Lo, P.W., Zhou, Q., 2003. Three-dimensional imaging of the human retina by high-speed optical coherence tomography. Opt Express 11, 2753–2761.

Hrebesh, M.S., Dabu, R., Sato, M., 2009. In vivo imaging of dynamic biological specimen by real-time single-shot full-field optical coherence tomography. Opt Commun 282, 674–683.

Huang, D., Swanson, E.A., Lin, C.P., Schuman, J.S., Stinson, W.G., Chang, W., et al., 1991. Optical coherence tomography. Science 254, 1178–1181.

Izatt, J.A., Hee, M.R., Owen, G.M., Swanson, E.A., Fujimoto, J.G., 1994a. Optical coherence microscopy in scattering media. Opt Lett 19, 590–592.

Izatt, J.A., Hee, M.R., Swanson, E.A., Lin, C.P., Huang, D., Schuman, J.S., et al., 1994b. Micrometer-scale resolution imaging of the anterior eye in vivo with optical coherence tomography. Arch Ophthalmol 112, 1584–1589.

Izatt, J.A., Kulkarni, M.D., Wang, H.W., Kobayashi, K., Sivak, M.V., 1996. Optical coherence tomography and microscopy in gastrointestinal tissues. IEEE J. Sel Topics Quantum Electron 2, 1017.

Kempe, M., Rudolph, J., 1996. Analysis of heterodyne and confocal microscopy. J Modern Opt 43, 2189–2204.

Kino, G.S., Chim, S.C., 1990. Mirau correlation microscope. Appl Opt 29, 3775–3783.

König, K., 2000. Multiphoton microscopy in life sciences. J Microsc 200, 83–104.

Larkin, K.G., 1996. Efficient nonlinear algorithm for envelope detection in white light interferometry. J Opt Soc Am A 13, 832–843.

Laude, B., De Martino, A., Drévillon, B., Benattar, L., Schwartz, L., 2002. Full-field optical coherence tomography with thermal light. Appl Opt 41, 6637–6645.

Leitgeb, R.A., Villiger, M., Bachmann, A.H., Steinmann, L., Lasser, T., 2006. Extended focus depth for fourier domain optical coherence microscopy. Opt Lett 31, 2450–2452.

Lexer, F., Hitzenberger, C.K., Drexler, W., Molebny, S., Sattmann, H., Sticker, M., Fercher, A.F., 1999. Dynamic coherent focus OCT with depth-independent transversal resolution. J Modern Opt 46, 541–553.

Moneron, G., Boccara, A.C, Dubois, A., 2005. Stroboscopic ultrahigh-resolution full-field optical coherence tomography. Opt Lett 30, 1351–1353.

Nassif, N., Cense, B., Park, B.H., Yun, S.H., Chen, T.C., Bouma, B.E., et al., 2004. In-vivo human retinal imaging by ultrahigh-speed spectral domain optical coherence tomography. Opt Lett 29, 480–482.

Oh, W.-Y., Bouma, B.E., Iftimia, N., Yelin, R., Tearney, G.J., 2006a. Spectrally-modulated full-field optical coherence microscopy for ultrahigh-resolution endoscopic imaging. Opt Express 14, 8675–8684.

Oh, W.Y., Bouma, B.E., Iftimia, N., Yun, S.H., Yelin, R., Tearney, G.J., 2006b. Ultrahigh-resolution full-field optical coherence microscopy using InGaAs camera. Opt Express 14, 726–735.

Perea-Gomez, A., Moreau, A., Camus, A., Grieve, K., Moneron, G., Dubois, A., et al., 2004. Initiation of gastrulation in the mouse embryo is preceded by an apparent shift in the orientation of the anterior-posterior axis. Curr Biol 14, 197–207.

Podoleanu, A.G., Dobre, G.M., Jackson, D.A., 1998. En-face coherence imaging using galvanometer scanner modulation. Opt Lett 23, 147–149.

Povazay, B., Bizheva, K., Unterhuber, A., Hermann, B., Sattmann, H., Fercher, A.F., et al., 2002. Submicrometer axial resolution optical coherence tomography. Opt Lett 27, 1800–1802.

Ralson, T.S., Marks, D.L., Kamalabadi, F., Boppart, S.A., 2005. Deconvolution methods for mitigation of transverse blurring in optical coherence tomography. IEEE Trans Image Proc Special Issue Molec Cell Bioimag 14, 1254.

Sacchet, D., Brzezinski, M., Moreau, J., Georges, P., Dubois, A., 2010. Motion artifact suppression in full-field optical coherence tomography. Appl Opt 49, 1480–1488.

Sato, M., Nagata, T., Niizuma, T., Neagu, L., Dabu, R., Watanabe, Y., 2007. Quadrature fringes wide-field optical coherence tomography and its applications to biological tissues. Opt Commun 271, 573–580.

Simpson, T, Fonn, D., 2008. Optical coherence tomography of the anterior segment. Ocul Surf 6, 117–127.

Swanson, E.A., Izatt, J.A., Hee, M.R., Huang, D., Lin, C.P., Schuman, J.S., et al., 1993. In-vivo retinal imaging by optical coherence tomography. Opt Lett 18, 1864–1866.

Tearney, G.J., Bouma, B.E., Boppart, S.A., Golubovic, B., Swanson, E.A., Fujimoto, J.G., 1996. Rapid acquisition of in-vivo biological images by use of optical coherence tomography. Opt Lett 21, 1408–1410.

Vabre, L., Dubois, A., Boccara, A.C., 2002. Thermal-light full-field optical coherence tomography. Opt Lett 27, 530–532.

Wang, Y., Zhao, Y., Nelson, J.S., Chen, Z., Windeler, R.S., 2003. Ultrahigh-resolution optical coherence tomography by broadband continuum generation from a photonic crystal fiber. Opt Lett 28, 182–184.

Watanabe, Y., Sato, M., 2008. Three-dimensional wide-field optical coherence tomography using an ultrahigh-speed CMOS camera. Opt Commun 281, 1889–1895.

Wiesauer, K., Pircher, M., Götzinger, E., Bauer, S., Engelke, R., Ahrens, G., et al., 2005. En-face scanning optical coherence tomography with ultra-high resolution for material investigation. Opt Express 13, 1015–1024.

Wojtkowski, M., Leitgeb, R., Kowalczyk, A., Bajraszewski, T., Fercher, A.F., 2002. In-vivo human retinal imaging by Fourier domain optical coherence tomography. J Biomed Opt 7, 457–463.

9

RETROVIRAL LABELING AND IMAGING OF NEWBORN NEURONS IN THE ADULT BRAIN

Kurt A. Sailor[1,2], Hongjun Song[1,2,3,4], and Guo-li Ming[1,2,3,4]

[1]Institute for Cell Engineering, [2]The Solomon H. Snyder Department of Neuroscience, and, [3]Department of Neurology, Johns Hopkins University School of Medicine, Baltimore, Maryland, USA, [4]Diana Helis Henry Medical Research Foundation, New Orleans, Louisiana, USA

CHAPTER OUTLINE

Techniques to Label and Detect Newborn Neurons in the Adult Brain   202
Retrovirus-mediated Labeling of Adult-born Neurons   203
Single-cell Genetic Manipulation in Adult-born Neurons   205
Retrovirus Production and Delivery   207
Viral-labeled Cell Toxicity and Physiological Changes   209
Imaging Newborn Neurons in the Adult Brain   210
In Vivo Live Animal Imaging of Adult Neurogenesis   210
In Vivo Window Preparation   211
In Vivo Imaging Setup and Acquisition   213
Postacquisition Image Processing and Analysis   213
Future Directions in Live Animal Imaging of Adult Neurogenesis   214
Acknowledgments   216
References   216

Adult neurogenesis is a novel system of new circuit formation and maintenance in the mature neuronal network and an extraordinary form of structural and functional plasticity. The existence of proliferative zones and newborn neurons in the adult brain was first suggested in 1965 with the discovery of mitotic cells and new neurons in the dentate gyrus of the rat hippocampus and subependymal layer of the lateral ventricles (Altman and Das, 1965). There is now mounting evidence confirming that adult neurogenesis occurs in almost all mammals studied, including humans (Eriksson et al., 1998; Ge et al., 2006; Zhao et al., 2006). However, the mechanisms underlying cell

Cellular Imaging Techniques for Neuroscience and Beyond.
DOI: http://dx.doi.org/10.1016/B978-0-12-385872-6.00009-X

fate determination of adult neural stem cells, activity-dependent regulation of neural development, and the function of ongoing neurogenesis in the adult brain are not fully understood (Zhao et al., 2006; Bonaguidi et al., 2011; Ming and Song, 2011).

In rodents, newly formed hippocampal dentate granule cells are derived from adult neural stem cells in the subgranular zone (SGZ), a thin neurogenic region between the hilus and granule cell layer of the dentate gyrus. New periglomerular neurons and olfactory granule cells are derived from neural stem cells in the subventricular zone (SVZ) lining the lateral ventricles and are integrated into the olfactory bulb after migrating anteriorly via the rostral migratory stream (Ming and Song, 2005, 2011). In both regions new neurons incorporate over a span of weeks, and once they reach maturity they are indistinguishable from the existing older neuronal population, as suggested by their morphological and electrophysiological properties (van Praag et al., 2002; Ge et al., 2006). Early approaches for labeling dividing cells and their progeny used radiolabeled thymidine or its analog, which incorporates into genomic DNA during the S-phase of cell cycle to birthdate cells. Immunohistochemistry (IHC) with cell type-specific antibodies was used in conjunction to determine the identity of labeled cells. With the advent of oncoretrovirus-mediated methods to exclusively label dividing cells with fluorescent proteins, such as green fluorescent protein (GFP), the complete structure of a newborn neuron can be directly visualized live, allowing for electrophysiological and morphological analysis to track their development (van Praag et al., 2002; Ge et al., 2006).

The combined evolution of molecular labeling and modern imaging techniques has driven rapid progress in the field of neuroscience. This chapter will cover retroviral techniques to probe adult neurogenesis, starting with methods to label and genetically manipulate the newborn neuronal population in adult rodents *in vivo*. We will then explore various imaging methods to visualize these neurons. Finally, we will provide a glimpse into the future of these combined techniques and questions that may be answered in the field. This will not be a review of neurogenesis (Gage, 2000; Ming and Song, 2005, 2011; Whitman and Greer, 2009) or a complete review of imaging techniques (Wilt et al., 2009); instead it is a specific review on approaches to retrovirally label and image new neurons in the adult brain.

Techniques to Label and Detect Newborn Neurons in the Adult Brain

In the 1960s, the incorporation of tritiated thymidine nucleotide into nuclei, which was systemically injected into an animal and detected by autoradiography, allowed birthdating of populations of dividing cells and led to the initial discovery of two proliferative zones in the adult brain (Altman and Das, 1965). In the early 1980s, a new synthetic thymidine analog, bromodeoxyuridine (BrdU), was developed to label dividing cells (Gratzner, 1982). BrdU can be systemically injected into an animal with the advantage of being detected with antibodies, which allows for coimmunostaining for other markers. With this approach, newborn cells in the adult proliferative zones were shown to eventually differentiate into mature neurons (Kuhn et al., 1996). Currently, antibodies for doublecortin and neuronal nuclei (NeuN) are commonly used as immature and mature neuronal markers, respectively, to determine the maturation stages of BrdU$^+$ neurons (Ge et al., 2006). In addition to marking the cell's birthdate, critical periods related to survival and plasticity during adult neurogenesis can be determined by injecting a cohort of animals and then analyzing the brain tissue at different time points (Mouret et al., 2008).

More recently, Iodo and Chloro analogs of uridine have been developed for staggered labeling, with the advantage of differentiating the cell proliferation versus cell survival contribution to adult neurogenesis within the same animal (Llorens-Martin and Trejo, 2011). The development of ethynyl deoxyuridine (EdU), another uridine analog, further simplified the labeling procedure by eliminating the harsh acid tissue pretreatment as required with halogenated uridines and the need to use an antibody to detect the modified nucleotide (Salic and Mitchison, 2008). Instead, a chemical reaction directly bonds a fluorophore to EdU, which greatly increases the signal with less background fluorescence (our laboratory observation).

While DNA analogs are easy to use and inexpensive, they have some disadvantages and limitations (Table 9.1). First, these analogs are teratogens, making them potentially harmful for the animal and labeled cells (Skalko and Packard, 1975). Second, the labeling procedure requires tissue fixation, making it infeasible for live imaging and electrophysiology analysis. Third, DNA analogs label the nucleus and thus rely on colocalization imaging with other cell-specific antibodies to determine the cell type and structures. As the criteria for colocalization of non-nuclear-labeled cell morphology markers with nuclear labeling can be uncertain even with confocal imaging, this approach has introduced controversy over the existence of neurogenesis in the adult primate cortex (Gould et al. 1999; Rakic, 2002).

Retrovirus-mediated Labeling of Adult-born Neurons

The adoption of lentiviral and adeno-associated viruses with re-engineered viral tropism in neuroscience research provides investigators with a powerful tool to infect live cells in neural tissue. Labeling of neurons with viral-mediated expression of a fluorescent reporter protein,

Table 9.1 Methods to Label and Visualize Newborn Adult Neurons: Advantages and Disadvantages

Method	Cost	Difficulty	Advantages	Disadvantages
LABELING NEWLY FORMED NEURONS				
BrdU DNA analogs	Inexpensive	Easy delivery to animal, simple IHC procedure	Bright labeling, can use fluorescent probe for multilabeling IHC with morphology markers	Teratogenic, require acid or chemical treatment, labels nucleus, does not show cell morphology
IHC proliferation markers	Moderate	Standard IHC protocols	No injection of animal, can be used in combination with morphology markers	Variable results, no long-term marker for tracking new cells
Retrovirus-mediated expression	Moderate to expensive	Moderate to produce and inject into region	Complete morphology of cell, knockdown and overexpression, quick production, *in vivo* image, electrophysiology, bright fluorescence with strong promoters	Biosafety issues, limited insert size, variable titer quality, invasive
Transgenic mice	Expensive to develop, moderate to maintain	Difficult to develop	Noninvasive, *in vivo* image or electrophysiology, complete morphology	Variable brightness, limited number of lines for adult neurogenesis
IMAGING NEWLY FORMED NEURONS				
Brightfield imaging	Inexpensive	Easy	DAB labeling allows amplification, large field of view imaging, ideal for stereology, fast imaging	Limited to no multilabeling capability, blurry images, not suitable for thick sections
Epifluorescent imaging	Moderate	More complex	Multilabel imaging using fluorescent antibodies/proteins, fast imaging, good for screening brain sections for fluorescent protein-positive cells	Blurry images, not suitable for accurate colocalization detection of proliferation and morphology markers, not suitable for thick sections
One-photon confocal imaging	Expensive	Moderate	Sharp, high contrast images, ideal for colocalization of proliferation and morphology markers	Bleaching in z-path, not suitable for *in vivo* imaging, depth limitation
Two-photon live animal imaging	Very expensive	Difficult	Multi time point deep tissue imaging of same neuron to detect dynamics, complete structure imaging, biologically safe/little bleaching, behavioral manipulation effects during imaging, genetic calcium indicators	Surgical preparation difficult/variable quality, restricted to olfactory bulb neurogenesis, surgical damage to region, blood vessels hinder view

IHC, immunohistochemistry; and DAB, diaminobenzidine.

such as GFP or its variants, reveals the complete structure of a cell. Thus morphological and electrophysiological analysis can be carried out to probe individual cell and neuronal network function, a distinct advantage over the use of DNA analogs (Yu and Schaffer, 2005). The viral labeling approach has a relatively low cost with a much reduced experimental time frame compared to developing a transgenic animal (Table 9.1).

While lenti- and adeno-associated viruses can infect both dividing and nondividing cells, murine leukemia retroviruses (MLVs) are ideally suited for studying adult neurogenesis since they only integrate into dividing cells (Lewis and Emerman, 1994). The expression of MLV genes requires nuclear entry during mitosis, therefore providing specific labeling of active proliferating neural progenitors in the adult brain with relatively accurate birthdating. The retrovirus incorporates into the host genome, resulting in stable expression throughout the life of the infected progenitor cells and their progeny. The retrovirus-mediated fluorescent protein expression in many cases is sufficiently bright such that the complete structure of a neuron derived from the infected progenitor, including complete dendrites, dendritic spines, axons, and synaptic boutons, can be visualized by confocal microscopy, even without the need of IHC to amplify the signal (Figure 9.1). Numerous groups have performed electrophysiological studies using retroviral labeling to determine the physiological properties of adult-generated neurons at progressive developmental stages in both the olfactory bulb and the dentate gyrus (Carleton et al., 2003; Ge et al., 2006), along with classifying the morphological development of their dendritic structures (Petreanu and Alvarez-Buylla, 2002; Zhao et al., 2006). In addition, retroviral-mediated labeling has been used to track the migration of neuroblasts in the SVZ and rostral migratory stream (Kakita and Goldman, 1999), and to determine the ultrastructure of newborn neurons and their neighbors by combining immunogold or chromagen labeling of GFP with electron microscopy (Faulkner et al., 2008; Toni et al., 2008).

Single-cell Genetic Manipulation in Adult-born Neurons

Modification of MLV vectors has further permitted genetic manipulations for mechanistic studies and elucidation of functions of specific genes. Strong promoters, such as ubiquitin-C (ubi), eukaryotic elongation factor 1-α (EF-1α), and cytomegalovirus , are used in combination within an MLV vector for simultaneous expression of a fluorescent protein for labeling the proliferating cell and a gene or an shRNA (short hairpin RNA under the U6 promoter) for genetic manipulation. Retrovirus-mediated shRNA knockdown of endogenous gene expression has led to the identification of several factors and molecular

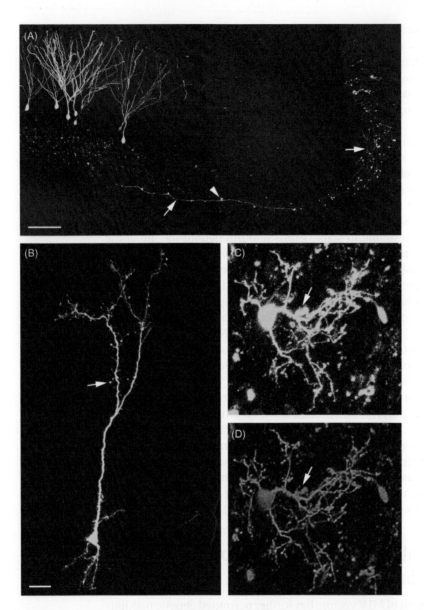

Figure 9.1 Retroviral labeling of adult-born neurons. (A) Adult-born dentate granule cells in the mouse hippocampus labeled by infection with GFP-expressing MLV retrovirus and sacrificed at 8 weeks after virus injection. Note that the complete dendritic structure of granule cells can be visualized along with their axons (arrows) and synaptic boutons (arrow head) projecting to the CA3 hippocampal subfield. (B) An adult-born granule cell in the mouse olfactory bulb labeled by infection with GFP-expressing MLV retrovirus and sacrificed 3 weeks after virus injection. The complete dendritic structure along with spines can be visualized. (C) An adult-born mouse olfactory bulb periglomerular neuron labeled by infection with GFP-expressing MLV retrovirus and two-photon *in vivo* imaged through an open skull cranial window at 5 weeks after viral injection. Also shown in D is the same neuron as in C with deconvolution. Note that with deconvolution, fine structures lost in the raw image (C, arrow) due to axial and lateral light scattering become defined (D, arrow) also with a significant decrease in overall noise. Scale bars: (A) 100 μm, (B–D) 15 μm. Please see color plates at the back of the book.

pathways involved in the regulation of adult neurogenesis (Ge et al., 2006; Duan et al., 2007; Jessberger et al., 2008; Kuwabara et al., 2009; Kim et al., 2009, 2010). There has also been some success in using MLV for overexpression of either wild-type or dominant-negative forms of a gene (Lie et al., 2005; Duan et al., 2007; Jessberger et al., 2008; Kang et al., 2011). However, gene delivery by retroviruses is limited by the insert size that can be packaged into the virus genome (up to 6.5 kb); therefore, MLV-based retroviruses are not suitable for expressing large genes. Furthermore, genetically modified animals, such as promoter-specific floxed mouse lines have been used in conjunction with retrovirus expressing Cre recombinase to achieve single-cell gene knockout in adult newborn neurons *in vivo* (Tashiro et al., 2006).

One general limitation of the retroviral approach in studying adult neurogenesis is its invasive method of delivery. High titers of MLVs are directly injected to the hilus region of the hippocampus or into the SVZ under stereotaxic guidance. The damage can be minimized by using a small needle (25 gauge or smaller) or using a pulled glass fine-tipped pipette (~10 μm).

Retrovirus Production and Delivery

A flowchart of the steps required for retrovirus production, injection, tissue processing, and imaging is illustrated in Figure 9.2. Replication-incompetent MLV viruses are created by transfecting a helper cell line (e.g., HEK293 cells) with three vectors: one containing gag-pol polypeptide sequences, one with an envelope protein, and one with the retroviral expression vector for expressing the gene(s) of interest, for example, ubi promoter driving GFP (Miller, 1990). The retroviral expression vector contains a retrovirus-packaging signal and is designed with the target gene containing both the promoter and gene of interest, which are flanked by long terminal repeat sequences demarcating the end of the retroviral genome, allowing stable integration into the host cell's genome (Figure 9.2). Upon assembly in the helper cells, only the retroviral expression vector is packaged, thus the virus is replication incompetent due to the virus lacking gag-pol and envelope protein sequences that are essential for viral reproduction. Integration of the retroviral expression vector into the host genome does not require viral protein synthesis. Without the production of these proteins in the host, the immune response is attenuated (Sliva and Schierle, 2010). The combination of these factors makes working with retroviruses much safer than working with native viruses, but biosafety protocols must be properly followed.

A laboratory that produces viruses requires access to fundamental equipment for molecular biology, mammalian cell culture, and ultracentrifugation. We normally use a two-plasmid approach where HEK293 cells with the *gag-pol* genes incorporated into their genome

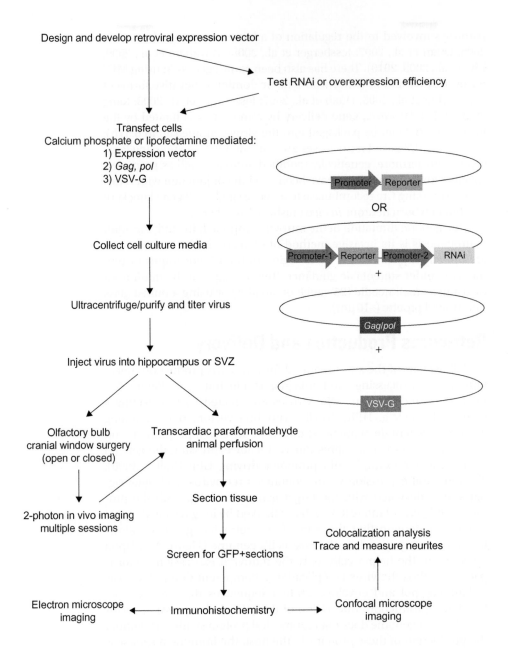

Figure 9.2 Flowchart of the production of retroviruses, *in vivo* labeling, and imaging.

(HEK293-GP2 cells) are transfected using the calcium phosphate method (Graham and van der Eb, 1973). The retroviral expression vector and a vector expressing the envelope glycoprotein of the vesicular stomatitis virus (VSV-G) are mixed with calcium chloride and HEPES phosphate buffered saline is added to produce suspended microscopic calcium phosphate crystals containing the plasmids. This precipitate solution is added to highly confluent HEK293-GP2 cells and incubated to allow for endocytosis of the crystals and subsequent incorporation of the plasmid (Jordan et al., 1996). The media is changed and the supernatant is collected multiple times after transfection and can be stored at −80°C for future use.

To isolate and concentrate the viral particles, supernatants are ultracentrifuged at 25,000 g and the pellet containing concentrated viral particles is then resuspended. A key to this protocol is the use of the VSV-G envelope protein, since it can withstand high g forces making the viral particle purification relatively quick and simple. A complete protocol similar to our approach has been published for expressing Cre-recombinase, which can be applied to any retroviral expression vector (Tashiro et al., 2006). In addition, a very general calcium phosphate transfection protocol video for making retrovirus is available (Gavrilescu and Van Etten, 2007). For laboratories incapable of producing their own virus, many commercial vendors and public institution cores can provide high titer virus using their own stock constructs or custom constructs provided by the investigator.

Numerous steps are involved to retrovirally label newborn neurons in the adult brain, and of great importance is care and practice in perfecting stereotaxic injections. A stereotaxic frame is required to securely position the animal, and a manipulator arm for holding the injection device is required for accurate injections. Several protocols have been published explaining the procedure for stereotaxic injection of retrovirus into the rodent brain for studying adult neurogenesis (Tashiro et al., 2006; Gu et al., 2011).

Viral-labeled Cell Toxicity and Physiological Changes

One concern with retroviral infection of cells is that high-titer viruses may result in multicopy insertion into the host genome and lead to potentially toxic effects due to a high level of expression. For example, GFP has been reported to be deleterious when highly expressed (Liu et al., 1999; Baens et al., 2006). The expression level of retroviral-labeled GFP under the ubi promoter is generally weaker than that of genetically encoded GFP in the Thy-1-GFP transgenic mouse lines (Feng et al., 2000). In the past decade, neurons in Thy-1-GFP mice have been thoroughly investigated with no apparent fluorescent protein-induced toxicity or alteration of neural activity. Several studies directly compared nonretroviral-labeled GFP⁻ mature neurons and GFP⁺ adult-born neurons older than 4 weeks of cell age in the same animal and found no discernible differences in electrophysiological properties (van Praag et al., 2002; Ge et al., 2006;

Overstreet-Wadiche and Westbrook, 2006). In addition, retroviral-labeled cells with GFP expression have been shown to survive up to 14 months, supporting GFP being relatively inert by retroviral mediated expression (Zhao et al., 2006).

Imaging Newborn Neurons in the Adult Brain

In the adult hippocampus and olfactory bulb, it is not uncommon to observe that retroviral-labeled neurons exhibit different levels of fluorescence within the same animal and variable numbers of cells are infected in different animals even when using the same batch of retrovirus. Many factors contribute to this variation and some are more difficult to control for, such as variability caused by differences in the copy number of retrovirus inserted into the host genome and the effect of the insert sequence on reporter expression. Other factors, such as the quality of stereotaxic injection, the titer of viral particles, and the efficiency of the promoter, can in general be improved or adjusted. If the retroviral-labeled brain sections are to be used for IHC, one practice is to screen the fixed brain sections for the presence of infected cells under microscope at a low magnification. If the intensity of fluorescent protein is weak, signal amplification with a matched wavelength secondary antibody may be necessary before confocal imaging.

Confocal microscopy is the standard tool for visualizing fluorescently labeled adult-born neurons. A confocal microscope uses an adjustable pinhole between a single point detector and the specimen. This pinhole is "confocal" to the focal plane of the tissue, thus any light in focus will pass through the pinhole while out-of-focus light will be blocked (Shotton and White, 1989). The imaging depth can be up to 100 μm, depending on the wavelength of the laser and the opacity of the tissue. If the brain sections are immunostained with more than one antibody with different fluorophores, multiple channels acquisition configuration should be used. To avoid cross talk between the fluorescent channels, the correct configuration needs to be set up where the excitation laser and matching emission filter (i.e., 488 nm laser and a bandpass 500–550 nm filter) should only be scanning and acquiring as separate tracks for each fluorophore. Cross talk or bleed-through is a common artifact associated with simultaneous configurations in which all the lasers scan while the separate filters for the fluorophores detect simultaneously.

In Vivo Live Animal Imaging of Adult Neurogenesis

With the advent of the two-photon microscope, high-resolution deep-brain imaging of live animals has become a reality, allowing for *in vivo* tracking of changes of neuronal structure and activity in response to various manipulations and behavioral tasks (Svoboda and Yasuda, 2006; Stutzmann, 2008). Fluorescent-labeled neurons in the intact animal can be detected using two-photon confocal imaging with sufficient resolution to discriminate individual neurites and spines. Moreover, the same neuron can be identified and imaged at multiple time points for tracking morphological changes, effectively removing any "blurring" of neurite changes that occur with averaged population analysis (Knott et al., 2006; Chen et al., 2011).

A distinct advantage of *in vivo* imaging is the visualization of dynamic structural changes in real time. This is impossible when using conventional histological methods where net changes in overall spine density or neurite length are measured at the population level, but the actual gain and loss of spines or neurite motility of the same cell cannot be detected. Furthermore, single-photon confocal imaging has limited tissue penetration whereas two-photon imaging can penetrate up to 1 mm allowing imaging of complete neuronal structures in a single or multiple fields of view. The same two-photon imaged neurons can be identified *ex vivo* and electron microscope imaged to find correlations in ultrastructural changes with *in vivo* dynamics (Knott et al., 2006).

The *in vivo* two-photon imaging of adult neurogenesis so far is limited to the olfactory bulb due to its superficial localization for accessibility. Adult hippocampal neurogenesis is still out of reach because of its deep brain localization, which is beyond the depth limit of two-photon imaging. Developmental changes of newly born GFP⁺ periglomerular cells and granule cells in the adult olfactory bulb have been characterized, which showed active structural plasticity even after 3 months (Mizrahi, 2007). In a later study, Livneh et al. (2009) used postsynaptic density-95 GFP fusion protein as a postsynaptic marker and examined the structural changes of periglomerular neurons. By using intrinsic signal imaging to track the blood flow dynamics on the olfactory surface, they were able to map the activity domains of specific odorants. Interestingly, when the animals were subjected to odorants for multiple days, the periglomerular cells beneath the active regions had larger dendritic projections compared to the control regions. Recently, this technique was further used to show that localized odor-induced activity stabilizes the spine structures of adult-born granule cells (Livneh and Mizrahi, 2011). These studies have demonstrated the feasibility of combining direct sensory manipulation and *in vivo* imaging to probe activity-induced dynamic changes of newborn neurons.

In Vivo Window Preparation

In general, there are two major types of cranial window preparations for *in vivo* imaging: using either an open skull craniotomy or a closed thinned skull approach. There are advantages and disadvantages to the two major types of cranial window preparations, but they both require practice to produce a clear long-term window and the success rate is quite variable between individuals and laboratories (Holtmaat et al., 2009).

The open skull preparation involves making a craniotomy over the region of interest and a coverglass is placed on the dura and secured with cyanoacrylate glue and dental cement (Holtmaat et al., 2009). A metal bar is also secured to the skull for reproducible secure mounting of the animal's head to a microscope stage. The first imaging session for chronic studies is performed at least 3 to 4 weeks after window implantation, since possible surgery-induced inflammation under the window could cause unpredicted results and a transient period of decreased image quality (Lee et al., 2008). The open skull preparation has superior optical clarity and allows imaging at great depths (500–1000 µm). With this approach, the animals can be imaged for months without further surgical procedures. Due to bone regeneration into the opening and dura thickening, the open window approach is on average limited to 3 months and rebuilding the open cranial window is difficult and often unsuccessful (Holtmaat et al., 2009). Recently, a modified form of this window was used to chronically image newly formed neurons in the olfactory bulb (Adam and Mizrahi, 2011; Livneh and Mizrahi, 2011).

The thinned-skull preparation does not completely expose the brain; instead, the skull is carefully shaved to within a thickness of 20 µm, making the bone transparent (Yang et al., 2010). A metal bar is attached to the skull to secure the animal's head to the microscope stage. Since this preparation is minimally invasive with little or no damage to the underlying brain tissue, injury-induced artifacts are reduced (Davalos et al., 2005; Nimmerjahn et al., 2005). The maximum imaging depth with a closed cranial window is 200–300 µm, thus imaging is limited to superficial neuronal structures. Since the skull regenerates rapidly it needs to be rethinned for subsequent imaging sessions, but this can only be performed for a maximum of five sessions, although the duration between sessions can be months to years (Yang et al., 2010).

Recently, a hybrid technique has been developed where a large area of the skull (up to 3 mm diameter) is thinned and a coverglass is glued to the thinned skull with transparent cyanoacrylate (Drew et al., 2010). This approach allows for a minimally invasive, large chronically open window with little or no bone regeneration, since the glue creates an impenetrable layer for osteoblast infiltration. Furthermore, there is no dura thickening, as can occur with the open window preparation, which gradually causes decreased imaging depth and quality (Holtmaat et al., 2009). The key to the hybrid window technique is to produce an even and sufficiently thinned skull

(10–15 μm). This is achieved by using a fine polishing grit slurry on the bone surface and swirling this solution until the desired thickness is achieved (Drew et al., 2010). Similar to the thinned-skull approach, the hybrid window imaging depth is restricted to 200–300 μm.

Irrespective of the choice of window preparation, proper control experiments need to be performed, including IHC for reactive astrocytes and activated microglia, since the window type may affect experimental results (Xu et al., 2007).

In Vivo Imaging Setup and Acquisition

To image deep brain structures, a multiphoton pulsed laser source is required to excite the fluorophores. Ti:sapphire pulsed lasers are the predominant two-photon source, which produces extremely brief pulses of infrared light at high frequency. Current models have tuning ranges from 650 to 1200 nm providing sufficient range for the excitation of most genetically encoded fluorescent proteins (Stutzmann, 2008). The employment of infrared light greatly reduces the scattering of photons, provides superior tissue penetration, and introduces much less phototoxicity. Fluorophore excitation is achieved by simultaneously absorbing two long wavelength photons, and this two-photon effect is confined to the focal plane causing limited out-of-focus signal; therefore, all emitted light can be collected, without the need of a pinhole, allowing for high sensitivity (Denk et al., 1994).

To minimize movement artifact during *in vivo* imaging, a translatable microscope stage securing the animal's skull is used to provide a stable imaging platform. A metal tab or hexagonal nut is secured to the animal's skull, which mates to an assembly attached to the stage to manually translate the animal under the objective (Holtmaat et al., 2009). Recently, an automatic alignment stage was designed using a "hexapod" positioning system for reproducible accurate positioning between imaging sessions, which makes alignment of the different time point images much easier (Scheibe et al., 2011).

Postacquisition Image Processing and Analysis

Due to its deep brain localization, most studies of adult hippocampal neurogenesis rely on confocal imaging of labeled newborn granule cells of brain sections after fixation. We have used single-photon confocal imaging to examine in detail the dendritic development of retroviral-labeled newly formed dentate gyrus granule cells with or without genetic manipulation (Ge et al., 2006; Duan et al., 2007; Kim et al., 2009). For these studies, brains were sectioned at

40 µm thickness and then the hippocampus was confocal imaged. After image acquisition, the dendritic structure was traced using ImageJ software (National Institutes of Health) with the NeuronJ plug-in (Meijering et al., 2004). NeuronJ uses an algorithm to find contiguous filamentous structure based on regional intensity, providing a semi-automatic guided interface for easy and accurate tracing. One caveat with this approach is that the dendritic structure of a neuron is often interrupted or cut during sectioning; therefore only granule cells located within the middle portion of a section with relatively complete dendritic structure can be analyzed.

To overcome this thin sectioning-induced loss of structure, we have used two-photon imaging of thicker hippocampal slices (300 µm) or of the whole dorsal blade of the hippocampus using a whole-mount preparation. We use Imaris scientific software (Bitplane) for image processing and data analysis. Imaris can handle very large datasets, has a complete 3D interface, and the "FilamentTracer" plug-in provides semi-automatic guided tracing in 3D, similar to the 2D NeuronJ plug-in for ImageJ. Another advantage of this software is its compatibility with newly developed high-frequency monitors and 3D glasses for complete 3D visualization, making the tracing of complex structures possible.

For *in vivo*-imaged datasets, the images can be noisy and the heterogeneity of brain tissue causes significant spreading of the captured light in the lateral, and to a greater degree, in the axial direction (Figure 9.1C). Postacquisition median filtering can be used to remove noise while preserving edges, but it also causes noticeable blurring. Since the acquired image is made up of multiple points of light and these points are spread due to non-ideal optical properties (tissue layers, coverglass thickness), the point spread functions (PSF) can be measured or mathematically estimated and "reassigned" to a point. Deconvolution image processing is based on this principle and produces enhanced sharpness and contrast images as illustrated in Figure 9.1D (Chia and Levene, 2009). Software packages provide blind deconvolution algorithms that do not require measuring experimental PSFs. When applying blind deconvolution, the objective, fluorophore, immersion medium, and aberration parameters are entered and the software computes a theoretical PSF to apply to the dataset.

Future Directions in Live Animal Imaging of Adult Neurogenesis

There are still significant unknowns in the field of adult neurogenesis and technological development undoubtedly has the potential to provide novel tools to probe these questions. Microscopy and multiphoton imaging technology is advancing in parallel with new genetic tools, such as longer wavelength fluorescent proteins, endoscopic imaging, head-mounted two-photon systems for freely moving animals, and genetically encoded calcium indicators. These advances open new doors to understand the function of this extraordinary form of neural plasticity in the adult brain. For example, *in vivo* imaging of newborn neurons in the adult hippocampus in rodents may soon be possible by adapting a recently developed microendoscope technique (Barretto et al., 2011). By implanting a micro-optical probe containing a gradient refractive index (GRIN) microlens into the tissue and using a standard two-photon microscope and objective, deep structures, such as the hippocampus, can be imaged through the exposed end of the probe on the skull surface. Since a portion of the cortex needs to be removed for microlens implantation, this approach is invasive in nature. The authors demonstrated that by placing the probe tip at a natural tissue border, between the corpus callosum and the hippocampus surface above CA1, they minimized the damage beneath the lens. The newly formed neurons in the dentate gyrus are localized even deeper within the hippocampus, thus it may require tunneling the probe through the CA1 to the hippocampal fissure, a natural border between the CA1 lacunosum molecular layer and the suprapyramidal dentate gyrus molecular layer.

Microprism implantation into the hippocampus could also be used for probing adult neurogenesis, although for chronic preparations, the injury effects may cause too much artifact (Chia and Levene, 2009). The advantage of this approach is that the acquired images are in the sagittal or coronal planes, which is an ideal orientation for granule cells in the dentate gyrus, whereas a GRIN microlens would image in the horizontal plane, perpendicular to the dendritic long axis of the neuron.

Future systems could also provide a mobile, stable imaging platform attached to the skull for relatively unhindered imaging in a freely moving animal (Helmchen et al., 2001; Ghosh et al., 2011). There are still significant technical hurdles with this technique, since the current systems are quite bulky. Even with very secure attachment to the skull, the behaving animal causes movement artifacts that blur images, which makes resolving fine structures like dendritic spines difficult.

Adult neurogenesis in the hippocampus has been probed with electrophysiology in brain slices *ex vivo* and the function of adult neurogenesis has been studied by ablating the new neuron population with irradiation or antimitotic drug delivery and looking at changes in behavior (Shors et al., 2001; Saxe et al., 2006; Garthe et al., 2009). While informative, the true network activity of newborn granule cells in their *in vivo* environment remains elusive. Calcium imaging, using retroviral-infected granule cells expressing the genetically encoded calcium indicator GCaMP3, for example (Tian et al., 2009), with a proper tunneled window could provide a method to finally probe the firing pattern of the newly formed neurons and how it may correlate with specific behaviors.

In summary, retroviral-specific labeling, in concert with striking advances in optical methods and computational power, provide powerful tools to manipulate and visualize newborn neurons in the mature brain. We now have current imaging approaches to address questions in adult neurogenesis and neuroscience in general at a level unimaginable a decade ago.

Acknowledgments

We thank K. Christian for comments. The work in the authors' laboratories was supported by the National Institutes of Health and Maryland Stem Cell Research Foundation. The authors acknowledge the joint participation by the Diana Helis Henry Medical Research Foundation through its direct engagement in the continuous active conduct of medical research in conjunction with The Johns Hopkins Hospital and the Johns Hopkins University School of Medicine and the Foundation's Parkinson's Disease Program No. H-1.

References

Adam, Y., Mizrahi, A., 2011. Long-term imaging reveals dynamic changes in the neuronal composition of the glomerular layer. J Neurosci 31, 7967–7973.

Altman, J., Das, G., 1965. Autoradiographic and histological evidence of postnatal hippocampal neurogenesis in rats. J Comp Neurol 124, 319–335.

Baens, M., Noels, H., Broeckx, V., Hagens, S., Fevery, S., Billiau, A., et al., 2006. The dark side of EGFP: defective polyubiquitination. PloS One 1, e54.

Barretto, R., Ko, T., Jung, J., Wang, T., Capps, G., Waters, A., et al., 2011. Time-lapse imaging of disease progression in deep brain areas using fluorescence microendoscopy. Nat Med 17, 223–228.

Bonaguidi, M., Wheeler, M., Shapiro, J., Stadel, R., Sun, G., Ming, G., et al., 2011. *In vivo* clonal analysis reveals self-renewing and multipotent adult neural stem cell characteristics. Cell 145, 1142–1155.

Carleton, A., Petreanu, L., Lansford, R., Alvarez-buylla, A., Lledo, P., 2003. Becoming a new neuron in the adult olfactory bulb. Nat Neurosci 6, 507–518.

Chen, J., Lin, W., Cha, J., So, P., Kubota, Y., Nedivi, E., 2011. Structural basis for the role of inhibition in facilitating adult brain plasticity. Nat Neurosci 14, 587–594.

Chia, T., Levene, M., 2009. Microprisms for *in vivo* multilayer cortical imaging. J Neurophysiol 102, 1310–1314.

Davalos, D., Grutzendler, J., Yang, G., Kim, J., Zuo, Y., Jung, S., et al., 2005. ATP mediates rapid microglial response to local brain injury *in vivo*. Nat Neurosci 8, 752–758.

Denk, W., Delaney, K., Gelperin, A., Kleinfeld, D., Strowbridge, B., Tank, D., et al., 1994. Anatomical and functional imaging of neurons using 2-photon laser scanning microscopy. J Neurosci Methods 54, 151–562.

Drew, P., Shih, A., Driscoll, J., Knutsen, P., Blinder, P., Davalos, D., et al., 2010. Chronic optical access through a polished and reinforced thinned skull. Nat Methods 7, 5–10.

Duan, X., Chang, J., Ge, S., Faulkner, R., Kim, J., Kitabatake, Y., et al., 2007. Disrupted-in-schizophrenia 1 regulates integration of newly generated neurons in the adult brain. Cell 130, 1146–1158.

Eriksson, P., Perfilieva, E., Bjork-Eriksson, T., Alborn, A., Nordborg, C., Peterson, D., et al., 1998. Neurogenesis in the adult human hippocampus. Nat Med 4, 1313–1317.

Faulkner, R., Jang, M., Liu, X., Duan, X., Sailor, K., Kim, J., et al., 2008. Development of hippocampal mossy fiber synaptic outputs by new neurons in the adult brain. Proc Natl Acad Sci USA 105, 14157–14162.

Feng, G., Mellor, R., Bernstein, M., Keller-Peck, C., Nguyen, Q., Wallace, M., et al., 2000. Imaging neuronal subsets in transgenic mice expressing multiple spectral variants of GFP. Neuron 28, 41–51.

Gage, F., 2000. Mammalian neural stem cells. Science 287, 1433–1438.

Garthe, A., Behr, J., Kempermann, G., 2009. Adult-generated hippocampal neurons allow the flexible use of spatially precise learning strategies. PloS One 4, e5464.

Gavrilescu, L., Van Etten, R., 2007. Production of replication-defective retrovirus by transient transfection of 293T cells. J Visualized Exp 10, 550.

Ge, S., Goh, E., Sailor, K., Kitabatake, Y., Ming, G., Song, H., 2006. GABA regulates synaptic integration of newly generated neurons in the adult brain. Nature 439, 589–593.

Ghosh, K., Burns, L., Cocker, E., Nimmerjahn, A., Ziv, Y., Gamal, A., et al., 2011. Miniaturized integration of a fluorescence microscope. Nat Methods 8, 871–878.

Gould, E., Reeves, A., Graziano, M., Gross, C., 1999. Neurogenesis in the neocortex of adult primates. Science 286, 548–552.

Graham, F., van der Eb, A., 1973. A new technique for the assay of infectivity of human adenovirus 5 DNA. Virology 52, 456–467.

Gratzner, H., 1982. Monoclonal antibody to 5-bromo- and 5-iododeoxyuridine: a new reagent for detection of DNA replication. Science 218, 474–475.

Gu, Y., Janoschka, S., Ge, S., 2011. Studying the integration of adult-born neurons. J Visualized Exp 49, 2548.

Helmchen, F., Fee, M., Tank, D., Denk, W., 2001. A miniature head-mounted two-photon microscope. high-resolution brain imaging in freely moving animals. Neuron 31, 903–912.

Holtmaat, A., Bonhoeffer, T., Chow, D., Chuckowree, J., De Paola, V., Hofer, S., et al., 2009. Long-term, high-resolution imaging in the mouse neocortex through a chronic cranial window. Nat Protoc 4, 1128–1144.

Jessberger, S., Aigner, S., Clemenson, G., Toni, N., Lie, D., Karalay, O., et al., 2008. Cdk5 regulates accurate maturation of newborn granule cells in the adult hippocampus. PLoS Biol 6, e272.

Jordan, M., Schallhorn, A., Wurm, F., 1996. Transfecting mammalian cells: optimization of critical parameters affecting calcium-phosphate precipitate formation. Nucleic Acids Res 24, 596–601.

Kakita, A., Goldman, J., 1999. Patterns and dynamics of SVZ cell migration in the postnatal forebrain: monitoring living progenitors in slice preparations. Neuron 23, 461–472.

Kang, E., Burdick, K., Kim, J., Duan, X., Guo, J., Sailor, K., et al., 2011. Interaction between FEZ1 and DISC1 in regulation of neuronal development and risk for schizophrenia. Neuron 72, 559–571.

Kim, J., Duan, X., Liu, C., Jang, M., Guo, J., Pow-anpongkul, N., et al., 2009. DISC1 regulates new neuron development in the adult brain via modulation of AKT-TOR signaling through KIAA1212. Neuron 63, 761–773.

Kim, P., Duan, X., Huang, A., Liu, C., Ming, G., Song, H., et al., 2010. Aspartate racemase, generating neuronal D-aspartate, regulates adult neurogenesis. Proc Natl Acad Sci USA 107, 3175–3179.

Knott, G., Holtmaat, A., Wilbrecht, L., Welker, E., Svoboda, K., 2006. Spine growth precedes synapse formation in the adult neocortex *in vivo*. Nat Neurosci 9, 1117–1124.

Kuhn, H., Dickinson-Anson, H., Gage, F., 1996. Neurogenesis in the dentate gyrus of the adult rat: age-related decrease of neuronal progenitor proliferation. J Neurosci 16, 2027–2033.

Kuwabara, T., Hsieh, J., Muotri, A., Yeo, G., Warashina, M., Lie, D., et al., 2009. Wnt-mediated activation of NeuroD1 and retro-elements during adult neurogenesis. Nat Neurosci 12, 1097–1105.

Lee, W., Chen, J., Huang, H., Leslie, J., Amitai, Y., So, P., et al., 2008. A dynamic zone defines interneuron remodeling in the adult neocortex. Proc Natl Acad Sci USA 105, 19968–19973.

Lewis, P., Emerman, M., 1994. Passage through mitosis is required for oncoretroviruses but not for the human immunodeficiency virus. J Virol 68, 510–516.

Lie, D., Colamarino, S., Song, H., Desire, L., Mira, H., Consiglio, A., et al., 2005. Wnt signalling regulates adult hippocampal neurogenesis. Nature 437, 1370–1375.

Liu, H., Jan, M., Chou, C., Chen, P., Ke, N., 1999. Is green fluorescent protein toxic to the living cells? Biochem Biophys Res Commun 260, 712–717.

Livneh, Y., Feinstein, N., Klein, M., Mizrahi, A., 2009. Sensory input enhances synaptogenesis of adult-born neurons. J Neurosci 29, 86–97.

Livneh, Y., Mizrahi, A., 2011. Experience-dependent plasticity of mature adult-born neurons. Nat Neurosci (Advanced online publication).

Llorens-Martín, M., Trejo, J., 2011. Multiple birthdating analyses in adult neurogenesis: a line-up of the usual suspects. Front Neurosci 5, 76.

Meijering, E., Jacob, M., Sarria, J., Steiner, P., Hirling, H., Unser, M., 2004. Design and validation of a tool for neurite tracing and analysis in fluorescence microscopy images. Cytometry A 58, 167–176.

Miller, A., 1990. Retrovirus packaging cells. Hum Gene Ther 1, 5–14.

Ming, G., Song, H., 2011. Adult neurogenesis in the Mammalian brain: significant answers and significant questions. Neuron 70, 687–702.

Ming, G., Song, H., 2005. Adult neurogenesis in the mammalian central nervous system. Annu Rev Neurosci 28, 223–250.

Mizrahi, A., 2007. Dendritic development and plasticity of adult-born neurons in the mouse olfactory bulb. Nat Neurosci 10, 444–452.

Mouret, A., Gheusi, G., Gabellec, M., de Chaumont, F., Olivo-Marin, J., Lledo, P., 2008. Learning and survival of newly generated neurons: when time matters. J Neurosci 28, 11511–11516.

Nimmerjahn, A., Kirchhoff, F., Helmchen, F., 2005. Resting microglial cells are highly dynamic surveillants of brain parenchyma *in vivo*. Science 308, 1314–1318.

Overstreet-Wadiche, L., Westbrook, G., 2006. Functional maturation of adult-generated granule cells. Hippocampus 16, 208–215.

Petreanu, L., Alvarez-Buylla, A., 2002. Maturation and death of adult-born olfactory bulb granule neurons: role of olfaction. J Neurosci 22, 6106–6113.

Rakic, P., 2002. Neurogenesis in adult primates. Prog Brain Res 138, 3–14.

Salic, A., Mitchison, T., 2008. A chemical method for fast and sensitive detection of DNA synthesis *in vivo*. Proc Natl Acad Sci USA 105, 2415–2420.

Saxe, M., Battaglia, F., Wang, J., Malleret, G., David, D., Monckton, J., et al., 2006. Ablation of hippocampal neurogenesis impairs contextual fear conditioning and synaptic plasticity in the dentate gyrus. Proc Natl Acad Sci USA 103, 17501–17506.

Scheibe, S., Dorostkar, M., Seebacher, C., Uhl, R., Lison, F., Herms, J., 2011. 4D in *in vivo* 2-photon laser scanning fluorescence microscopy with sample motion in 6 degrees of freedom. J Neurosci Methods 200, 47–53.

Shors, T., Miesegaes, G., Beylin, A., Zhao, M., Rydel, T., Gould, E., 2001. Neurogenesis in the adult is involved in the formation of trace memories. Nature 410, 372–376.

Shotton, D., White, N., 1989. Confocal scanning microscopy: three-dimensional biological imaging. Trends Biochem Sci 14, 435–439.

Skalko, R., Packard, D., 1975. Mechanisms of halogenated nucleoside embryotoxicity. Ann NY Acad Sci 255, 552–558.

Sliva, K., Schnierle, B., 2010. Selective gene silencing by viral delivery of short hairpin RNA. Virol J 7, 248.

Stutzmann, G., 2008. Seeing the brain in action: how multiphoton imaging has advanced our understanding of neuronal function. Microsc Microanal 14, 482–491.

Svoboda, K., Yasuda, R., 2006. Principles of two-photon excitation microscopy and its applications to neuroscience. Neuron 50, 823–839.

Tashiro, A., Zhao, C., Gage, F., 2006. Retrovirus-mediated single-cell gene knockout technique in adult newborn neurons *in vivo*. Nat Protoc 1, 3049–3055.

Tian, L., Hires, S., Mao, T., Huber, D., Chiappe, M., Chalasani, S., et al., 2009. Imaging neural activity in worms, flies and mice with improved GCaMP calcium indicators. Nat Methods 6, 875–881.

Toni, N., Laplagne, D., Zhao, C., Lombardi, G., Ribak, C., Gage, F., et al., 2008. Neurons born in the adult dentate gyrus form functional synapses with target cells. Nature Neurosci 11, 901–907.

van Praag, H., Schinder, A., Christie, B., Toni, N., Palmer, T., Gage, F., 2002. Functional neurogenesis in the adult hippocampus. Nature 415, 1030–1034.

Whitman, M., Greer, C., 2009. Adult neurogenesis and the olfactory system. Prog Neurobiol 89, 162–175.

Wilt, B., Burns, L., Wei, H., Ghosh, K., Mukamel, E., Schnitzer, M., 2009. Advances in light microscopy for neuroscience. Annu Rev Neurosci 32, 435–506.

Xu, H., Pan, F., Yang, G., Gan, W., 2007. Choice of cranial window type for *in vivo* imaging affects dendritic spine turnover in the cortex. Nat Neurosci 10, 549–551.

Yang, G., Pan, F., Parkhurst, C., Grutzendler, J., Gan, W., 2010. Thinned-skull cranial window technique for long-term imaging of the cortex in live mice. Nat Protoc 5, 201–208.

Yu, J., Schaffer, D., 2005. Advanced targeting strategies for murine retroviral and adeno-associated viral vectors. Adv Biochem Eng/Biotechnol 99, 147–167.

Zhao, C., Teng, E., Summers, R., Ming, G., Gage, F., 2006. Distinct morphological stages of dentate granule neuron maturation in the adult mouse hippocampus. J Neurosci 26, 3–11.

References

10

STUDY OF MYELIN SHEATHS BY CARS MICROSCOPY

Chun-Rui Hu[1], Bing Hu[1], and Ji-Xin Cheng[2]

[1]CAS Key Laboratory of Brain Function and Disease, and School of Life Sciences, University of Science and Technology of China, Hefei, China, [2]Weldon School of Biomedical Engineering, Purdue University, West Lafayette, Indiana, USA

CHAPTER OUTLINE
Traditional Myelin Imaging Methods   221
Principle and History of CARS Microscopy   224
Technical Characteristics of CARS Microscopy   226
CARS Microscopy for *Ex Vivo* and *In Vivo* Myelin Imaging   227
    The CARS Microscope   227
    Sample Preparation for CARS Imaging   228
    CARS Images of Myelin   229
    In Vivo CARS Imaging of the Spinal Cord   229
Mechanistic Understanding of Demyelination and Remyelination Enabled by CARS Imaging   231
    Lysophosphatidylcholine-induced Demyelination   231
    Glutamate Toxicity   234
    Electrical Stimulation   234
    Experimental Autoimmune Encephalomyelitis Model   235
Other Methods for *In Vivo* Imaging of Myelin   236
    THG Microscopy   236
    TPEF Microscopy and Transgenic Animal Models   236
    Nuclear MRI for Diagnosis of Myelin-related Diseases   237
    PET   238
Outlook for Myelin Imaging by CARS Microscopy   238
Acknowledgments   240
References   240

Traditional Myelin Imaging Methods

In the past several decades our understanding of the fundamental characteristics of myelin has been established largely due

Cellular Imaging Techniques for Neuroscience and Beyond.
DOI: http://dx.doi.org/10.1016/B978-0-12-385872-6.00010-6

to electron microscopy, as it is the classical and "gold standard" method for revealing the ultrastructure of myelin sheaths (Hirano and Dembitzer, 1967; Hartline, 2008). In the vertebrate central nervous system (CNS), myelin is formed by oligodendrocytes with one oligodendrocyte wrapping around several axons. In the peripheral nervous system, the myelin sheaths of axons are formed by Schwann cells, where each Schwann cell only wraps on one axon (Sherman and Brophy, 2005; Simons and Trotter, 2007; Aggarwal et al., 2011). The gaps of consecutive myelin segments are called "nodes of Ranvier." These short segments of bare axons on a fiber are responsible for the saltatory conduction of nerve impulses (Landon and Williams, 1963; Peters, 1966; Waxman, 2006).

Myelin contains 70–85% lipids and 15–30% proteins. The lipids provide the insulation for the axons while the proteins are necessary for myelin functionality and integrity (Baumann and Pham-Dinh, 2001). Even today, electron microscopy is indispensable and being widely used in scientific and medical research of myelin and myelinated nerve fibers. Electron microscopy has revealed the intrinsic molecules and extrinsic environmental factors that influence the development and organization of myelin sheaths. However, the use of this technique in clinical diagnosis is limited due to the highly expensive equipment and the time- and resource-consuming, delicate sample preparation. Alternative histomorphological methods have been developed for visualization of myelin, such as staining with Luxol Fast Blue (Margolis and Pickett, 1956; Scholtz, 1977), Sudan Black (Gerrits et al., 1992), periodic acid-Schiff, oil red O, osmium (Stoeckenius, 1957; Di Scipio et al., 2008), Loyez (Kelemen and Becus, 1969), and so forth. Compared with electron microscopy, histomorphology provides a convenient method for the rapid diagnosis in the pathology lab of damaged myelin. Yet, having broadly the same limitations as electron microscopy, the histomorphology methods can only examine fixed samples. Nuclear MR-related techniques have provided a strategy to observe myelin in living organisms (Neema et al., 2007). MRI is widely applied all around the world for clinical diagnosis of human tissues, including CNS diseases like multiple sclerosis (MS) (Oda and Udaka, 2008; Haller et al., 2009). However, the resolution of MRI is about 2 mm, which severely limits this imaging instrument to act as a reliable tool for ultrastructural analysis of myelin sheaths. Other myelin imaging technologies like diffusion tensor imaging (DTI) (Goldberg-Zimring et al., 2005; Yu et al., 2006), magnetization transfer imaging (MTI) (McGowan et al., 1998; Filippi and Rovaris, 2000), and positron emission tomography (PET) (Schiepers et al., 1997; Kiferle et al., 2011), have improved the resolution to about 1 mm. Yet these imaging techniques are still insufficient for detailed observation of subcellular structures, and they

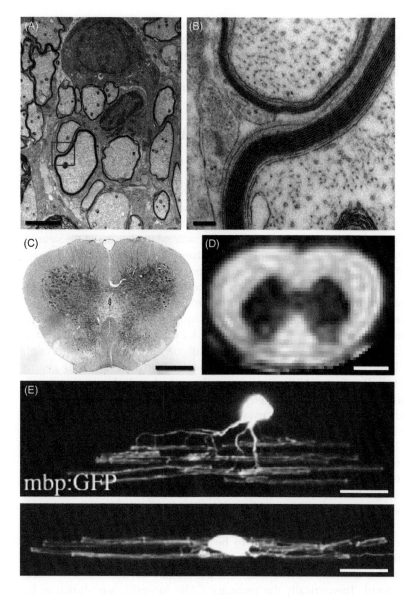

Figure 10.1 Traditional method for myelin observation. (A) EM data show an oligodendrocyte that exhibits a process and wraps on two axons (*). Scale bar, 5 μm. (B) High-magnification imaging of the box in (A), showing the compact multilaminar membrane of myelin sheath. Scale bar, 100 nm. (C) Transverse section of spinal cord stained by Luxol Fast Blue. Myelin shows as light blue structure around the white matter. (D) DTI imaging showing myelin structure in the spinal cord of the rat. The strong myelin signal rises from the water diffusion and the anisotropy in the highly regulated myelin structure. Scale bars: (in C, D) 1 mm. (E) Two-photon fluorescence imaging of the Tg (mbp:EGFP) zebrafish reveals the *bis*-membrane morphology of the myelin sheath formed by the single oligodendrocyte. Scale bar, 10 μm. E is adapted from Almeida et al., 2011. Please see color plates at the back of the book.

only generate a general morphological picture of myelin architecture. Recently, several lines of transgenic zebrafish have been developed for *in vivo* myelin imaging. The expression of green fluorescent protein (GFP) is driven by the promoter of myelin special protein, such as nkx 2.2 (Shin et al., 2003), olig 2 (Kirby et al., 2006), mbp (Jung et al., 2010; Almeida et al., 2011), and plp (Yoshida and Macklin, 2005). With two-photon excited fluorescence (TPEF) microscopy, the myelin in this transparent organism can be observed. Figure 10.1 shows myelin seen by the traditional methods. In recent years, coherent anti-Stokes Raman scattering (CARS) microscopy has provided a novel strategy to perform *in vivo* high-resolution imaging of myelin sheaths. In this chapter we describe the principle, history, and technical features of CARS microscopy and its application in myelin research.

Principle and History of CARS Microscopy

CARS microscopy is based on the nonlinear optical process of coherent anti-Stokes Raman scattering. CARS is a four-wave mixing process in which a pump field E_p (ω_p) and a Stokes field E_S (ω_s) interacts with the sample, and generates an anti-Stokes field (E_{as}) at the frequency of $2\omega_p-\omega_s$. When the energy difference ($\omega_p-\omega_s$) between the pump and Stokes fields matches the Raman-active molecular vibration of the sample, the anti-Stokes signal is significantly enhanced (Cheng et al; 2002; Cheng, 2007). The energy diagram of a CARS process is shown in Figure 10.2A.

CARS as a nonlinear optical process was first studied in 1965 by Maker and Terhune (1965) at the Ford Motor Company. CARS spectroscopy emerged in the early 1970s and has been an efficient technology for chemical analysis (Tolles et al., 1977; Clark and Hester, 1988). The first CARS microscope was constructed by Duncan et al. (1982) using two visible dye laser beams and non-collinear geometry. The first CARS microscope showed chemical-selective imaging capability and was used to observe the epidermis cells of onions. However, the resolution provided by the non-collinear beam geometry appeared insufficient. In a 1999 paper, Zumbusch et al. revisited CARS microscopy by using two tightly focused near-IR laser beams in a collinear geometry. The collinear beam geometry significantly improved the imaging quality. Moreover, the use of near-IR laser reduced nonresonant background by avoiding two-photon electronic resonance with molecules.

In the 1999 paper of Zumbusch et al. the spectral width of a femtosecond pulse is assumed to be much broader than most Raman line widths, which limited the spectral selectivity and devoted most of the excitation energy to the generation of nonresonant background. Theoretically the resonant CARS intensity was shown to be saturated by femtosecond pulse excitation, whereas the nonresonant background increases in proportion to the square of the pulse spectral width (Cheng et al., 2001a). As most Raman line widths are comparable with the spectral width of a picosecond laser beam, using picosecond lasers not only improved the spectral resolution but also reduced the background (Cheng et al., 2001a). During the early development of CARS microscopy, major efforts focused on decreasing the nonresonant background and on improving the detection sensitivity.

Two important developments were epi-detection and polarization-sensitive detection. In an epi-detected CARS microscope, the solvent nonresonant background can be fully rejected as the solvent CARS signal primarily goes forward in the same direction as the incident beams and cannot be detected backward (Volkmer et al., 2001; Cheng et al., 2001a). As the depolarization properties of the CARS signal and nonresonant background are different, using an analyzer whose polarization is perpendicular to that of nonresonant CARS signal may help. Thus, polarization CARS efficiently eliminates the background

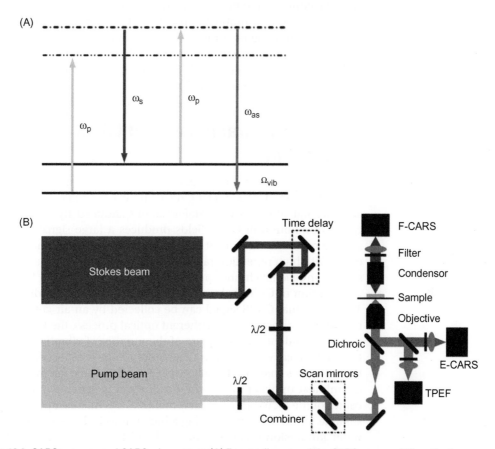

Figure 10.2 CARS process and CARS microscope. (A) Energy diagram of the CARS process. When the frequency difference between the pump and Stokes signals $(\omega_p - \omega_s)$ matches the Raman-active molecular vibrational frequency, Ω_{vib}, the anti-Stokes signal is coherently generated at a frequency $\omega_{as} = 2\omega_p - \omega_s$. (B) Schematic of a CARS microscope. The synchronized pump and Stokes picoseconds-pulse trains are collinearly combined and polarized. The combined beams are directed to a confocal microscope. A high-NA objective focuses the beams to the sample. The CARS signal is detected simultaneously in the forward and backward direction. The TPEF signal is spectrally separated in the same optical pathway from the E-CARS signal. Please see color plates at the back of the book.

from both the sample scatters and the solvent (Cheng et al., 2001b). Other methods for suppressing the nonresonant background include time-resolved CARS (Volkmer et al., 2002), phase control and shaping CARS (Oron et al., 2003), heterodyne interference CARS (Evans et al., 2004; Potma et al., 2006), and frequency modulation CARS (Ganikhanov et al., 2006; Chen et al., 2010). Along with these developments, CARS microscopy was applied in the fields of chemistry, materials sciences, and biology and medicine fields (Cheng, 2007; Evans and Xie, 2008). Today, CARS microscopy has become a platform for multimodal, nonlinear optical imaging, including CARS, TPEF, second harmonic generation (SHG), third-harmonic generation (THG) imaging, and pump-probe microscopy. Like other modalities of nonlinear optical imaging probing-specific molecular structure, multimodal nonlinear optical microscopy allows the study of complex biological tissues (Le et al., 2007; Huff et al., 2008; Chen et al., 2009; Wang et al., 2009; Yue et al., 2011).

Technical Characteristics of CARS Microscopy

CARS microscopy uses signals from inherent molecular vibrations. Therefore, CARS imaging needs no exogenous tags and avoids any internal ultrastructural deformations induced by labeling. The coherent addition of CARS fields produces a large signal, thus permitting real-time imaging at the same speed as fluorescence microscopy. With high repetition rate (in MHz) lasers, the image acquisition speed can reach 20 frames per second (Evans et al., 2005). The coherent addition of CARS fields also makes the signal highly directional. The directional CARS signal can be collected by an air condenser in the forward direction. As a coherent optical process, the CARS intensity is dependent on the square of the number of vibrational oscillators in the focal volume. Thus CARS microscopy is especially sensitive for molecules that are densely packed, such as lipid membranes. As a nonlinear optical process, the CARS effect is generated only at the center of the focus. This property makes CARS microscopy inherently suitable for 3D sectioning. Using a high numerical aperture (NA) 60× water immersion objective, the axial and lateral resolution of CARS can reach values of 0.23 and 0.75 µm, respectively (Cheng et al., 2002). Such high resolution allows the detection of subcellular structure in thick tissue. By using near-IR excitation beams, CARS microscopy can penetrate tissue as deep as 0.4 mm (Evans and Xie, 2008). Unlike Raman scattering, the CARS signal is generated at the frequency of the anti-Stokes field $(2\omega_p - \omega_s)$, which is blue shifted from the two excitation frequencies of pump and Stokes field. Thus a CARS signal can be easily spectrally separated from the excitation field and the one photon fluorescence background can be rejected (Cheng and Xie, 2004). The CARS signal can be detected by photomultipliers of high quantum efficiency.

CARS Microscopy for *Ex Vivo* and *In Vivo* Myelin Imaging

As a key component in the vertebrate nervous system, myelin is a lamellar membranous structure containing 70% lipid and 30% protein by dry weight (Baumann and Pham-Dinh, 2001). The high density of CH_2 bonds in the fatty acid chains contribute to a large CARS intensity when $(\omega_p-\omega_s)$ matches the characteristic frequency of CH_2 stretch vibrations at $2840\,cm^{-1}$ (Wang et al., 2005). In the following section, we describe the CARS microscope configuration and the procedure of myelin sample preparation, and we show some typical CARS images of myelin in different nervous systems.

The CARS Microscope

A schematic drawing of the CARS microscope used for myelin imaging (Wang et al., 2005; Fu et al., 2007; Huff et al., 2011) is shown in Figure 10.2B. The pump and Stokes laser beams are provided by two Ti: sapphire oscillators (Mira 900; Coherent Inc.). Both lasers are tunable from 700 to 1000 nm. One laser operating at 706 nm served as the master and provided the clock for synchronization with the other laser. The pulse durations of both lasers were 2.5 ps; the corresponding spectral widths of the laser light matches the line width of the CH_2 symmetric stretch Raman band. The frequency difference of two laser beams was tuned to $2840\,cm^{-1}$ corresponding to the CH_2 symmetric stretching mode. A half-wave plate was used to control the polarization of both laser beams. The repetition rate of both lasers was reduced from 78 MHz to a few MHz by means of a Pockels cell (Model 350-160; Conoptics) to reduce the average power while maintaining pulse peak intensity. The two lasers were tightly synchronized (Sync-Lock; Coherent Inc.) and collinearly combined using a dichroic combiner (LWP-45-R720-7850-PW-1004-UV; CVI Laser LLC). The collinearly combined beams were directed into the scanning unit of a confocal microscope (Olympus America Inc.) and focused into a sample through a 60× water immersion objective lens with a numerical aperture (NA) of 1.2 or a 20× air objective lens where NA = 0.75. Images were acquired by scanning a pair of galvanometer mirrors. The epi-detected CARS (E-CARS) signal was collected by the same objective and detected by an external photomultiplier (PMT; H7422, Hamamatsu Photonics). TPEF imaging was achieved by the same picosecond laser beams. A dichroic splitter was used to spectrally separate the epi-detected TPEF signal from the E-CARS signal. The TPEF signal was detected with the same type of PMT. For *in vivo* imaging, the pump and Stokes beams were directed into an upright laser scanning confocal microscope (FV300/BX51WI; Olympus America Inc.), which was equipped with a dipping mode 40/60× water immersion objective lens or a miniature microscope objective (MPO), focusing the beams into a sample. All of the imaging experiments were conducted at the optics lab temperature, 22°C.

Sample Preparation for CARS Imaging

Preparation of Spinal Cord Samples

The fresh spinal cord samples for myelin imaging were surgically isolated from guinea pigs. The spinal cord was first split into two halves by sagittal division and then cut radially to separate the ventral white matter from the gray matter. The isolated ventral white matter strip was mounted on a chambered glass coverslip and kept in oxygen-bubbled Krebs' solution (NaCl 124 mM, KCl 2 mM, KH_2PO_4 1.2 mM, $MgSO_4$ 1.3 mM, $CaCl_2$ 2 mM, dextrose 10 mM, $NaHCO_3$ 26 mM, and sodium ascorbate 10 mM, equilibrated with 95% O_2, 5% CO_2). The samples for CARS imaging to visualize the axonal myelin were not labeled, and the axons in the sample maintained good morphology for at least 10 hours in the chamber.

Preparation of Brain Slices

BALB/c mice were anesthetized with 0.4–0.6 ml of a mixture containing 0.625 mg/ml xylazine and 6.25 mg/ml ketamine. During deep anesthesia, they were perfused transcardially with 10 ml of cold phosphate buffer solution (PBS) (pH, 7.4), and then with 10 ml of PBS containing 4% paraformaldehyde (PFA) for *in vivo* fixation. The whole brain was immediately excised and kept in 4% PFA for an additional period of 3 days. The whole brain was then transferred to PBS and sectioned into 300 μm horizontal slices using a tissue slicer (OTS-4000; Electron Microscopy Sciences). The slices were stored in PBS prior to imaging.

Animal Preparation for In Vivo Imaging

Long-Evans or Sprague Dawley rats were anesthetized and held in a stereotaxic apparatus. An incision was surgically opened through the skin and muscle tissue at the T-10 segment of the spinal cord. After exposure of the vertebral arch, a 2 mm diameter concave perforation was made with a small drill to expose the spinal cord for imaging with the miniature microscope objective. The hole was filled with sterile saline to keep the tissue hydrated. An alternative approach consisted of exposing the spinal cord, which was exposed by dorsal half-laminectomy at T-10 for imaging with a 40× water immersion objective.

CARS Images of Myelin

Typical CARS images of myelin sheath in the white matter of guinea pig spinal cord and brain are shown in Figures 10.3A and B, respectively. The myelin wrapping on the parallel axons displays a high contrast, with a resonant signal to nonresonant background ratio of 10:1. The alignment of myelin fibers in the spinal cord is more regular than that in the brain. A node of Ranvier, which has been included in a myelinated axon, was visualized by high-resolution CARS imaging (Figure 10.3C). Three-dimensional reconstructed images show the tubular structure of myelin sheath in the spinal cord (Figure 10.3D). Z-stacks of CARS images of myelin provide spatial distribution information of brain myelin fibers. The high-density myelinated axons orient randomly in brain cortex (Figure 10.3E). In the corpus callosum, myelin fiber bundles aggregate tightly in an alignment pattern like in spinal cord (Figure 10.3F). The more detailed substructures including the node of Ranvier and the Schmidt-Lanterman incisure are shown in Figure 10.3G–I. Figure 10.3G shows a node of Ranvier between two adjacent segments of the myelin sheath. In the paranodal region, the myelin cytoplasmic surface forms several fluffy loops, which are comparable with those seen in electron microscopic images. Figure 10.3H shows a Schmidt-Lanterman incisure in which the myelin layers are separated by the cytoplasm of oligodendrocyte cells and the lateral loops are clearly recognized. At the top or bottom of Schmidt-Lanterman incisures, an annular arrangement winds through the myelin sheath (Figure 10.3I).

The whole mouse brain was mapped by 9579 CARS images acquired with a 20× air objective to illustrate the myelinated fibers in different regions (Figure 10.3J). In the cerebellum, the white matter region displays a high-intensity CARS signal. The white matter tracts associated with fiber bundles in the genu of corpus callosum (gcc) and fimbria of the hippocampus (fi) are also clearly visualized by bright CARS contrast. The cingulum and the external capsule (ec) connected to the genu of corpus callosum were identified. A large number of small axonal bundles in the caudate putamen (CPu) was displayed as a distributed CARS plaque morphology.

In Vivo CARS Imaging of the Spinal Cord

Recently Shi et al. (2011) carried out *in vivo* CARS imaging of demyelination and remyelination in live rats. After surgical exposure of the spinal cord, a miniature microscope objective or a water immersion 40× objective was used for imaging the myelin sheath in

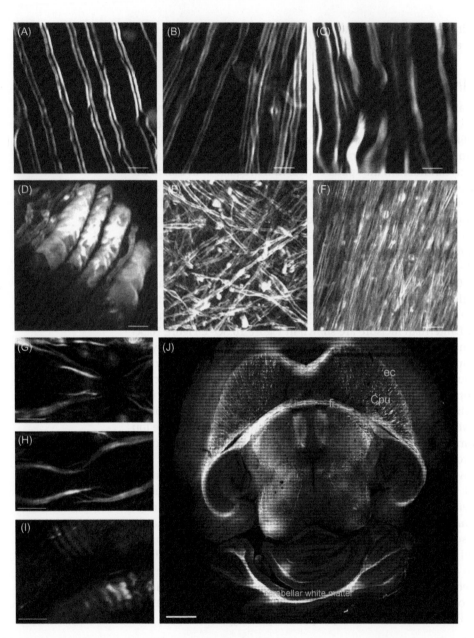

Figure 10.3 Imaging myelin *ex vivo* by CARS microscopy. Typical CARS imaging shows parallel myelin fibers in the white matter of spinal cord (A) and brain slice (B). Myelin appears as a high vibrational contrast, whereas other places only show a low nonresonant background. (C) A node of Ranvier connecting two myelin segments. (D) Three-dimensional reconstruction of CARS imaging showing the fistular structure of myelin sheath in the spinal cord. (E) Many axons with myelin appear randomly oriented in cerebral cortex. (F) Bundle of high-density myelin sheath with regular orientation in corpus callosum. A–F are adapted from Fu et al., 2008b. (G) Typical substructure of a node of Ranvier;

spinal cord. Figure 10.4A and B shows the *in vivo* imaging setup. The *in vivo* myelin images acquired by the miniature microscope objective and by a 40× objective are shown in Figure 10.4C and D, respectively. After secretory phospholipase A2 (sPLA2-III) injection into the spinal cord, *in vivo* CARS imaging was performed to follow the myelin degradation and regeneration process. One day after sPLA2-III injection, the myelin was destroyed as many vesicles appeared (Figure 10.4E). One week later, myelin debris started to be removed by macrophages and/or microglia activity (Figure 10.4F). Three weeks later normal myelin morphology had returned. The TPEF image of ethidium bromide-labeled Schwann cells implicated that remyelination had been done by Schwann cells (Figure 10.4G). CARS microscopy has also been used to monitor demyelination and remyelination in peripheral nerves (Belanger et al., 2011).

Mechanistic Understanding of Demyelination and Remyelination Enabled by CARS Imaging

Demyelination, the loss of normal myelin, accounts for long-term neurologic disability. MS is the most commonly acquired demyelinating disease in humans, affecting around 2.5 million people worldwide (Compston and Coles, 2002). Focal areas of intensive demyelination and white matter infiltration by lymphocytes and mononuclear cells are two of the pathological hallmarks of the disease. Over the past, CARS microscopy has been employed to explore the mechanism of demyelination in chemical and animal models of MS, and also has been used to monitor the remyelination of *in vivo* animals.

Lysophosphatidylcholine-induced Demyelination

Demyelination induced by lysophosphatidylcholine (LPC) has been widely used for the study of myelin diseases (Degaonkar et al., 2002a,b;

Figure 10.3 (Continued) with paranodal myelin appearing as several incompact loops that are formed by the cytoplasmic surface of oligodendrocytes. (H) Schmidt-Lanterman incisure. (I) Schmidt-Lanterman incisure morphology represented as the annular arranged wrap on the surface of a myelin sheath. G–I are adapted from Wang et al., 2005. (J) A whole mosaic brain of CARS imaging showing the high contrast of myelin in different regions. Adapted from Fu et al., 2008a. Gcc, genu of corpus callosum; CPu, caudate putamen; ec, external capsule; fi, fimbria hippocampus. Scale bars: (in A, C–F) 10 μm; (in B) 5 μm; (in G–I) 5 μm; (in J) 1 mm.

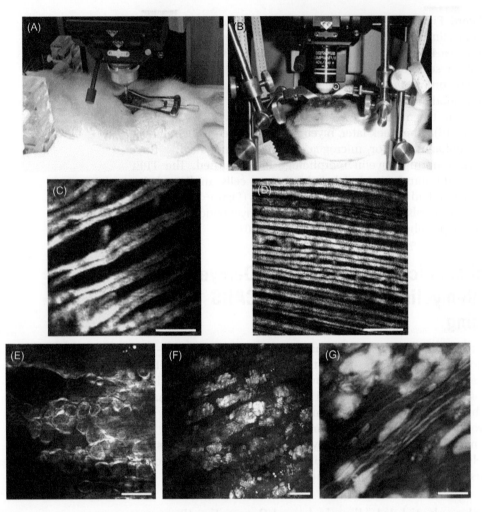

Figure 10.4 Imaging myelin *in vivo* by CARS microscopy. (A). The *in vivo* imaging setup equipped with an MPO. (B) A 40× dipping objective also can used to conduct *in vivo* spinal cord imaging. Myelin sheath morphology in exposed spinal cord was revealed via an MPO (C) and 40× dipping objective (D), respectively. (E–G) Myelin sheath repair progress after sPLA2-III injection. (E) Broken-down myelin one day after sPLA2-III injection, including many vesicles. (F) Myelin vesicles appearing to be engulfed by macrophages and/or microglia one week later. (G) Three weeks later, normal myelin morphology began to re-emerge, and the TPEF imaging of ethidium bromide-labeled Schwann cells' nuclei implicates that the naked axons are remyelinated by Schwann cells (G). Scale bars: (in A) 10 μm; (in B) 20 μm; (in C) 20 μm; (in D); 20 μm; (in E), 10 μm. Adapted from Shi et al., 2011.

Vereyken et al., 2009; Nagai et al., 2010). Yet, the mechanism underlying this model is unclear. Using CARS microscopy, Fu and coworkers (2007) found that LPC-induced myelin degradation is Ca^{2+} dependent. In their experiment, the ventral white matter in the spinal cord of the female guinea pig was isolated and cultured *ex vivo*. Five minutes

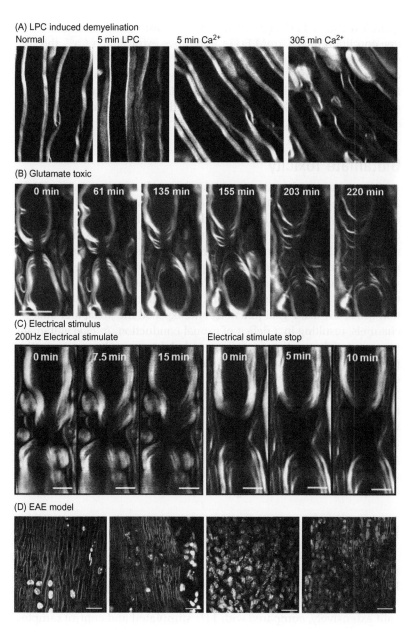

(A) LPC induced demyelination

Normal | 5 min LPC | 5 min Ca²⁺ | 305 min Ca²⁺

(B) Glutamate toxic

0 min | 61 min | 135 min | 155 min | 203 min | 220 min

(C) Electrical stimulus

200Hz Electrical stimulate | Electrical stimulate stop

0 min | 7.5 min | 15 min | 0 min | 5 min | 10 min

(D) EAE model

Figure 10.5 Demyelination visualized by CARS microscopy. (A) After injection of LPC, the compact myelin sheath loosens, starting from the outside of the myelin, and the demyelination region encroaches gradually into the inside. After being treated with LPC for 5 min, the intense CARS signal decreases significantly. The thickness of myelin increases because of the relaxation of the multilaminar compact myelin sheath. Scale bar, 10 μm. Adapted from Fu et al., 2007 with permission. (B) After the incubation of 1 mM glutamate, the paranodal myelin sheath retracts from the node of Ranvier, and the length of node significantly increases. The myelin loops split severely and the interval between paranodal myelin increases. Scale bar, 10 μm. (C) During electrical stimulation at 200 Hz, paranodal myelin retracts and the node of Ranvier lengthens. Retraction continues even after the electrical stimulation has been stopped. Scale bar, 5 μm. (D) In the naive whiter matter, myelin fibers are regularly distributed, with visible myelin structure and with only a few glial cells. At the onset stage of EAE, the myelin does not display any obvious abnormality, but the number of glial cells increases. In the acute stage of EAE, myelin loses its regular alignment and many glial cells aggregate at the lesion site. At the remission state, myelin restores and the parallel structures reappear. Scale bar, 20 μm. Adapted from Fu et al., 2011. Please see color plates at the back of the book.

after injection of LPC the myelin sheath started to swell, as revealed by decrease of CARS intensity (Figure 10.5A). Around the LPC injected site, the degraded myelin sheath formed a local lesion. The same demyelination process was also observed *in vivo*, in which the LPC was injected into the sciatic nerve of a mouse. The Ca²⁺ ionophore A23187-treated spinal cord manifested the same demyelinating phenomenon as LPC treated tissue. If LPC injected spinal cords were incubated in a Ca^{2+}-free condition plus 0.5 mM EGTA to remove extracellular Ca^{2+}, the size of the lesion was considerably reduced. Furthermore, two calcium-dependent enzymes, $cPLA_2$ and calpain, were shown to participate in LPC-induced demyelination (Fu et al., 2007). Collectively these data revealed that the LPC-induced demyelination process is caused by Ca^{2+} overload.

Glutamate Toxicity

Glutamate toxicity occurs commonly in white matter injury (Platt, 2007; Sivakumar et al., 2010), but the exact molecular and cellular events are still not very clear. Fu and coworkers (2009) monitored the paranodal myelin degeneration process induced by glutamate toxicity through real-time CARS imaging of myelin in *ex vivo* spinal cord stripes. After a 60 minute glutamate incubation, the typical and compact lamellar myelin sheath degraded and the paranodal myelin sheath retracted away from the node of Ranvier (Figure 10.5B). Furthermore, glutamate application broke the axon–glial junctions and exposed the juxtaparanodal K^+ channels, resulting in a deficit of axonal conduction. *In vivo* glutamate treatment of the sciatic nerve also revealed the same phenomenon. The paranodal myelin retraction was quantitatively measured by the ratio of nodal length to nodal diameter. The glutamate receptor agonists, NMDA and kainate, significantly increased the nodal length to diameter ratio. The glutamate-induced paranodal myelin splitting and retraction were similar with the observation of Ca^{2+} ionophore A23187 treatment. Calcium-free culture medium or calpain inhibitor coincubation with glutamate rescued the degenerated paranodal myelin structure, suggesting that glutamate toxicity-induced myelin degeneration is mediated by Ca^{2+} influx and subsequent calpain activation (Fu et al., 2009).

Electrical Stimulation

CARS microscopy has been used to monitor the response of myelin to high-frequency electrical stimulation (Huff et al., 2011). In spinal cord subjected to 15 minutes of 200 Hz electrical stimulation, the paranodal myelin retracted from the node of Ranvier and split extensively, compared with 1 Hz stimulated and control samples (Figure 10.5C). The ratio of nodal length to axon diameter was used for the quantitative indication of the paranodal myelin retraction. Normal tissue had a "healthy" ratio of less than 1.0, whereas 200 Hz stimulated tissue significantly increased this ratio to 9.8. Moreover, the paranodal myelin retraction was dependent on the stimulation frequency. Paranodal myelin retraction during electrical stimulation was correlated with the diameter of the axon. The myelin wrapping on thicker axons was more likely to retract under the electrical stimulation than myelin wrapping on thinner axons. After electrical stimulation, K^+ channels were exposed outside to the myelin, and the junction protein Caspr departed from the paranodal domain and diffused along the axon. Simultaneously, neurofascin lengthened and deformed severely. Calcium staining demonstrated that electrical stimulus caused entry of Ca^{2+} into the axon, and the Ca^{2+} overloading activated the calpain. The calpain activation, in turn, caused the degradation of cytoskeletal proteins, and disrupted the linkages of myelin-anchoring proteins to the axon (Huff et al., 2011). As high-frequency stimulation is widely used for the treatment of many CNS diseases like epilepsy and Parkinson's disease (Ogura et al., 2004; Oommen et al., 2005; Alarcon and Valentin, 2011), the CARS observations may help in understanding the mechanism of this therapy.

Experimental Autoimmune Encephalomyelitis Model

Experimental autoimmune encephalomyelitis (EAE) is a T-cell-mediated autoimmune disease of the CNS that causes the CNS myelin to be recognized as an exogenous immunogen and subsequently be attacked by T-cells. This eventually induces an irreversible CNS demyelination process (Gold, 2005; Jurynczyk et al., 2008). The EAE model has been employed extensively to explore the mechanism underlying myelin loss and therapies for myelin repair (Piraino et al., 2005; Merkler et al., 2006; Wang et al., 2006). Using CARS microscopy, Fu et al. (2011) observed a thorough demyelination process in the EAE model and provided new insight of into the intermolecular deformation in the necrosis of myelin. These data (Figure 10.5D) showed that, at EAE onset stage, the spinal cord manifests a high density of glial cell, but no apparent myelin debris compared with the naive mouse. In the peak acute stage, the glial cells sharply increase four times, and the normal parallel myelin fibrous structure disappears, with irregular myelin debris being displayed. Finally, in the remission stage of EAE, the glial cell and myelin fiber recover partially compared with the normal level. In the demyelinating lesion, the intermolecular chain becomes more disorderly than normal myelin, as reflected by the increase of CH_3 CARS signal compared to asymmetric CH_2 CARS signal. The unsaturation of the fatty acid chain in myelin debris also increased as evaluated from a high ratio of C=C strength in vibration bands to H-C-H. Detailed CARS imaging revealed that the length of the node of Ranvier increased and paranodal myelin retracted, and paired K^+ channel Kv1.2 was exposed and deformed as the breakdown of the axon–glial junctions occurred. Further, the demyelination of the EAE model revealed a characteristic neurofascin elongation in the paranodal myelin domain (Fu et al., 2011).

CARS microscopy has also been combined with reflectance confocal microscopy to observe myelin and axon damage in spinal cords extracted from EAE mice (Imitola et al., 2011). The myelin structure in the corpus callosum, cerebellum, and spinal cord was imaged. In EAE mice, the CARS and reflectance signals from corpus callosum decreased significantly, compared to naive mice. Moreover, in the demyelination lesion of EAE spinal cord, the presence of microglia increased, indicating the involvement of microglia in demyelination and/or remyelination.

Other Methods for *In Vivo* Imaging of Myelin

THG Microscopy

Besides CARS microscopy, THG microscopy has been employed for noninvasive, label-free imaging of myelin (Farrar et al., 2011). THG is a third-order nonlinear optical process that requires three photons (ω_1) and generates one photon at the tripled frequency ($3\omega_1$). As the Gouy phase shift of excitation beam crosses the focus, the THG signal is produced at the optical interfaces of an inhomogeneous tissue. With this contrast mechanism, THG microscopy has been used for imaging axons and dendrites of neurons in the brain (Witte et al., 2011). The THG signals of myelin sheaths in fixed mouse spinal cord and brain are comparable to the TPEF signal from FluoroMyelin Red-labeled structures (Farrar et al., 2011). *In vivo* THG imaging of yellow fluorescent protein transgenic mouse dorsal thoracic spinal cord and Texas Red-labeled zebrafish Mauthner axon allowed visualization of myelin sheath wrapped on a single axon. Yet compared to CARS imaging, THG microscopy needs to be further improved to obtain higher resolution myelin images.

TPEF Microscopy and Transgenic Animal Models

TPEF microscopy has also been widely applied in myelin imaging (Fitzner et al., 2006; Micu et al., 2007; Rinholm et al., 2011; Sobottka et al., 2011). By using various antibodies to recognize the proteins distributed on the myelin sheath, TPEF has allowed a better understanding of the protein composition of myelin sheath and node of Ranvier (Fitzner et al., 2006; Arnaud et al., 2009). TPEF and confocal laser scanning microscopy have been used for the discovery of the unknown proteins affecting myelination, remyelination, and the assembly of the node of Ranvier (Pedraza and Colman, 2000; Tiwari-Woodruff et al., 2001; Cid et al., 2005; Eftekharpour et al., 2007). Combined with transgenic animals, TPEF also allows *in vivo* imaging of myelin (Kirby et al., 2006; Almeida et al., 2011). As a specific example, using the transgenic zebrafish model, a recent study revealed that, after specific ablation of the myelinated oligodendrocytes, the new oligodendrocyte precursor cells migrate to the lesions and differentiate into a mature oligodendrocyte and then succeed to wrap on the demyelinated naked axons (Kirby et al., 2006).

Nuclear MRI for Diagnosis of Myelin-related Diseases

MRI, MTI, and DTI all possess the ability to image white matter *in vivo* and *in situ*. These technologies are mostly used in clinical applications for diagnosis or for prolonged studies of myelin-related diseases.

MRI is the most frequently used technology for the detection of myelin damage in the clinical field. The MRI signal arises from the magnetization change of the different protons in the tissues. In a powerful external magnetic field the proton magnetization direction is forcibly aligned parallel or anti-parallel with the strong external magnetic field. A brief ratio-frequency wave is used to disturb the direction of the aligned protons. After the ratio-frequency wave turns off, the disturbed direction of protons returns to their previous state, with a decaying signal generated in the local radio-frequency coils (Runge et al., 1984). In the human body, innumerable protons are present in water, carbohydrates, proteins, lipids, and other macromolecules. Thus the relatively high presence of protons in the molecular architecture of the myelin provides a comparative contrast compared to edema, inflammation, demyelination, and axonal damage in MS lesions. MRI includes conventional T1 hypointense and T2 hyperintense signals; however, both approaches have a limited ability to assess myelin loss or repair processes.

MTI is based on the measure of magnetization transfer ratio (MTR). In animal tissues, the unbound protons in free water and bound protons in macromolecules continuously interact and exchange, resulting in a special MTR. Myelin damage always induces a loss of MTR, while MTR increase indicates the repair of myelin (Neema et al., 2007). DTI is based on the diffusion anisotropy effect and motion characteristic of water molecules in the tissues (Goldberg-Zimring et al., 2005; Yu et al., 2006). The direction, randomness, and velocity of water molecules will decrease, as the molecular motility has been limited in a highly organized tissue-like myelin sheath. Thus, the measure of the diffusion and anisotropy of water molecules will provide a molecular composition contrast of the myelin structure. In the MS patient, demyelination will increase the water diffusion or decrease the anisotropy. MTI and DTI are more sensitive and specific than MRI in detecting the myelin lesion, but their spatial resolution is still insufficient for ultrastructural observations.

PET

PET is based on the electronic detection of short-lived positron-emitting radiotracers in the body. PET is widely used to measure the metabolic and functional characteristic of the living tissue of humans, and is a powerful technology for disease diagnosis, as reviewed previously (Kumar et al., 2005). PET has been employed for evaluation of myelin damage among MS patients (Pozzilli et al., 1992), and is still a favorable tool for myelin imaging and diagnosis. Many radiotracers for MS diagnosis have been developed that are safe and can cross the blood–brain barrier (Stankoff et al., 2006, 2011). The biggest disadvantage of PET is the high radiation hazard generated by the radiotracers. Also, the resolution of PET is approximately several micrometers (Moses, 2011), which is not sufficient for observations on single axons.

Outlook for Myelin Imaging by CARS Microscopy

During the course of demyelinating diseases of the mammalian CNS, remyelination always occurs as a very complex process involving many intrinsic and extrinsic factors as well as interaction with the immune system. During remyelination, oligodendrocyte precursor cells are recruited and migrate to the demyelinating site. These cells then differentiate into mature oligodendrocytes and reestablish contact with the newly generated or the demyelinated axons (Franklin and ffrench-Constant, 2008). Then, the oligodendrocyte processes wrap on the axon and form the compacted myelin sheath. At the same time, many myelin-related proteins are generated and assembled into the myelin membrane and the nodes of Ranvier. Remyelination is important for recovery of function of the demyelinating nervous system, as it restores the saltatory conduction of nerve impulse and provides protection to the naked axons from damage by glutamate and calcium ion. Yet, the mammalian CNS often fails in complete remyelination, resulting in thinner myelin sheaths and internodes of shorter length, especially in MS (Blakemore, 1981; Franklin and Hinks, 1999; Franklin and ffrench-Constant, 2008; Franklin and Kotter, 2008).

Study of myelin formation is important for a complete understanding of remyelination and may provide fundamental knowledge for treating myelin-associated diseases. The developmental myelination and the remyelination processes are thought to follow the same biological events: oligodendrocyte precursor cell proliferation, migration, differentiation, and mature oligodendrocyte wrapping on axons. Thus the recapitulation hypothesis indicates that remyelination should recapitulate the developmental myelination process (Fancy et al., 2011). However, remyelination occurs together with an inflammatory reaction, in which many inhibitory factors prevent remyelination and the instructive factors fail to promote remyelinization (Fancy et al., 2011).

Many demyelination models have been developed to investigate the demyelination and remyelination processes and to explore potential treatments for demyelination diseases. The mammalian animal models are not well suited for remyelination studies, due to their long-term repair (even partially), their expensive cost, and the amount of time involved performing experiments. As the CNS is embedded in the skull and vertebra, it is hard for *in vivo* imaging of the demyelinating or remyelinating process in real time. Thus, to more efficiently study potential therapies to stop demyelination or promote remyelination, more appropriate animal models, as well as the powerful tools like CARS microscopy, are critically needed to monitor how demyelination occurs and why remyelination often fails. In the last two decades, the zebrafish has been established as an excellent model organism for the study of vertebrate biology. Its transparent characteristic makes zebrafish suitable for performing *in vivo* nervous system imaging. As previously mentioned, several research groups have developed transgenic zebrafish lines to observe myelin sheaths. The general morphology of myelin architecture can be revealed by TPEF microscopy. Another very fascinating characteristic of zebrafish is its strong regeneration capability (LeClair and Topczewski, 2010) of organs such as the heart (Raya et al., 2004; Major and Poss, 2007), liver (Sadler et al., 2007; Kan et al., 2009), and fin (Whitehead et al., 2005; Shao et al., 2011) as well as the nervous system (Becker and Becker, 2008). Our data reveal that zebrafish possess a robust capability of remyelination (Bing Hu, unpublished). After the adult zebrafish optic nerves were completely transected, the axons regenerated. Oligodendrocytes formed a relatively compact myelin sheath on the regenerated axons. Three months later, the g-ratio of regenerated myelin sheath almost reached the normal level with no significant difference. Therefore, zebrafish not only suit *in vivo* myelin imaging, but they also act as an excellent model for demyelination and remyelination research (Kazakova et al., 2006; Buckley et al., 2008; Monk and Talbot, 2009).

In summary, CARS microscopy is a powerful tool with high sensitivity and contrast, is label-free and non-invasive, and it provides the high resolution necessary to detect nodal myelin. Taking advantage of the transparent characteristic of zebrafish, live myelin imaging by CARS microscopy will shed light on the most real processes of myelin formation, demyelination, and remyelination. Such investigations are expected to promote myelin therapy in the future.

Acknowledgments

Several photos are from our previous published and unpublished work. We thank our coworkers Professor Wutian Wu, Professor E.D. Wu, Professor Kwok-fai So, and Connie Wong. The myelin research work has been granted by NIH R01 EB7243 to C.J.X. and National Key Basic Research Program 2011CB504402, NSFC 31070950 to H.B.

References

Aggarwal, S., Yurlova, L., Simons, M., 2011. Central nervous system myelin: structure, synthesis and assembly. Trends Cell Biol 21, 585–593.

Alarcon, G., Valentin, A., 2011. Cortical stimulation with single electrical pulses in human epilepsy. Clin Neurophysiol 123, 223–224.

Almeida, R.G., Czopka, T., Ffrench-Constant, C., Lyons, D.A., 2011. Individual axons regulate the myelinating potential of single oligodendrocytes in vivo. Development 138, 4443–4450.

Arnaud, E., Zenker, J., de Preux Charles, A.S., Stendel, C., Roos, A., Medard, J.J., et al., 2009. SH3TC2/KIAA1985 protein is required for proper myelination and the integrity of the node of Ranvier in the peripheral nervous system. Proc Natl Acad Sci USA 106, 17528–17533.

Baumann, N., Pham-Dinh, D., 2001. Biology of oligodendrocyte and myelin in the mammalian central nervous system. Physiol Rev 81, 871–927.

Becker, C.G., Becker, T., 2008. Adult zebrafish as a model for successful central nervous system regeneration. Restor Neurol Neurosci 26, 71–80.

Belanger, E., Henry, F.P., Vallee, R., Randolph, M.A., Kochevar, I.E., Winograd, J.M., et al., 2011. In vivo evaluation of demyelination and remyelination in a nerve crush injury model. Biomed Opt Express 2, 2698–2708.

Blakemore, W.F., 1981. Remyelination in the CNS. Prog Clin Biol Res 59A, 105–109.

Buckley, C.E., Goldsmith, P., Franklin, R.J., 2008. Zebrafish myelination: a transparent model for remyelination? Dis Model Mech 1, 221–228.

Chen, B.C., Sung, J., Lim, S.H., 2010. Chemical imaging with frequency modulation coherent anti-Stokes Raman scattering microscopy at the vibrational fingerprint region. J Phys Chem B 114, 16871–16880.

Chen, H., Wang, H., Slipchenko, M.N., Jung, Y., Shi, Y., Zhu, J., et al., 2009. A multimodal platform for nonlinear optical microscopy and microspectroscopy. Opt Express 17, 1282–1290.

Cheng, J.X., 2007. Coherent anti-Stokes Raman scattering microscopy. Appl Spectrosc 61, 197–208.

Cheng, J.X., Xie, X.S., 2004. Coherent anti-Stokes Raman scattering microscopy: instrumentation, theory and applications. J Phys Chem B 108, 827–840.

Cheng, J.X., Book, L.D., Xie, X.S., 2001b. Polarization coherent anti-Stokes Raman scattering microscopy. Opt Lett 26, 1341–1343.

Cheng, J.X., Jia, Y.K., Zheng, G., Xie, X.S., 2002. Laser-scanning coherent anti-Stokes Raman scattering microscopy and applications to cell biology. Biophys J 83, 502–509.

Cheng, J.X., Volkmer, A., Book, L.D., Xie, X.S., 2001a. An epi-detected anti-Stokes Raman scattering (E-CARS) microscope with high spectral resolution and high sensitivity. J Phys Chem B 105, 1277.

Cid, C., Alvarez-Cermeno, J.C., Salinas, M., Alcazar, A., 2005. Anti-heat shock protein 90beta antibodies decrease pre-oligodendrocyte population in perinatal and adult cell cultures. Implications for remyelination in multiple sclerosis. J Neurochem 95, 349–360.

Clark, R.J.H., Hester, R.E., 1988. Advances in Nonlinear Spectroscopy. John Wiley, New York.

Compston, A., Coles, A.., 2002. Multiple sclerosis. Lancet 359, 1221–1231.

Degaonkar, M.N., Jayasundar, R., Jagannathan, N.R., 2002a. Sequential diffusion-weighted magnetic resonance imaging study of lysophosphatidyl choline-induced experimental demyelinating lesion: an animal model of multiple sclerosis. J Magn Reson Imaging 16, 153–159.

Degaonkar, M.N., Khubchandhani, M., Dhawan, J.K., Jayasundar, R., Jagannathan, N.R., 2002b. Sequential proton MRS study of brain metabolite changes monitored during a complete pathological cycle of demyelination and remyelination in a lysophosphatidyl choline (LPC)-induced experimental demyelinating lesion model. NMR Biomed 15, 293–300.

Di Scipio, F., Raimondo, S., Tos, P., Geuna, S., 2008. A simple protocol for paraffin-embedded myelin sheath staining with osmium tetroxide for light microscope observation. Microsc Res Tech 71, 497–502.

Duncan, M.D., Reintjes, J., Manuccia, T.J., 1982. Scanning coherent anti-Stokes Raman microscope. Opt Lett 7, 350–352.

Eftekharpour, E., Karimi-Abdolrezaee, S., Wang, J., El Beheiry, H., Morshead, C., Fehlings, M.G., 2007. Myelination of congenitally dysmyelinated spinal cord axons by adult neural precursor cells results in formation of nodes of Ranvier and improved axonal conduction. J Neurosci 27, 3416–3428.

Evans, C.L., Xie, X.S., 2008. Coherent anti-stokes Raman scattering microscopy: chemical imaging for biology and medicine. Annu Rev Anal Chem 1, 883–909.

Evans, C.L., Potma, E.O., Xie, X.S., 2004. Coherent anti-stokes raman scattering spectral interferometry: determination of the real and imaginary components of nonlinear susceptibility chi(3) for vibrational microscopy. Opt Lett 29, 2923–2925.

Evans, C.L., Potma, E.O., Puoris'haag, M., Cote, D., Lin, C.P., Xie, X.S., 2005. Chemical imaging of tissue in vivo with video-rate coherent anti-Stokes Raman scattering microscopy. Proc Natl Acad Sci USA 102, 16807–16812.

Fancy, S.P., Chan, J.R., Baranzini, S.E., Franklin, R.J., Rowitch, D.H., 2011. Myelin regeneration: a recapitulation of development? Annu Rev Neurosci 34, 21–43.

Farrar, M.J., Wise, F.W., Fetcho, J.R., Schaffer, C.B., 2011. In vivo imaging of myelin in the vertebrate central nervous system using third harmonic generation microscopy. Biophys J 100, 1362–1371.

Filippi, M., Rovaris, M., 2000. Magnetisation transfer imaging in multiple sclerosis. J Neurovirol 6 (Suppl. 2), S115–S120.

Fitzner, D., Schneider, A., Kippert, A., Mobius, W., Willig, K.I., Hell, S.W., et al., 2006. Myelin basic protein-dependent plasma membrane reorganization in the formation of myelin. EMBO J 25, 5037–5048.

Franklin, R.J., ffrench-Constant, C., 2008. Remyelination in the CNS: from biology to therapy. Nat Rev Neurosci 9, 839–855.

Franklin, R.J., Hinks, G.L., 1999. Understanding CNS remyelination: clues from developmental and regeneration biology. J Neurosci Res 58, 207–213.

Franklin, R.J., Kotter, M.R., 2008. The biology of CNS remyelination: the key to therapeutic advances. J Neurol 255 (Suppl. 1), 19–25.

Fu, Y., Huff, T.B., Wang, H.W., Wang, H., Cheng, J.X., 2008a. Ex vivo and in vivo imaging of myelin fibers in mouse brain by coherent anti-Stokes Raman scattering microscopy. Opt Express 16, 19396–19409.

Fu, Y., Talavage, T.M., Cheng, J.X., 2008b. New imaging techniques in the diagnosis of multiple sclerosis. Expert Opin Med Diagn 2, 1055–1065.

Fu, Y., Sun, W., Shi, Y., Shi, R., Cheng, J.X., 2009. Glutamate excitotoxicity inflicts paranodal myelin splitting and retraction. PLoS One 4, e6705.

Fu, Y., Wang, H., Huff, T.B., Shi, R., Cheng, J.X., 2007. Coherent anti-Stokes Raman scattering imaging of myelin degradation reveals a calcium-dependent pathway in lyso-PtdCho-induced demyelination. J Neurosci Res 85, 2870–2881.

Fu, Y., Frederick, T.J., Huff, T.B., Goings, G.E., Miller, S.D., Cheng, J.X., 2011. Paranodal myelin retraction in relapsing experimental autoimmune encephalomyelitis visualized by coherent anti-Stokes Raman scattering microscopy. J Biomed Opt 16, 106006.

Ganikhanov, F., Evans, C.L., Saar, B.G., Xie, X.S., 2006. High-sensitivity vibrational imaging with frequency modulation coherent anti-Stokes Raman scattering (FM CARS) microscopy. Opt Lett 31, 1872–1874.

Gerrits, P.O., Brekelmans-Bartels, M., Mast, L., Gravenmade, E.J., Horobin, R.W., Holstege, G., 1992. Staining myelin and myelin-like degradation products in the spinal cords of chronic experimental allergic encephalomyelitis (Cr-EAE) rats using Sudan black B staining of glycol methacrylate-embedded material. J Neurosci Methods 45, 99–105.

Gold, R., 2005. Overcoming failure to repair demyelination in EAE: an important step towards neuroprotection in multiple sclerosis. J Neuroimmunol 170, 1–2.

Goldberg-Zimring, D., Mewes, A.U., Maddah, M., Warfield, S.K., 2005. Diffusion tensor magnetic resonance imaging in multiple sclerosis. J Neuroimaging 15 (4 Suppl.), 68S–81S.

Haller, S., Pereira, V.M., Lalive, P.H., Chofflon, M., Vargas, M.I., Lovblad, K.O., 2009. Magnetic resonance imaging in multiple sclerosis. Topics in magnetic resonance imaging: TMRI 20, 313–323.

Hartline, D.K., 2008. What is myelin? Neuron Glia Biol 4, 153–163.

Hirano, A., Dembitzer, H.M., 1967. A structural analysis of the myelin sheath in the central nervous system. J Cell Biol 34, 555–567.

Huff, T.B., Shi, Y., Fu, Y., Wang, H., Cheng, J.X., 2008. Multimodal nonlinear optical microscopy and applications to central nervous system imaging. IEEE J Sel Top Quantum Electron 14, 4–9.

Huff, T.B., Shi, Y., Sun, W., Wu, W., Shi, R., Cheng, J.X., 2011. Real-time CARS imaging reveals a calpain-dependent pathway for paranodal myelin retraction during high-frequency stimulation. PLoS One 6, e17176.

Imitola, J., Cote, D., Rasmussen, S., Xie, X.S., Liu, Y., Chitnis, T., et al., 2011. Multimodal coherent anti-Stokes Raman scattering microscopy reveals microglia-associated myelin and axonal dysfunction in multiple sclerosis-like lesions in mice. J Biomed Opt 16, 021109.

Jung, S.H., Kim, S., Chung, A.Y., Kim, H.T., So, J.H., Ryu, J., et al., 2010. Visualization of myelination in GFP-transgenic zebrafish. Dev Dyn 239, 592–597.

Jurynczyk, M., Jurewicz, A., Bielecki, B., Raine, C.S., Selmaj, K., 2008. Overcoming failure to repair demyelination in EAE: gamma-secretase inhibition of Notch signaling. J Neurol Sci 265, 5–11.

Kan, N.G., Junghans, D., Izpisua Belmonte, J.C., 2009. Compensatory growth mechanisms regulated by BMP and FGF signaling mediate liver regeneration in zebrafish after partial hepatectomy. FASEB J 23, 3516–3525.

Kazakova, N., Li, H., Mora, A., Jessen, K.R., Mirsky, R., Richardson, W.D., et al., 2006. A screen for mutations in zebrafish that affect myelin gene expression in Schwann cells and oligodendrocytes. Dev Biol 297, 1–13.

Kelemen, J., Becus, M., 1969. Rapid myelin sheath staining in paraffin, frozen and celloidin sections. A new modification of the Weigert-Loyez staining method. Acta Morphol Acad Sci Hung 17, 105–112.

Kiferle, L., Politis, M., Muraro, P.A., Piccini, P., 2011. Positron emission tomography imaging in multiple sclerosis–current status and future applications. Eur J Neurol 18, 226–231.

Kirby, B.B., Takada, N., Latimer, A.J., Shin, J., Carney, T.J., Kelsh, R.N., et al., 2006. In vivo time-lapse imaging shows dynamic oligodendrocyte progenitor behavior during zebrafish development. Nat Neurosci 9, 1506–1511.

Kumar, S., Rajshekher, G., Prabhakar, S., 2005. Positron emission tomography in neurological diseases. Neurology India 53, 149–155.

Landon, D.N., Williams, P.L., 1963. Ultrastructure of the node of Ranvier. Nature 199, 575–577.

Le, T.T., Langohr, I.M., Locker, M.J., Sturek, M., Cheng, J.X., 2007. Label-free molecular imaging of atherosclerotic lesions using multimodal nonlinear optical microscopy. J Biomed Opt 12, 054007.

LeClair, E.E., Topczewski, J., 2010. Development and regeneration of the zebrafish maxillary barbel: a novel study system for vertebrate tissue growth and repair. PLoS One 5, e8737.

Major, R.J., Poss, K.D., 2007. Zebrafish heart regeneration as a model for cardiac tissue repair. Drug Discov Today Dis Models 4, 219–225.

Maker, P.D., Terhune, R.W., 1965. Study of optical effects due to an induced polarization third order in the electric field strength. Phys Rev 137, A801–A818.

Margolis, G., Pickett, J.P., 1956. New applications of the Luxol fast blue myelin stain. Lab Invest 5, 459–474.

McGowan, J.C., Filippi, M., Campi, A., Grossman, R.I., 1998. Magnetisation transfer imaging: theory and application to multiple sclerosis. J Neurol Neurosurg Psychiatry 64 (Suppl. 1), S66–S69.

Merkler, D., Ernsting, T., Kerschensteiner, M., Bruck, W., Stadelmann, C., 2006. A new focal EAE model of cortical demyelination: multiple sclerosis-like lesions with rapid resolution of inflammation and extensive remyelination. Brain 129, 1972–1983.

Micu, I., Ridsdale, A., Zhang, L., Woulfe, J., McClintock, J., Brantner, C.A., et al., 2007. Real-time measurement of free Ca^{2+} changes in CNS myelin by two-photon microscopy. Nat Med 13, 874–879.

Monk, K.R., Talbot, W.S., 2009. Genetic dissection of myelinated axons in zebrafish. Curr Opin Neurobiol 19, 486–490.

Moses, W.W., 2011. Fundamental limits of spatial resolution in PET. Nucl Instrum Methods Phys Res A 648 (Suppl. 1), S236–S240.

Nagai, J., Uchida, H., Matsushita, Y., Yano, R., Ueda, M., Niwa, M., et al., 2010. Autotaxin and lysophosphatidic acid1 receptor-mediated demyelination of dorsal root fibers by sciatic nerve injury and intrathecal lysophosphatidylcholine. Mol Pain 6, 78.

Neema, M., Stankiewicz, J., Arora, A., Guss, Z.D., Bakshi, R., 2007. MRI in multiple sclerosis: what's inside the toolbox? Neurotherapeutics 4, 602–617.

Oda, M., Udaka, F., 2008. [Magnetic resonance imaging in multiple sclerosis]. Nihon Rinsho 66, 1098–1102.

Ogura, M., Nakao, N., Nakai, E., Uematsu, Y., Itakura, T., 2004. The mechanism and effect of chronic electrical stimulation of the globus pallidus for treatment of Parkinson disease. J Neurosurg 100, 997–1001.

Oommen, J., Morrell, M., Fisher, R.S., 2005. Experimental electrical stimulation therapy for epilepsy. Curr Treat Options Neurol 7, 261–271.

Oron, D., Dudovich, N., Silberberg, Y., 2003. Femtosecond phase-and-polarization control for background-free coherent anti-Stokes Raman spectroscopy. Phys Rev Lett 90, 213902.

Pedraza, L., Colman, D.R., 2000. Fluorescent myelin proteins provide new tools to study the myelination process. J Neurosci Res 60, 697–703.

Peters, A., 1966. The node of Ranvier in the central nervous system. Q J Exp Physiol Cogn Med Sci 51, 229–236.

Piraino, P.S., Yednock, T.A., Messersmith, E.K., Pleiss, M.A., Freedman, S.B., Hammond, R.R., et al., 2005. Spontaneous remyelination following prolonged inhibition of alpha4 integrin in chronic EAE. J Neuroimmunol 167, 53–63.

Platt, S.R.., 2007. The role of glutamate in central nervous system health and disease—a review. Vet J 173, 278–286.

Potma, E.O., Evans, C.L., Xie, X.S., 2006. Heterodyne coherent anti-Stokes Raman scattering (CARS) imaging. Opt Lett 31, 241–243.

Pozzilli, C., Fieschi, C., Perani, D., et al., 1992. Relationship between corpus callosum atrophy and cerebral metabolic asymmetries in multiple sclerosis. J Neurol Sci 12, 51–57.

Raya, A., Consiglio, A., Kawakami, Y., Rodriguez-Esteban, C., Izpisua-Belmonte, J.C., 2004. The zebrafish as a model of heart regeneration. Cloning Stem Cells 6, 345–351.

Rinholm, J.E., Hamilton, N.B., Kessaris, N., Richardson, W.D., Bergersen, L.H., Attwell, D., 2011. Regulation of oligodendrocyte development and myelination by glucose and lactate. J Neurosci 31, 538–548.

Runge, V.M., Price, A.C., Kirshner, H.S., Allen, J.H., Partain, C.L., James Jr., A.E., 1984. Magnetic resonance imaging of multiple sclerosis: a study of pulse-technique efficacy. AJR Am J Roentgenol 143, 1015–1026.

Sadler, K.C., Krahn, K.N., Gaur, N.A., Ukomadu, C., 2007. Liver growth in the embryo and during liver regeneration in zebrafish requires the cell cycle regulator, uhrf1. Proc Natl Acad Sci USA 104, 1570–1575.

Schiepers, C., Van Hecke, P., Vandenberghe, R., Van Oostende, S., Dupont, P., Demaerel, P., et al., 1997. Positron emission tomography, magnetic resonance imaging and proton NMR spectroscopy of white matter in multiple sclerosis. Mult Scler 3, 8–17.

Scholtz, C.L., 1977. Quantitative histochemistry of myelin using Luxol Fast Blue MBS. Histochem J 9, 759–765.

Shao, J., Chen, D., Ye, Q., Cui, J., Li, Y., Li, L., 2011. Tissue regeneration after injury in adult zebrafish: the regenerative potential of the caudal fin. Dev Dyn 240, 1271–1277.

Sherman, D.L., Brophy, P.J., 2005. Mechanisms of axon ensheathment and myelin growth. Nat Rev Neurosci 6, 683–690.

Shi, Y., Zhang, D., Huff, T.B., Wang, X., Shi, R., Xu, X.M., et al., 2011. Longitudinal in vivo coherent anti-Stokes Raman scattering imaging of demyelination and remyelination in injured spinal cord. J Biomed Opt 16, 106012.

Shin, J., Park, H.C., Topczewska, J.M., Mawdsley, D.J., Appel, B., 2003. Neural cell fate analysis in zebrafish using olig2 BAC transgenics. Methods Cell Sci 25, 7–14.

Simons, M., Trotter, J., 2007. Wrapping it up: the cell biology of myelination. Curr Opin Neurobiol 17, 533–540.

Sivakumar, V., Ling, E.A., Lu, J., Kaur, C., 2010. Role of glutamate and its receptors and insulin-like growth factors in hypoxia induced periventricular white matter injury. Glia 58, 507–523.

Sobottka, B., Ziegler, U., Kaech, A., Becher, B., Goebels, N., 2011. CNS live imaging reveals a new mechanism of myelination: The liquid croissant model. Glia 59, 1841–1849.

Stankoff, B., Freeman, L., Aigrot, M.S., Chardain, A., Dolle, F., Williams, A., et al., 2011. Imaging central nervous system myelin by positron emission tomography in multiple sclerosis using [methyl-(1)(1)C]-2-(4'-methylaminophenyl)-6-hydroxybenzothiazole. Ann Neurol 69, 673–680.

Stankoff, B., Wang, Y., Bottlaender, M., Aigrot, M.S., Dolle, F., Wu, C., et al., 2006. Imaging of CNS myelin by positron-emission tomography. Proc Natl Acad Sci USA 103, 9304–9309.

Stoeckenius, W., 1957. [Osmium tetroxide staining of intracellular myelin patterns]. Exp Cell Res 13, 410–414.

Tiwari-Woodruff, S.K., Buznikov, A.G., Vu, T.Q., Micevych, P.E., Chen, K., Kornblum, H.I., et al., 2001. OSP/claudin-11 forms a complex with a novel member of the tetraspanin super family and beta1 integrin and regulates proliferation and migration of oligodendrocytes. J Cell Biol 153, 295–305.

Tolles, W.M., Nibler, J.W., McDonald, J.R., Harvey, A.B., 1977. A review of the theory and application of coherent anti-Stokes Raman spectroscopy (CARS). Appl Spectrosc 31, 253–272.

Vereyken, E.J., Fluitsma, D.M., Bolijn, M.J., Dijkstra, C.D., Teunissen, C.E., 2009. An in vitro model for de- and remyelination using lysophosphatidyl choline in rodent whole brain spheroid cultures. Glia 57, 1326–1340.

Volkmer, A., Book, L.D., Xie, X.S., 2002. Time-resolved coherent anti-Stokes Raman scattering microscopy: imaging based on Raman free induction decay. Appl Phys Lett 80, 1505–1507.

Volkmer, A., Cheng, J.X., Xie, X.S., 2001. Vibrational imaging with high sensitivity via epi-detected coherent anti-Stokes Raman scattering microscopy. Phys Rev Lett 87, 023901.

Wang, C., Gold, B.G., Kaler, L.J., Yu, X., Afentoulis, M.E., Burrows, G.G., et al., 2006. Antigen-specific therapy promotes repair of myelin and axonal damage in established EAE. J Neurochem 98, 1817–1827.

Wang, H., Fu, Y., Zickmund, P., Shi, R., Cheng, J.X., 2005. Coherent anti-stokes Raman scattering imaging of axonal myelin in live spinal tissues. Biophys J 89, 581–591.

Wang, H.W., Langohr, I.M., Sturek, M., Cheng, J.X., 2009. Imaging and quantitative analysis of atherosclerotic lesions by CARS-based multimodal nonlinear optical microscopy. Arterioscler Thromb Vasc Biol 29, 1342–1348.

Waxman, S.G., 2006. Axonal conduction and injury in multiple sclerosis: the role of sodium channels. Nat Rev Neurosci 7, 932–941.

Whitehead, G.G., Makino, S., Lien, C.L., Keating, M.T., 2005. fgf20 is essential for initiating zebrafish fin regeneration. Science 310, 1957–1960.

Witte, S., Negrean, A., Lodder, J.C., de Kock, C.P., Testa Silva, G., Mansvelder, H.D., et al., 2011. Label-free live brain imaging and targeted patching with third-harmonic generation microscopy. Proc Natl Acad Sci USA 108, 5970–5975.

Yoshida, M., Macklin, W.B., 2005. Oligodendrocyte development and myelination in GFP-transgenic zebrafish. J Neurosci Res 81, 1–8.

Yu, C.S., Li, K.C., Lin, F.C., Jiang, T.Z., Sun, H., Chen, B., 2006. [Diffusion tensor imaging of the normal-appearing brain tissue in relapsing-remitting multiple sclerosis]. Zhonghua Yi Xue Za Zhi 86, 1260–1264.

Yue, S.H., Slipchenko, M.N., Cheng, J.X, 2011. Multimodal nonlinear optical microscopy. Laser Photons Rev 5, 496–512. doi: 10.1002/lpor.201000027.

Zumbusch, A., Holtom, G.R., Xie, X.S., 1999. Three-dimensional vibrational imaging by coherent anti-Stokes Raman scattering. Phys Rev Lett 82, 4142–4145.

HIGH-RESOLUTION APPROACHES TO STUDYING PRESYNAPTIC VESICLE DYNAMICS USING VARIANTS OF FRAP AND ELECTRON MICROSCOPY

Kevin Staras[1], and Tiago Branco[2,3]

[1]*School of Life Sciences, University of Sussex, Brighton, UK,* [2]*Wolfson Institute for Biomedical Research,* [3]*Department of Neuroscience, Physiology, and Pharmacology, University College London, London, UK*

CHAPTER OUTLINE

Introduction   248
Quantifying Dynamic Events at the Macromolecular Scale   249
FRAP for Studying Mobility   251
    FRAP to Study Intrasynaptic Vesicle Mobility   251
    FRAP to Study Intersynaptic Vesicle Mobility   253
    FRAP and Phototoxicity   254
Variations on FRAP Using Photoswitchable Fluorophores   255
    Application of Dendra2 to the Tracking of Synaptic Vesicle Dynamics   257
    Considerations for Dendra2 Photoswitching   258
Linking Fluorescence and Ultrastructure: Correlative Approaches for
    Assaying Presynaptic Function   259
    Photoconversion and Correlative Microscopy Using FM1-43   260
    Effective Photoconversion of FM-dye with DAB in Hippocampal Cultures   262
    Correlative Approach   263
Structure–Function Relationships of Vesicle Pools in Hippocampal
    Synapses   263
Combining FRAP with Correlative Electron Microscope   265
    FRAP-CLEM for Detailing Synaptic Vesicle Dynamics   266
    Inverse FRAP-CLEM   268

Cellular Imaging Techniques for Neuroscience and Beyond.
DOI: http://dx.doi.org/10.1016/B978-0-12-385872-6.00011-8

Concluding Remarks 269
Acknowledgments 269
References 270

Introduction

In the central nervous system, the transfer of information between neurons relies mainly on chemical transmission at individual synapses, which are ultrastructurally specialized neuron–neuron apposition points (Murthy and De Camilli, 2003; Sudhof, 2004; Rizzoli and Betz, 2005). A synapse comprises a presynaptic compartment, defined by a cluster of neurotransmitter-containing synaptic vesicles, and a specialized postsynaptic structure. Synaptic transmission proceeds with the exocytic fusion of vesicles with the plasma membrane at a site termed the active zone, and released neurotransmitter diffuses across extracellular space to activate receptors located at the postsynaptic terminal. In trying to understand the mechanisms that regulate information flow within neuronal networks, the flexibility of this process associated with learning and memory, as well as synaptic dysfunction events that underlie some neurological pathologies, researchers have long since recognized the value of having reliable readouts of synaptic function. Investigations using electrophysiological approaches have proved highly informative for this purpose, but they are mainly focused on characterizing neuronal connections—the compound output of all participating synapses (Branco and Staras, 2009). In the last two decades, however, major additional insights have come with the advent of imaging approaches, which permit functional readouts at the level of individual synapses.

In broad terms, three main factors have driven this change. First, there have been technical developments in imaging systems (microscopes, cameras, light sources, etc.) that allow individual synapses to be imaged with higher spatial resolution, higher sensitivity, and higher temporal acquisition rates than previously achievable. Second, there is the revolution that has occurred in the development of chemical and genetically encoded optical probes (Lippincott-Schwartz and Patterson, 2003), permitting targeted readouts of many different aspects of synaptic function in living neurons. The third factor is the development of increasingly sophisticated correlative microscopy approaches that allow optical signals to be directly coupled to postfix ultrastructural information facilitating detailed investigations of structure–function relationships. Individually, these developments have proved to be highly informative for elucidating aspects of synaptic function. However, as we outline below, exploiting these approaches in combination has proved particularly advantageous, allowing researchers to gain a new appreciation of the dynamic nature of synapses and providing insights into the highly orchestrated molecular events that underlie information transmission.

Clearly a complete review of this field is beyond the scope of one chapter, but we will restrict our discussion to a variety of methods for characterizing synaptic processes, and, specifically, synaptic vesicle dynamics at presynaptic terminals using relatively affordable and nonspecialist approaches. A particular focus will be on illustrating the benefits of combining light-based and ultrastructural approaches for detailing structure–function relationships. We will first discuss conventional methods such as fluorescence recovery after photo-bleaching (FRAP) for monitoring and quantifying dynamic synaptic processes, and then introduce more recent variants of this strategy utilizing newly developed fluorophores, which can provide additional information about dynamic events. We then consider approaches that combine the specific advantages of light-based imaging methods and ultrastructural characterization using correlative approaches. We outline functional parameters amenable to electron microscopy (EM) and discuss how dynamic events recorded in light microscopy can also be reported at the nanoscale. Although the emphasis of this chapter will be on the presynaptic terminal and dynamics of synaptic vesicles, which is the major focus in our laboratory, these techniques are broadly applicable to a wide range of cell biological questions.

Quantifying Dynamic Events at the Macromolecular Scale

The development of modern time-lapse imaging systems heralded a new view of cell biological processes in living tissue. With this advance, researchers gained a further appreciation of the dynamic nature of cellular function and, in particular, the realization that the mobility of subcellular structures was often intimately linked to specific biological events. Additional insight came with the development of small fluorescent probes that could be tagged to structures of interest or even individual molecular components. In this way, dynamic properties of specific target elements could be tracked effectively over time. Although this combination of a targeted fluorescent label and time-lapse imaging methods is highly informative for some applications, gaining detailed quantitative readouts can still be challenging. This is particularly true in experimental examples where the aim is to quantify movements of numerous macromolecular-scale elements that are tightly clustered within a compartment. In this case, the movement of individual elements is effectively obscured by the fluorescence and dynamics of the whole population. Synaptic vesicles at presynaptic terminals are an illustrative example of such an arrangement. Methods to fluorescently label synaptic vesicles in living neurons have been widely adopted in the last 25 years (Ryan, 2001). These

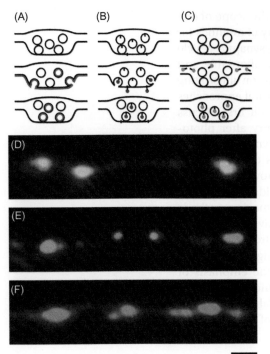

(A)   (B)   (C)

(D)

(E)

(F)

1 μm

Figure 11.1 Example methods for labeling synaptic vesicle pools. (A) Schematic of FM-dye labeling method. Vesicles are labeled by the addition of FM-dye to extracellular solution followed by neuronal stimulation. Dye binds to the outer leaflet of plasma membrane and is inserted onto the lumenal membrane of the vesicle during endocytosis. Washing to remove extracellular FM-dye reveals selective labeling of recycled vesicles. (B) Schematic of synaptic vesicle labeling using antibodies directed to vesicular proteins. Native vesicular proteins with lumenal epitopes move to the extracellular membrane surface during exocytosis. Fluorescently tagged antibodies target proteins and are taken into vesicles during endocytosis. Washing to remove surface antibody reveals selective labeling of recycled vesicles. (C) Schematic of genetically encoded labeling of vesicles. cDNA coding for synaptic vesicle protein and a fluorescent protein tag is transfected into neurons. Protein is trafficked to target synapses and expressed in vesicles. (D) FM1-43 labeling with presynaptic terminals appearing as discrete fluorescent punctum. (E) Live antibody labeling with synaptotagmin I Oyster550. (F) Expression of Synaptophysin I-mCherry. Please see color plates at the back of the book.

include FM-styryl-dyes (Figure 11.1A and D), specific markers of recycling vesicles that are taken up onto vesicle lumenal membrane after exo-endocytic recycling (Betz and Bewick, 1992; Ryan et al., 1993), live antibody labels targeted to specific vesicular proteins (Figure 11.1B and E) (Kraszewski et al., 1995; Willig et al., 2006; Hua et al., 2010), genetically encoded constructs expressing fluorescently tagged (e.g., green fluorescent protein) synaptic vesicle proteins (Figure 11.1C and F) (Ahmari et al., 2000; Sankaranarayanan and Ryan, 2000) and in elegant recent work, nanoscale fluorescent particles called q-dots (Zhang et al., 2007a, 2009). All of these methods can label populations of synaptic vesicles so that, when imaged, each presynaptic terminal is visualized as a discrete fluorescent punctum (Figure 11.1D–F). This approach has been hugely successful in allowing the detailed characterization of vesicle fusion and recycling kinetics during transmission at individual terminals. However, gaining a sense of vesicle mobility within a terminal or, for example, quantifying levels of fluorescent traffic passing through a synapse, is more problematic, because individual vesicle movements are masked by the vesicle population as a whole. Technically sophisticated approaches such as fluorescence correlation spectroscopy (Jordan et al., 2005; Shtrahman et al., 2005), single particle tracking (Lemke and Klingauf, 2005), superresolution microscopy (Westphal et al., 2008; Kamin et al., 2010), and q-dot-labeling methods (Zhang et al., 2007a, 2009) have provided important information on these kinds of issue, but they represent more advanced and specialist solutions that still come with limitations. For many applications, variations on an approach termed FRAP (Lippincott-Schwartz et al., 2003) offers an elegant, relatively simple and economical approach to tackling this problem. We first briefly review the main features of the FRAP approach and then consider its value for monitoring presynaptic vesicle dynamics.

FRAP for Studying Mobility

FRAP is an important and widely used approach for assessing mobility characteristics of small intracellular structures or proteins in living cells. First developed in the mid-1970s (Axelrod et al., 1976; Elson et al., 1976; Jacobson et al., 1976; Schlessinger et al., 1976), the applications for FRAP substantially increased with the development of affordable laser-based imaging systems and GFP-based protein tagging approaches (Lippincott-Schwartz et al., 2003). The basic principle is that a defined compartment within a larger structure containing many individual fluorescent components is irreversibly photobleached using high-power focused laser light. Subsequently this region is imaged over time and the temporal recovery of fluorescence signal provides a measure of the repopulation of the compartment with mobile fluorescent components from neighboring nonphotobleached regions (Figure 11.2A). Specifically, the FRAP signal gives a quantitative readout of the average rate of mobility of fluorescent components and also any immobile fraction corresponding to the residual unrecovered signal (reviewed in Lippincott-Schwartz et al., 2003). A major strength of the FRAP approach is that it requires relatively little in the way of specialized equipment: most modern commercially available confocal systems, for example, are able to perform semi-automated multisite FRAP experiments.

FRAP to Study Intrasynaptic Vesicle Mobility

The FRAP approach has been elegantly exploited in a wide variety of different systems to address all manner of issues related to dynamics of small structures (Lippincott-Schwartz et al., 2003). Here we specifically consider its application for examining synaptic vesicle dynamics. One excellent illustrative example comes from research into vesicle mobility within lizard cone photoreceptor terminals (Figure 11.2B and C; Rea et al., 2004). In this system, individual terminals house a substantial vesicle population that researchers could label effectively with FM-dye. Moreover, the large size of the

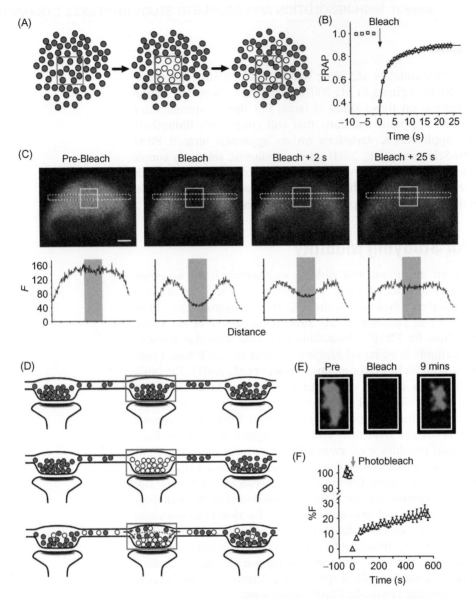

Figure 11.2 Using FRAP to monitor the fluorescence mobility. (A) Schematic illustrating principle of FRAP for quantifying mobility in a defined compartment. An ROI within a compartment containing mobile fluorescent elements (left) is photobleached (middle). Recovery of fluorescence in the ROI occurs as fluorescent elements move in from outside the ROI. (B, C) Example of FRAP at a cone photoreceptor terminal where fluorescence corresponds to synaptic vesicles labeled with the dye FM1-43. (B) Typical fluorescence recovery curve at an ROI following photobleach at time 0. (C) Sample images for plot in (B). Top, terminal labeled with FM1-43 is photobleached in square ROI and shows rapid fluorescence recovery over 25 s. Bottom, fluorescence intensity cross-section through terminal (dashed rectangle in top images) showing profile of fluorescence change. (D). Schematic illustrating principle of FRAP for quantifying mobility between compartments. Top, synaptic vesicles are labeled with FM-dye at presynaptic terminals. Middle, a target synapse is photobleached. Bottom, lateral mobility of vesicles arising from other nonphotobleached synapses leads to recovery of fluorescence at the bleached terminal. (E) Sample images from time-lapse sequence for FRAP experiment showing prebleaching, bleaching, and recovery stages. (F) Typical plot of fluorescence recovery into target terminal after bleaching at time 0. (B and C are adapted from Rea et al., 2004, with permission. E and F are adapted from Darcy et al., 2006b.)

photoreceptor terminal permitted the specific photobleaching of a subregion in the center of the synapse while fluorescence in the terminal surrounding this subregion was preserved (Figure 11.2C). The recovery of fluorescence into the bleached region provides a measure of the mobility rate, and the extent of recovery reports the fraction of immobilized vesicles in the terminal (Figure 11.2B). The findings of this study revealed that in these photoreceptor terminals, which lack the vesicle-anchoring protein, synapsin, vesicles move at rates approaching free diffusion with almost all vesicles participating in movement (Rea et al., 2004). Using similar strategies, FRAP has been exploited in a wide range of systems to characterize features of vesicle dynamics, including retinal bipolar cell terminals from goldfish (Holt et al., 2004) and mouse (LoGiudice et al., 2008), hippocampal synapses (Kraszewski et al., 1996; Li and Murthy, 2001; Darcy et al., 2006a,b; Stan et al., 2010), mouse neuromuscular junction (NMJ; Gaffield and Betz, 2007), *Drosophila* NMJ (Shakiryanova et al., 2006), and frog NMJ (Henkel et al., 1996b; Gaffield et al., 2006).

FRAP to Study Intersynaptic Vesicle Mobility

The FRAP approach outlined earlier, relies on localized photobleaching of a subcompartment of a larger fluorescent structure. However, in a variation of this approach FRAP is also a powerful method for assaying mobile traffic *between* connected but discrete compartments. In our laboratory, we employ FRAP in this way to investigate the lateral movement of synaptic vesicles along axons between presynaptic terminals (Figure 11.2D). Functionally recycling vesicles are labeled with FM-dye by maximal stimulation of cultures [e.g., 600 action potentials (APs) 10 Hz via field stimulation or with a hyperkalemic solution] in the presence of the dye (10 μM). After a washing step using Advasep-7 (1 mM; Biotium), an FM-dye chelating agent that aids the removal of dye from the membrane surface (Kay et al., 1999), punctate fluorescent staining is readily observed (Figure 11.1A and 11.2E), corresponding to the labeled clusters of functionally recycling vesicle pools at individual terminals.

To monitor intersynaptic exchange of vesicles between terminals, we employ a standard FRAP protocol that is easily implemented on most modern confocal imaging systems. In essence, the experiment is simply a long time-lapse sequence punctuated by a single, short photobleaching protocol that eliminates the fluorescence signal at a single target synapse. The prebleach imaging period provides information on the baseline fluorescence and shows the extent of imaging-related photobleaching. The postbleach imaging period is used to monitor fluorescence recovery. To limit signal degradation, the experimenter is strongly advised to image with the minimal sample illumination required to assay fluorescence at the target terminal. For the photobleaching step, effective removal of fluorescence at the target synapse involves irradiation using a wavelength of light which efficiently excites the fluorophore. Imaging systems typically offer a fast spot-bleach mode or support a user-defined region-of-interest (ROI) facility. Within this ROI function, a "zoom-in" mode permits a tighter region of illumination for the photobleach process with less "bleed-out" to areas outside the target zone (Chudakov et al., 2007).

Once the region(s) for photobleaching is defined, the properties of the bleach protocol must be set. For a given system and specific experimental demand, the conditions need to be individually determined. A number of factors are particularly important including the fluorophore type, the objective lens numerical aperture (NA), the laser power, the illumination wavelength, the area of the ROI, the dwell time, and the number of scans over the ROI. Once correct settings are established, most systems run a FRAP protocol in an automated way and generate a time-lapse sequence from which information about fluorescence intensity values at the target terminal can be extracted. For this type of experiment, the assumption is that any fluorescence recovery observed at the photobleached terminal should arise only as a consequence of the lateral movement of synaptic vesicles along axons from neighboring nonphotobleached synapses. As such, the rate of recovery provides a measure of the turnover of recycling vesicles at a target synapse arising from extrasynaptic sites (Darcy et al., 2006a,b). An example of the time course of fluorescence recovery is shown in Figure 11.2E and F. Strikingly, the recovery curve reveals a complex multiphase process — the rapid phase presumably corresponding to the fast transport of vesicles along the axon and a slower incorporation step where mobilized vesicles become stably integrated into the synaptic terminal (Figure 11.2F; Darcy et al., 2006b). FRAP experiments such as these demonstrate the lateral movement of vesicles between synapses in cultured hippocampal neurons and provide quantitative information on the rate of vesicle turnover. They also provide direct evidence for the idea that mobile vesicles can become stably incorporated into presynaptic terminals, contributing to the stable fluorescence recovery signal (Darcy et al., 2006b).

FRAP and Phototoxicity

One concern with FRAP experiments is the possibility that photobleaching steps could have detrimental effects on normal physiological functions of the target structure. It is well established that excessive photo-irradiation of biological structures can have damaging consequences for biological processes and this is directly observable at the ultrastructural level (Deerinck et al., 1994; Teng and Wilkinson, 2000;

Harata et al., 2001; Tozer et al., 2007). It is imperative, therefore, to limit photo-irradiation to the minimum level required to achieve sufficient photobleaching. Alongside this, the experimenter should use functional assays of the target structure to determine that physiological functions are not impaired. The choice of assays naturally depends on the process under study. For the FRAP application outlined previously, we exploit FM-dye loading to test for normal vesicle recycling capabilities after photobleaching (Henkel et al., 1996b; Darcy et al., 2006a,b). In control experiments, synapses are photobleached and then subsequently reloaded with FM-dye to ensure that labeling of recycling vesicles after bleaching is achievable at a level comparable with the initial load. Moreover, reloaded synapses can subsequently be challenged with a further stimulation protocol to evoke an additional round of exocytosis (Darcy et al., 2006a). This process can be monitored as FM-dye loss in time-lapse imaging sequences and comparisons between the kinetics of destaining at previously bleached synapses can be directly compared with the same process at neighboring unbleached regions.

Variations on FRAP Using Photoswitchable Fluorophores

Most studies looking at the dynamic behavior of subcellular components rely on a combination of photo-illumination of a single-color fluorescence moiety associated with a target structure of interest and time-lapse imaging. While informative, this strategy ultimately provides limited information about the origins and fates of target elements over time. This is illustrated by the example of a time-lapse imaging sequence where the objective is to track a mobile fluorescent element within an area containing other fluorescent structures (Figure 11.3A), which is a typical situation in many cell biological experiments. In this case if the mobile element merges with one or several other similarly fluorescent components, then information about the fate of the target element is lost. This problem arises because the target element and the other structures are labeled with the same fluorescent tag. FRAP experiments based on single-color fluorophores are similarly limited: once photobleaching has occurred, the spatiotemporal dynamics of the bleached subpopulation can no longer be tracked (Figure 11.3B). Recently, however, the development of a new class of fluorophores has offered a solution to these limitations by allowing the fate of subpopulations of fluorescently tagged proteins to be tracked by assigning them a unique spectral signature.

This family of fluorophores, termed photoswitchable proteins (Chapman et al., 2005; Chudakov et al., 2007; Lippincott-Schwartz and

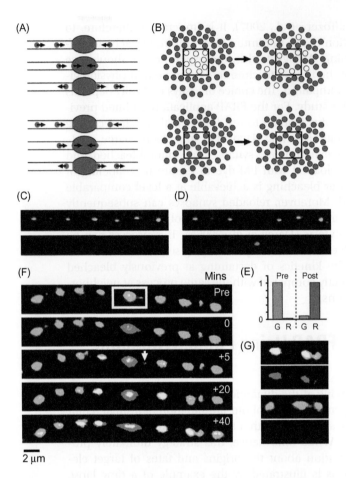

Figure 11.3 Using a photoswitchable fluorophore to monitor the fate of mobile elements. (A) Schematic showing how photoswitchable fluorophores provide additional information for tracking movement of fluorescent elements between compartments in time-lapse sequences. The fate of trafficking fluorescent particles (top) is unclear when they encounter a larger fluorescent structure. However, when photoswitched (bottom), the movement of a mobile element can be tracked, revealing a feature of the trafficking (a direction reverse) obscured using single-color imaging. (B) Schematic illustrating additional information gained by photoswitching method when used in a defined compartment. Top, conventional FRAP provides no information on fate of bleached elements. Using photoswitching (bottom), the fate of elements at photoswitched sites can be monitored alongside recovery of fluorescent elements into target region. (C, D) Experiment illustrating a critical requirement for a photoswitching experiment: that a discrete target structure can be photoswitched without influencing the fluorescence of neighbors. Images show a population of presynaptic terminals along an axon in hippocampal cultures expressing SytI-Dendra2 before (C) and after (D) photoswitching of the central synapse. Top shows green channel and bottom shows red channel. (E) Fluorescence intensity changes in a target synapse for green and red channels pre- and post-switching. (F) Example of photoswitching experiment to examine lateral vesicle mobility. A target synapse along an axon is selectively photoswitched and the spread of new red fluorescence to synaptic neighbors is monitored over time. Panels show the merge of green and red channels where overlap appears yellow. (G) Illustrates an important requirement for this experiment: that synapses in the same field of view but on different axons do not photoswitch. Top shows synapse in a target region 40 minutes after photoswitching (upper, red/green overlay; lower, red alone). Bottom show synapses in same field of view but on different axon (upper, overlay; lower, red alone). (F and G are adapted from Staras et al., 2010, with permission.) Please see color plates at the back of the book.

Patterson, 2009), undergoes a change in its spectral emission properties in response to a specific "switching" wavelength of light. Over recent years a number of these fluorophores have become available, of which a subset that is monomeric including Dendra2 (Gurskaya et al., 2006), PS-CFP2 (Chudakov et al., 2004), and mEosFP (Wiedenmann et al., 2004), are particularly applicable to studies tracking mobile proteins (Chudakov et al., 2007). Other related fluorophores whose spectral properties can be altered by activating wavelengths of light, such as PA-GFP (a photoactivatable compound whose fluorescence increases ~100-fold in response to illumination at 488 nm; Patterson and Lippincott-Schwartz, 2002) and Dronpa (Ando et al., 2004), rsFastLime (Stiel et al., 2007), and mTFP0.7 (Henderson et al., 2007), which show reversible photoconversion between dark and bright states (Chudakov et al., 2007), also have considerable value for certain experimental demands. Discussion of the detailed properties of these fluorophores is outlined at length elsewhere by Chapman et al. (2005), Chudakov et al. (2007), and Lippincott-Schwartz and Patterson (2009) and is beyond the scope of this chapter. However, we highlight one of these, Dendra2, from the octocoral *Dendronephthya* sp., which has several useful characteristics for tracking the dynamics of subpopulations of proteins within cells and neurons (Gurskaya et al., 2006; Chudakov et al., 2007). Dendra2 is photoswitchable with two distinct excitation-emission spectra and a specific switching wavelength (~488 nm), which irreversibly changes its spectral properties from one form to the other. For many studies, the blue light wavelength for switching Dendra2, rather than the potentially more damaging UV illumination required for other switchable fluorophores, is particularly appealing where concerns about phototoxicity effects are paramount (Gurskaya et al., 2006).

Application of Dendra2 to the Tracking of Synaptic Vesicle Dynamics

Exploitation of photoswitchable proteins can reveal new kinds of information about dynamic events in cells. Here we outline an experimental approach from our lab using Dendra2 to track movements of synaptic vesicles within axons of hippocampal neurons. Our previous FRAP work (Darcy et al., 2006b; see earlier sections) highlighted the labile nature of synaptic vesicles between stable presynaptic terminals, suggesting that the vesicular composition of individual terminals is highly dynamic. Nonetheless, using conventional single-color fluorophores it is difficult to gain a full sense of trafficking dynamics, and, in particular, the eventual fate of a complement of vesicles initially residing at one terminal. To address this issue we generated a novel construct based on the fusion of the vesicle-associated protein, synaptophysin I (SypI), with Dendra2 tagged to the cytoplasmic end (Staras et al., 2010). SypI is a synaptic vesicle protein with one of the highest copy numbers (Takamori et al., 2006) and is widely employed as a marker of vesicles, in large part because it is mainly restricted to vesicles rather than surface membrane (Granseth et al., 2006). When expressed in hippocampal neurons, unswitched SypI-Dendra2 excited with 488 nm laser light emits green fluorescence, which could be efficiently collected by a 515/30 emission filter set, and had a punctate appearance (Figure 11.3C; Staras et al., 2010). At the same time, excitation with 543 nm light and a 590/70 emission filter did not produce a signal above background (Figure 11.3C). Photoswitching of a single presynaptic terminal with 488 nm laser light was highly effective and resulted in a large (>40-fold) increase in red signal and a ~12-fold decrease in green emission (Figure 11.3D and E). Using this method it became readily possible to track the fate of a population of vesicles initially residing at one synapse over time (Figure 11.3F). We observed that vesicles were highly mobile and shared between multiple synapses along an axonal process, supporting the idea that many synaptic vesicles belong to a synapse-spanning vesicle superpool (Staras et al., 2010). Thus, photoswitching approaches targeted at synaptic vesicle pools represent an important strategy for tracking vesicle dynamics and offer some specific advantages over conventional single-color fluorescent probes.

Considerations for Dendra2 Photoswitching

Experimental plans using a photoswitchable fluorophore, such as Dendra2, should take into account a number of technical points (see also Chudakov et al., 2007). Dendra2, in particular, is highly sensitive to photoswitching, and as such, it is critical to guard against unnecessary illumination of the sample; for example, great care should be taken during primary visualization. Identification of target structures should be carried out using highly attenuated and brief illumination with widefield fluorescence (e.g., a mercury lamp or LED) or attenuated laser light using fast speed scans with short pixel dwell times and preferably no zoom. Once a target region is chosen, a red channel image should first be collected and the red fluorescence level analyzed (Figure 11.3C). Red signal significantly above background levels suggests that a photoswitching reaction has already occurred, invalidating the experiment. For photoswitching, calibration for each specific experimental setup is critical to achieve reliable and robust results. Several approaches are possible. Dendra2 is efficiently photoswitched using unattenuated widefield illumination, but this method is not suitable for highly spatially restricted photoswitching. For applications where it is important to selectively target a small and specific region, laser illumination is required. As with FRAP experiments, effective photoswitching can be achieved by spot irradiation, which may be beneficial for applications examining fast dynamics (Chudakov et al., 2007), or by using a defined ROI. For experiments on presynaptic terminals, efficient switching of a ~1 μm synapse can be readily performed without significant fluorescence changes occurring in neighboring terminals located a few microns away (Figure 11.3C and D). During the bleach phase, it is recommended that pixel dwell time is increased to ensure effective photoswitching (Chudakov et al., 2007). Also, note that typically there is no critical requirement for high power laser illumination to achieve photoswitching (Staras et al., 2010); in fact, excessive illumination has been reported to photobleach both protein forms and cause problems related to phototoxicity (Chudakov et al., 2007). Concerns about unintentional image-related photoswitching also extend into the post-photoswitch period. In general, the experimental design should limit image acquisition to the slowest acquisition rate that can effectively reveal the dynamic process under study. Acquisition at higher rates could lead to image acquisition-related photoswitching, contaminating the signal with unintentionally switched fluorescence. This potential problem can be monitored by assessing the extent of photoswitching at sites within the imaging field of view but remote from the target region, for example, on a different axon (Figure 11.3G) (Staras et al., 2010). A lack of red fluorescence at these sites suggests that nonspecific fluorescence is unlikely to be a contaminating factor.

Linking Fluorescence and Ultrastructure: Correlative Approaches for Assaying Presynaptic Function

As outlined above, fluorescence approaches can be highly informative for reading out dynamic properties of target structures. Nonetheless, the information they yield is ultimately limited by the spatial resolution of light microscopy. As such, detailed information on structural specializations that might underlie or heavily influence dynamic events remains unavailable. Diffraction-unlimited fluorescence microscopes provide powerful new alternative strategies for detailing subcellular events (Huang et al., 2009), but still fall short of providing a full structural description of the target subcellular environment. This resolution still remains the preserve of electron microscopes, and in recent years considerable investment has gone into the development of strategies to "link" light- and EM-based imaging (Stell, 1964; Blackstad, 1965; reviewed in Modla and Czymmek, 2011). The correlative approach typically exploits a visible light or fluorescence signal that can be traceable, often via some kind of conversion process, into EM. Primary signals include histochemical or immunocytochemical markers, genetically encoded constructs, and various kinds of tracers or dye (Modla and Czymmek, 2011). Alongside a careful procedure to trace the target area down to ultrastructural detail, these markers can be used to produce a correlated readout of the fluorescence and EM signals so that high-resolution studies can be performed. This is a broad and well-developed field that has been extensively reviewed elsewhere (Leapman, 2004; Huang et al., 2010; Modla and Czymmek, 2011). One particularly appealing aspect of this approach is that if the fluorescence-based marker provides any functional information, and a trace of this marker is preserved into EM, then nanoscale readouts of functional assays become attainable. Here, continuing with the theme of presynaptic function, we specifically consider approaches that relate functional and dynamic properties of vesicle pools in hippocampal synapses and combine FM-dye and ultrastructural information arising from correlative EM. We first outline the general steps for conversion of FM-dye signal to an electron-dense product and to correlate with structures observed in light microscopy with those in EM. Next, we discuss examples of structure–function relationships attainable using this method. Finally, we consider how dynamic readouts of synaptic function attained at fluorescence microscopy level, such as FRAP, can also be assayed at an ultrastructural level through correlative approaches.

Photoconversion and Correlative Microscopy Using FM1-43

Presynaptic terminals are highly appealing subjects for correlative ultrastructural analysis. The major reason is that they have complex organization, which pertains directly to functional synaptic performance, but many of these parameters are not directly resolvable by conventional light microscopy. These include, for example, the number of synaptic vesicles, the number of docked vesicles, the number of release sites, the size of the active zone, and the size of the postsynaptic density. By performing a postfix ultrastructural assessment, the magnitude of these parameters can be determined and set in the context of functional data gained at a live fluorescence level, offering a powerful approach to characterize structure–function relationships.

As described earlier, the family of FM-styryl-dyes (including spectral variants such as FM1-43, FM2-10, FM4-64, and FM5-95) has proved to be a highly informative optical tool for readout of synaptic function (Betz and Bewick, 1992; Ryan et al., 1993; Klingauf et al., 1998). An added advantage of FM1-43 is that it is an effective agent for driving the photo-oxidation of diaminobenzidine (DAB) to yield an electron-dense product (Henkel et al., 1996a; Harata et al., 2001; Schikorski and Stevens, 2001; de Lange et al., 2003; Rea et al., 2004; Rizzoli and

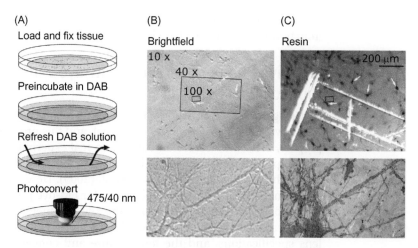

Figure 11.4 Steps in photoconversion and correlative EM for FM-dye-labeled hippocampal-cultured neurons. (A) Diagram illustrating basic steps in DAB photoconversion. Synapses are labeled with FM-dye and fixed. The sample is subsequently preincubated in DAB solution and then refreshed prior to photoconversion. The photoconversion process proceeds with illumination of the target region using the appropriate wavelength of light at the excitation peak for the fluorophore. (B) Illustrates correlative process. Top, at Brightfield level, multiple images of the target region at different wavelengths are taken to generate a montage. Bottom, detail of 100× field of view in top. (C) Top, equivalent area to (B), top after sample is embedded in resin. Bottom, equivalent to (B), bottom after thin sectioning and viewed in EM. Again, landmarks can be used to identify structures of interest. (B and C are adapted from Darcy et al., 2006a,b.) Please see color plates at the back of the book.

Betz, 2004; Darcy et al., 2006b; Branco et al., 2008, 2010; Welzel et al., 2011). In this way, translation of fluorescence-based information to the electron microscope becomes readily achievable and, specifically, provides a direct high-resolution readout of functional vesicle pools in individual synapses. The basic FM1-43 photoconversion methodology was first described in elegant work by the Tsien lab (Harata et al., 2001) in hippocampal cultures and the Betz lab in frog NMJ (Henkel et al., 1996a). Similar studies have since been performed in a variety of other systems including calyx of Held terminals (de Lange et al., 2003), neuroendocrine cells (Bauer et al., 2004), dorsal root ganglion neurons (Zhang et al., 2007b), cone photoreceptors (Rea et al., 2004), and *Drosophila* NMJ (Akbergenova and Bykhovskaia, 2010). The general approach (see Harata et al., 2001; Darcy et al., 2006a; Opazo and Rizzoli, 2010) involves FM-dye loading and tissue fixation, followed by photoillumination with blue light for ~10–50 minutes (see next section for further discussion) in the presence of extracellular DAB (1 mg/ml in phosphate-buffered saline; Figure 11.4A). This process drives the production of oxygen radicals leading to the photo-oxidation of DAB to form a dark insoluble precipitate (Meisslitzer-Ruppitsch et al., 2009). After processing, ultrastructural visualization reveals photoconverted vesicles as those with dark lumen, which are readily distinguishable from nonphotoconverted vesicles (Harata et al., 2001).

Effective Photoconversion of FM-dye with DAB in Hippocampal Cultures

Using the correct procedure for photoconversion of FM-dye is critical for yielding an appropriate EM signal, and this process should be properly calibrated before embarking on real experiments. There are several parameters that need to be considered, and they are reviewed at length by Meisslitzer-Ruppitsch et al. (2009). Here we briefly outline some of the specific considerations for FM-dye photoconversion in hippocampal cultured neurons. The key parameters of photoconversion are the intensity and duration of illumination, the objective lens specifications, and the temperature and concentration of dissolved oxygen in the extracellular solution. In the case of cultured hippocampal neurons, which effectively grow in a monolayer, concerns about tissue thickness are not a key consideration, but must be accounted for in thicker preparations, such as tissue slices (de Lange et al., 2003; Rea et al., 2004). Of course, as earlier, setting parameters for effective photoconversion must be considered in relation to the general requirement to keep the period of photo-illumination relatively short to prevent unwanted tissue photodamage.

For fixation, a number of points should be considered. The FM-dye should be of a fixable variety. There are a number of commercially available forms (e.g., FM1-43FX; Molecular Probes and AM1-43; Biotium) that include an aliphatic amine group permitting effective cross-linking when chemically fixed. The two main goals in fixing tissue are that it should be fast and result in excellent ultrastructural preservation. For hippocampal cultures we typically use 2% paraformaldehyde:2% glutaraldehyde. For thick tissue, the experimenter should consider microwave fixation, which can accelerate fixation times and improve fixation penetration (Jensen and Harris, 1989). To prevent the possibility of driving unwanted DAB reactions with glutaraldehyde autofluorescence, the sample is then washed in glycine and ammonium chloride, which acts to neutralize the fixative solution. Prior to photoconversion, we preincubate the sample for 10 minutes in DAB (1 mg/ml in phosphate-buffered saline) and then replace with fresh DAB before illuminating the sample. For cultures, we photoconvert with a 40× 0.8 NA objective on an upright Olympus BX51WI microscope using 470 nm light at 100% intensity using a 100 W mercury burner. Photoconversion, which results in the formation of a dark, electron-dense precipitate, is achieved in 12–14 minutes using this system, without the need for bubbling oxygen. Progression of the photoconversion reaction can be followed by monitoring the intensity of the DAB reaction under brightfield illumination as outlined previously (Harata et al., 2001).

Correlative Approach

In experiments aimed at relating light microscopy and EM information, the ideal approach is to employ a correlative strategy where specific fluorescence structures can be directly examined at the ultrastructural level. In this way, functional information gleaned from imaging assays can be specifically linked to underlying structural organization. There are many possible ways to achieve this objective (discussed in Modla and Czymmek, 2011), but for our work in dissociated cultures, a "roadmap" approach works well (Figure 11.4B). A major reason for this is that neuronal cultures grow to form individual 2D structures with uniquely identifiable landmarks. By taking a range of low and higher magnification differential interference contrast images combined with corresponding fluorescence images, a montage map that relates a target structure to other landmarks can be readily produced (Figure 11.4B). This map then provides the reference point for all subsequent stages where the target region needs to be re-identified. Initially, this is used in the postembedding stage when the region of interest needs to be marked for sectioning in the resin-encapsulated sample (Figure 11.4C). Subsequently it becomes important during ultrastructural visualization to identify specific target structures. In this way, in principle, any discrete structure appearing in a fluorescence image should be readily amenable to re-finding in electron micrographs.

Structure–Function Relationships of Vesicle Pools in Hippocampal Synapses

The family of FM-dyes and genetically encoded reporters that target vesicular components has proved very informative, for example, in defining and characterizing functional vesicle pools and subpools and detailing their role in synaptic transmission. In parallel, synapses have also been heavily studied at an ultrastructural level with detailed characterization of anatomical features of synaptic structures, including extensive organizational descriptions of the overall vesicle cluster. An approach combining both methodologies would thus allow the ultrastructural characterization of the functional subpools in the same way. The combination of FM-dyes and DAB photoconversion reaction has proved very informative for this purpose: researchers have detailed the characteristics of various subpools of presynaptic terminals in a range of systems. In particular, when combined with complete serial section EM, it has become possible to

Figure 11.5 Visualization of FM-dye labeling in electron micrographs. (A, B) Brightfield image and equivalent fluorescence image. (C) Overlay showing detail of rectangular region. (D, E) Equivalent region in (C) after serial sectioning. Fluorescence regions correspond to synaptic structures. (F–H). Detail on synapses from (E). Recycling vesicles appear with dark lumen after DAB photoconversion reaction. (I) Consecutive section series for synapse from (H). (J, K) Ultrastructural measurement of release probability. Electron micrograph of a synapse that was FM-dye labeled with 20 APs at 1 Hz prior to fixation and DAB photoconversion. The number of photoconverted vesicles correlates with synaptic release probability. (K) Three-dimensional reconstruction of synapse from (J). (A–E are adapted from Darcy et al., 2006a. F–I are adapted from Darcy et al., 2006b. J and K are adapted from Branco et al., 2008, with permission.)

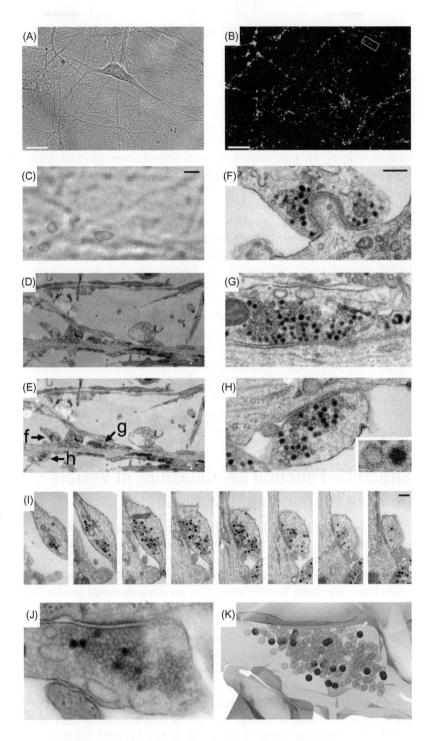

identify subpopulations in the context of the whole synaptic architecture, addressing questions about preferential organization and mixing of recycling vesicles (Schikorski and Stevens, 2001; Rizzoli and Betz, 2004; Darcy et al., 2006b; Branco et al., 2008; 2010; Staras and Branco, 2010). Figure 11.5 illustrates the approach and the nature of information available when studying the total recycling pool at hippocampal neurons: serial section EM reveals that after a saturating loading stimulus, vesicles of the recycling pool, visualized as those with dark lumen, are homogenously spatially distributed, implying that recently recycled vesicles are not compartmentalized within the pool as a whole (Figure 11.5F–I; Darcy et al., 2006b).

An extension of the same basic approach using fluorescence imaging, FM-dye photoconversion and correlative EM have been employed recently in our laboratory to provide an ultrastructural readout of a key parameter of synaptic performance, release probability (Figure 11.5J and K; Branco et al., 2008; 2010). The rationale for this is based on the assumption that, for a given synaptic terminal, triggering a small defined number of action potentials leads to a level of vesicle recycling that is directly related to the probability of vesicle fusion upon stimulation. This assumes a tight exocytic–endocytic coupling (Fernandez-Alfonso and Ryan, 2006) and relies on a limited stimulus-loading protocol (10–20 APs, 1 Hz) so that vesicle reuse is limited. As above, the level of vesicle turnover can be accurately assessed by performing the stimulation protocol in the presence of FM-dye and then fixing, photoconverting, and serially sectioning for EM. In this way, the number of photoconverted vesicles observed in a presynaptic terminal provides a direct readout of release probability (pr=number of photoconverted vesicles/number of APs applied) (Figure 11.5J and K; Branco et al., 2008; 2010). This requires a full serial reconstruction since a count of the total number of photoconverted vesicles is directly related to pr. Also, controls should be performed to ensure that the stimulation protocol used for loading is effective in evoking the correct number of action potentials. With these controls established, pr measurements can then be made in synapses where detailed ultrastructural information is also available (Figure 11.5J and K) (Branco et al., 2008; 2010).

Combining FRAP with Correlative Electron Microscope

FRAP provides a simple and informative approach for quantifying dynamic events occurring in populations of fluorescent structures. However, one important limitation is that an understanding of the recovered signal is ultimately restricted by the resolution limits of conventional microscopy. As such, nanoscale detail on the nature of the FRAP signal remains unknown. Moreover, information that might explain characteristics of the recovery process, such as structural specializations within the photobleached area, is also unavailable. One strategy to tackle these limitations is to combine a FRAP experiment with an ultrastructural investigation of the same region.

This type of approach is relatively straightforward, but places a number of specific demands on the experimental protocol. First, a correlative approach must be achievable so that the target photobleached region can be definitively re-identified at an ultrastructural level. Second, the fluorescence signal used to perform the FRAP experiment must be convertible into a form that can be readily visualized in ultrastructure. Third, the fluorescence signal must be sufficiently photobleached at the level of light microscopy so that it does not give rise to a fluorescence-associated product in EM. This latter point is absolutely critical since a central assumption of the experimental approach is that fluorescence-associated signal present at a previously photobleached site is a consequence of dynamic events and not simply residual signal arising from insufficient photobleaching (Darcy et al., 2006a,b; Staras et al., 2010). With this requirement comes a further demand: that the high level of photobleaching does not give rise to phototoxic effects. As discussed previously, this must be ascertained with appropriate tests of physiological function and the absence of ultrastructural damage (Darcy et al., 2006a,b).

Here we outline a FRAP-correlative electron microscope (CLEM) protocol aimed at uncovering ultrastructural detail on synaptic vesicle mobility in hippocampal cultures, combining FM-dye loading, synapse-specific photobleaching, DAB photoconversion, and CLEM. The main methods have been described individually earlier, but here we detail the additional steps for this experiment and provide suggestions on appropriate controls. A step-by-step description is given in (Darcy et al., 2006a). We also outline a variation on the conventional FRAP-CLEM approach — "inverse"-FRAP (Staras et al., 2010) — which permits the specific characterization of vesicle migration from a target synapse to its surroundings.

FRAP-CLEM for Detailing Synaptic Vesicle Dynamics

FRAP-CLEM is an ideal approach for gaining detailed insight into synaptic vesicle dynamics. Presynaptic terminals have a complex ultrastructure that is not directly resolvable by conventional light microscopy, and yet parameters of this structural organization are often intimately related to aspects of functional performance. As mentioned previously, FM-dye labeling offers a powerful and convenient tool for assaying recycling vesicle pools and when combined with DAB-based photoconversion and correlative EM, this functional

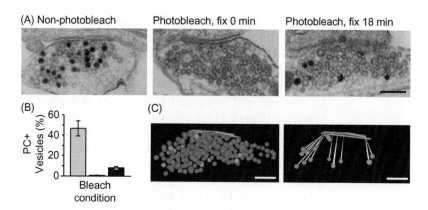

Figure 11.6 Readout of FRAP at the ultrastructural level — FRAP-CLEM. (A) Sample electron micrographs for synapses from samples labeled with FM-dye and treated in three different ways prior to fixing, photoconverting, serially sectioning, and viewing in EM. Left, not treated; middle, photobleached and immediately fixed; and right, photobleached and left for recovery period prior to fixing. Photoconverted vesicles have dark lumen. (B) Histogram summary of fraction of photoconverted vesicles in synapses for each condition from (A). (C) Three-dimensional reconstruction of synapse photobleached, fixed, and left for recovery period (right, A): the equivalent of the bleach-recovery signal shown in Figure 11.2E. Recycling vesicles are orange and nonrecycling vesicles appear blue. Vesicles are shown relative to the active zone (green). (A–C are adapted from Darcy et al., 2006a,b.) Please see color plates at the back of the book.

assay can be read out with ultrastructural detail. Adding a FRAP step to this protocol provides the opportunity to gain high-resolution dynamic information related to functional vesicle trafficking.

We used this approach to examine the turnover of synaptic vesicles at presynaptic terminal in the hippocampus, building on our observations from fluorescence-based experiments showing that vesicles are highly motile along axons between individual terminals (Darcy et al., 2006a,b). In short, the method begins with a conventional FRAP experiment in which a target terminal selected from a population of synapses loaded with FM-dye is individually photobleached and its fluorescence recovery monitored over time (Figure 11.6A). To gain an ultrastructural perspective, the preparation is then rapidly fixed so that procedures for photoconversion and EM preparation can be carried out. This technique utilizes many of the steps outlined in previous sections, but there are a number of additional concerns to consider. A key point is that the photobleaching step must completely eradicate the fluorescence signal so that none is present to drive the formation of a photoconverted product. If photobleaching is not fully effective, then the experiment is invalidated since it cannot be assumed that, at EM level, fluorescence-related photoconverted product corresponds to recovered fluorescence arising from trafficking events rather than residual fluorescence associated with incomplete photobleaching (Darcy et al., 2006a,b). It is critical, therefore, that appropriate controls are performed to establish the completeness of photobleaching: for example, an experiment in which recycling vesicles at a target synapse are FM-dye labeled and photobleached and then immediately fixed (Figure 11.6B and C). After photoconversion and preparation for EM, the number of photoconverted vesicles under these conditions should be counted and compared to synapses at loaded but unbleached synapses (Figure 11.6B and C). The expectation is that they should approach zero (Figure 11.6B and C). Again, it is imperative to ensure that the bleach protocol does not impair normal synaptic function. As discussed previously, one approach is to test the capability for subsequent rounds of FM-dye recycling after bleaching at a synapse.

Once it can be established that a fluorescence bleach protocol translates to an absence of photoconverted EM product, a FRAP experiment can proceed as normal. A target region is bleached followed by a period of recovery and then fixed, photoconverted, and prepared for ultrastructure. Recovered fluorescence signal at the light microscope level can then be read out with ultrastructural detail (Figure 11.6B). In the case of synaptic vesicles, important information about the relationship between vesicles and other parameters of the synapses, such as active zone size, docked vesicle number, and total pool size, becomes directly amenable to analysis (Darcy et al., 2006a,b). Also, with serial sectioning, 3D information about the FRAP signal can be established; for example, the spatial organization of photoconverted vesicles within the synaptic architecture (Figure 11.6D). In our laboratory, this approach was used to show that laterally mobile vesicles became newly incorporated into a target synapse over a short time period (~18 minutes). Reconstruction of serial sections allowed 3D organization of these non-native vesicles to be determined (Figure 11.6D).

Inverse FRAP-CLEM

A limitation of the basic FRAP approach is that it reports only the incoming fluorescence signal to a target site. For some experimental questions, information about the mobilization of fluorescence arising from a single source point is also important. To address this kind of issue, a version of the basic FRAP protocol can be utilized where the bleach protocol is directed at all fluorescent structures flanking a single target site, the fluorescence of which is preserved (Figure 11.7; Dundr et al., 2002; Lippincott-Schwartz et al., 2003). In this way, mobile elements arising from that target structure should be observed as the recovery of fluorescence at flanking sites. Used in conjunction with fluorescence photoconversion and correlative EM, high-resolution information can be collected. As is the case for the conventional FRAP approach, a critical requirement is to ensure that photobleaching is sufficient to remove fluorescence-related photoconversion product.

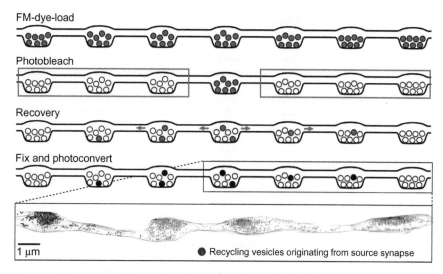

FM-dye-load

Photobleach

Recovery

Fix and photoconvert

1 µm   ● Recycling vesicles originating from source synapse

Figure 11.7 "Inverse" FRAP-CLEM for pursuing fates of mobile vesicles in hippocampal axons. Diagram illustrates the principle of the experiment. FM-dye-labeled synapses flanking a target terminal (first diagram) are irreversibly photobleached (second diagram). Recovery of fluorescence (third diagram) at bleached sites occurs as vesicles traffic from target terminal. (fourth diagram) preparation is fixed, photoconverted, and prepared for EM. Photoconverted signal appears electron dense. Bottom, sample reconstruction from a real experiment showing a segment of axon indicated with rectangles. The "source" synapse contains many photoconverted vesicles. Neighboring terminals show progressively fewer photoconverted vesicles, illustrating that movement of vesicles originates from source synapse. (Bottom diagram is adapted from Staras et al., 2010, with permission.) Please see color plates at the back of the book.

We have successfully used this kind of approach to examine the number of vesicles contributed by a single synaptic terminal to the synaptic neighborhood over time (Figure 11.7; Staras et al., 2010).

Concluding Remarks

This chapter outlines accessible approaches for assaying dynamic properties of neurons utilizing conventional fluorescence imaging methods and correlative EM. While the individual techniques are relatively simple to employ, when used together they provide high-resolution insights into structure–function relationships. Importantly, these powerful approaches will continue to be relevant as the armory of fluorescence-based probes expands further, offering nanoscale information on new aspects of neuronal function.

Acknowledgments

This work was supported by Wellcome Trust (WT084357MF) and BBSRC (BB/F018371) grants to K.S.

References

Ahmari, S.E., Buchanan, J., Smith, S.J., 2000. Assembly of presynaptic active zones from cytoplasmic transport packets. Nat Neurosci 3, 445–451.

Akbergenova, Y., Bykhovskaia, M., 2010. Synapsin regulates vesicle organization and activity-dependent recycling at drosophila motor boutons. Neurosci 170, 441–452.

Ando, R., Mizuno, H., Miyawaki, A., 2004. Regulated fast nucleocytoplasmic shuttling observed by reversible protein highlighting. Science 306, 1370–1373.

Axelrod, D., Koppel, D.E., Schlessinger, J., Elson, E., Webb, W.W., 1976. Mobility measurement by analysis of fluorescence photobleaching recovery kinetics. Biophys J 16, 1055–1069.

Bauer, R.A., Overlease, R.L., Lieber, J.L., Angleson, J.K., 2004. Retention and stimulus-dependent recycling of dense core vesicle content in neuroendocrine cells. J Cell Sci 117, 2193–2202.

Betz, W.J., Bewick, G.S., 1992. Optical analysis of synaptic vesicle recycling at the frog neuromuscular junction. Science 255, 200–203.

Blackstad, T.W., 1965. Mapping of experimental axon degeneration by electron microscopy of Golgi preparations. Z Zellforsch Mikrosk Anat 67, 819–834.

Branco, T., Staras, K., 2009. The probability of neurotransmitter release: variability and feedback control at single synapses. Nat Rev Neurosci 10, 373–383.

Branco, T., Marra, V., Staras, K., 2010. Examining size-strength relationships at hippocampal synapses using an ultrastructural measurement of synaptic release probability. J Struct Biol 172, 203–210.

Branco, T., Staras, K., Darcy, K.J., Goda, Y., 2008. Local dendritic activity sets release probability at hippocampal synapses. Neuron 59, 475–485.

Chapman, S., Oparka, K.J., Roberts, A.G., 2005. New tools for in vivo fluorescence tagging. Curr Opin Plant Biol 8, 565–573.

Chudakov, D.M., Lukyanov, S., Lukyanov, K.A., 2007. Tracking intracellular protein movements using photoswitchable fluorescent proteins PS-CFP2 and Dendra2. Nat Protoc 2, 2024–2032.

Chudakov, D.M., Verkhusha, V.V., Staroverov, D.B., Souslova, E.A., Lukyanov, S., Lukyanov, K.A., 2004. Photoswitchable cyan fluorescent protein for protein tracking. Nat Biotechnol 22, 1435–1439.

Darcy, K.J., Staras, K., Collinson, L.M., Goda, Y., 2006a. An ultrastructural readout of fluorescence recovery after photobleaching using correlative light and electron microscopy. Nat Protoc 1, 988–994.

Darcy, K.J., Staras, K., Collinson, L.M., Goda, Y., 2006b. Constitutive sharing of recycling synaptic vesicles between presynaptic boutons. Nat Neurosci 9, 315–321.

de Lange, R.P., de Roos, A.D., Borst, J.G., 2003. Two modes of vesicle recycling in the rat calyx of held. J Neurosci 23, 10164–10173.

Deerinck, T.J., Martone, M.E., Lev-Ram, V., Green, D.P., Tsien, R.Y., Spector, D.L., et al., 1994. Fluorescence photooxidation with eosin: a method for high resolution immunolocalization and in situ hybridization detection for light and electron microscopy. J Cell Biol 126, 901–910.

Dundr, M., Hoffmann-Rohrer, U., Hu, Q., Grummt, I., Rothblum, L.I., Phair, R.D., et al., 2002. A kinetic framework for a mammalian RNA polymerase in vivo. Science 298, 1623–1626.

Elson, E.L., Schlessinger, J., Koppel, D.E., Axelrod, D., Webb, W.W., 1976. Measurement of lateral transport on cell surfaces. Prog Clin Biol Res 9, 137–147.

Fernandez-Alfonso, T., Ryan, T.A., 2006. The efficiency of the synaptic vesicle cycle at central nervous system synapses. Trends Cell Biol 16, 413–420.

Gaffield, M.A., Betz, W.J., 2007. Synaptic vesicle mobility in mouse motor nerve terminals with and without synapsin. J Neurosci 27, 13691–13700.

Gaffield, M.A., Rizzoli, S.O., Betz, W.J., 2006. Mobility of synaptic vesicles in different pools in resting and stimulated frog motor nerve terminals. Neuron 51, 317–325.

Granseth, B., Odermatt, B., Royle, S.J., Lagnado, L., 2006. Clathrin-mediated endocytosis is the dominant mechanism of vesicle retrieval at hippocampal synapses. Neuron 51, 773–786.

Gurskaya, N.G., Verkhusha, V.V., Shcheglov, A.S., Staroverov, D.B., Chepurnykh, T.V., Fradkov, A.F., et al., 2006. Engineering of a monomeric green-to-red photoactivatable fluorescent protein induced by blue light. Nat Biotechnol 24, 461–465.

Harata, N., Ryan, T.A., Smith, S.J., Buchanan, J., Tsien, R.W., 2001. Visualizing recycling synaptic vesicles in hippocampal neurons by FM 1-43 photoconversion. Proc Natl Acad Sci USA 98, 12748–12753.

Henderson, J.N., Ai, H.W., Campbell, R.E., Remington, S.J., 2007. Structural basis for reversible photobleaching of a green fluorescent protein homologue. Proc Natl Acad Sci USA 104, 6672–6677.

Henkel, A.W., Lubke, J., Betz, W.J., 1996a. FM1-43 dye ultrastructural localization in and release from frog motor nerve terminals. Proc Natl Acad Sci USA 93, 1918–1923.

Henkel, A.W., Simpson, L.L., Ridge, R.M., Betz, W.J., 1996b. Synaptic vesicle movements monitored by fluorescence recovery after photobleaching in nerve terminals stained with FM1-43. J Neurosci 16, 3960–3967.

Holt, M., Cooke, A., Neef, A., Lagnado, L., 2004. High mobility of vesicles supports continuous exocytosis at a ribbon synapse. Curr Biol 14, 173–183.

Hua, Y., Sinha, R., Martineau, M., Kahms, M., Klingauf, J., 2010. A common origin of synaptic vesicles undergoing evoked and spontaneous fusion. Nat Neurosci 13, 1451–1453.

Huang, B., Babcock, H., Zhuang, X., 2010. Breaking the diffraction barrier: super-resolution imaging of cells. Cell 143, 1047–1058.

Huang, B., Bates, M., Zhuang, X., 2009. Super-resolution fluorescence microscopy. Annu Rev Biochem 78, 993–1016.

Jacobson, K., Derzko, Z., Wu, E.S., Hou, Y., Poste, G., 1976. Measurement of the lateral mobility of cell surface components in single, living cells by fluorescence recovery after photobleaching. J Supramol Struct 5, 565(417)–576(428).

Jensen, F.E., Harris, K.M., 1989. Preservation of neuronal ultrastructure in hippocampal slices using rapid microwave-enhanced fixation. J Neurosci Methods 29, 217–230.

Jordan, R., Lemke, E.A., Klingauf, J., 2005. Visualization of synaptic vesicle movement in intact synaptic boutons using fluorescence fluctuation spectroscopy. Biophys J 89, 2091–2102.

Kamin, D., Lauterbach, M.A., Westphal, V., Keller, J., Schonle, A., Hell, S.W., et al., 2010. High- and low-mobility stages in the synaptic vesicle cycle. Biophys J 99, 675–684.

Kay, A.R., Alfonso, A., Alford, S., Cline, H.T., Holgado, A.M., Sakmann, B., et al., 1999. Imaging synaptic activity in intact brain and slices with FM1-43 in C. elegans, lamprey, and rat. Neuron 24, 809–817.

Klingauf, J., Kavalali, E.T., Tsien, R.W., 1998. Kinetics and regulation of fast endocytosis at hippocampal synapses. Nature 394, 581–585.

Kraszewski, K., Daniell, L., Mundigl, O., De Camilli, P., 1996. Mobility of synaptic vesicles in nerve endings monitored by recovery from photobleaching of synaptic vesicle-associated fluorescence. J Neurosci 16, 5905–5913.

Kraszewski, K., Mundigl, O., Daniell, L., Verderio, C., Matteoli, M., De Camilli, P., 1995. Synaptic vesicle dynamics in living cultured hippocampal neurons visualized with CY3-conjugated antibodies directed against the lumenal domain of synaptotagmin. J Neurosci 15, 4328–4342.

Leapman, R.D., 2004. Novel techniques in electron microscopy. Curr Opin Neurobiol 14, 591–598.

Lemke, E.A., Klingauf, J., 2005. Single synaptic vesicle tracking in individual hippocampal boutons at rest and during synaptic activity. J Neurosci 25, 11034–11044.

Li, Z., Murthy, V.N., 2001. Visualizing postendocytic traffic of synaptic vesicles at hippocampal synapses. Neuron 31, 593–605.

Lippincott-Schwartz, J., Patterson, G.H., 2003. Development and use of fluorescent protein markers in living cells. Science 300, 87–91.

Lippincott-Schwartz, J., Patterson, G.H., 2009. Photoactivatable fluorescent proteins for diffraction-limited and super-resolution imaging. Trends Cell Biol 19, 555–565.

Lippincott-Schwartz, J., Altan-Bonnet, N., Patterson, G.H., 2003. Photobleaching and photoactivation: following protein dynamics in living cells. Nat Cell Biol Suppl., S7–S14.

LoGiudice, L., Sterling, P., Matthews, G., 2008. Mobility and turnover of vesicles at the synaptic ribbon. J Neurosci 28, 3150–3158.

Meisslitzer-Ruppitsch, C., Rohrl, C., Neumuller, J., Pavelka, M., Ellinger, A., 2009. Photooxidation technology for correlated light and electron microscopy. J Microsc 235, 322–335.

Modla, S., Czymmek, K.J., 2011. Correlative microscopy: A powerful tool for exploring neurological cells and tissues. Micron 42, 773–792.

Murthy, V.N., De Camilli, P., 2003. Cell biology of the presynaptic terminal. Annu Rev Neurosci 26, 701–728.

Opazo, F., Rizzoli, S.O., 2010. Studying synaptic vesicle pools using photoconversion of styryl dyes. J Vis Exp 15 pii. doi: 10.3791/1790.

Patterson, G.H., Lippincott-Schwartz, J., 2002. A photoactivatable GFP for selective photolabeling of proteins and cells. Science 297, 1873–1877.

Rea, R., Li, J., Dharia, A., Levitan, E.S., Sterling, P., Kramer, R.H., 2004. Streamlined synaptic vesicle cycle in cone photoreceptor terminals. Neuron 41, 755–766.

Rizzoli, S.O., Betz, W.J., 2004. The structural organization of the readily releasable pool of synaptic vesicles. Science 303, 2037–2039.

Rizzoli, S.O., Betz, W.J., 2005. Synaptic vesicle pools. Nat Rev Neurosci 6, 57–69.

Ryan, T.A., 2001. Presynaptic imaging techniques. Curr Opin Neurobiol 11, 544–549.

Ryan, T.A., Reuter, H., Wendland, B., Schweizer, F.E., Tsien, R.W., Smith, S.J., 1993. The kinetics of synaptic vesicle recycling measured at single presynaptic boutons. Neuron 11, 713–724.

Sankaranarayanan, S., Ryan, T.A., 2000. Real-time measurements of vesicle-SNARE recycling in synapses of the central nervous system. Nat Cell Biol 2, 197–204.

Schikorski, T., Stevens, C.F., 2001. Morphological correlates of functionally defined synaptic vesicle populations. Nat Neurosci 4, 391–395.

Schlessinger, J., Koppel, D.E., Axelrod, D., Jacobson, K., Webb, W.W., Elson, E.L., 1976. Lateral transport on cell membranes: mobility of concanavalin A receptors on myoblasts. Proc Natl Acad Sci USA 73, 2409–2413.

Shakiryanova, D., Tully, A., Levitan, E.S., 2006. Activity-dependent synaptic capture of transiting peptidergic vesicles. Nat Neurosci 9, 896–900.

Shtrahman, M., Yeung, C., Nauen, D.W., Bi, G.Q., Wu, X.L., 2005. Probing vesicle dynamics in single hippocampal synapses. Biophys J 89, 3615–3627.

Stan, A., Pielarski, K.N., Brigadski, T., Wittenmayer, N., Fedorchenko, O., Gohla, A., et al., 2010. Essential cooperation of N-cadherin and neuroligin-1 in the transsynaptic control of vesicle accumulation. Proc Natl Acad Sci USA 107, 11116–11121.

Staras, K., Branco, T., 2010. Sharing vesicles between central presynaptic terminals: implications for synaptic function. Front Synapt Neurosci 2, 20.

Staras, K., Branco, T., Burden, J.J., Pozo, K., Darcy, K.J., Marra, V., et al., 2010. A vesicle superpool spans multiple presynaptic terminals in hippocampal neurons. Neuron 66, 37–44.

Stell, W.K., 1964. Correlated light and electron microscope observations on Golgi preparations of goldfish retina. J Cell Biol 23, 89A.

Stiel, A.C., Trowitzsch, S., Weber, G., Andresen, M., Eggeling, C., Hell, S.W., et al., 2007. 1.8 A bright-state structure of the reversibly switchable fluorescent protein dronpa guides the generation of fast switching variants. Biochem J 402, 35–42.

Sudhof, T.C., 2004. The synaptic vesicle cycle. Annu Rev Neurosci 27, 509–547.

Takamori, S., Holt, M., Stenius, K., Lemke, E.A., Gronborg, M., Riedel, D., et al., 2006. Molecular anatomy of a trafficking organelle. Cell 127, 831–846.

Teng, H., Wilkinson, R.S., 2000. Clathrin-mediated endocytosis near active zones in snake motor boutons. J Neurosci 20, 7986–7993.

Tozer, J.T., Henderson, S.C., Sun, D., Colello, R.J., 2007. Photoconversion using confocal laser scanning microscopy: A new tool for the ultrastructural analysis of fluorescently labeled cellular elements. J Neurosci Methods 164, 240–246.

Welzel, O., Henkel, A.W., Stroebel, A.M., Jung, J., Tischbirek, C.H., Ebert, K., et al., 2011. Systematic heterogeneity of fractional vesicle pool sizes and release rates of hippocampal synapses. Biophys J 100, 593–601.

Westphal, V., Rizzoli, S.O., Lauterbach, M.A., Kamin, D., Jahn, R., Hell, S.W., 2008. Video-rate far-field optical nanoscopy dissects synaptic vesicle movement. Science 320, 246–249.

Wiedenmann, J., Ivanchenko, S., Oswald, F., Schmitt, F., Rocker, C., Salih, A., et al., 2004. EosFP, a fluorescent marker protein with UV-inducible green-to-red fluorescence conversion. Proc Natl Acad Sci USA 101, 15905–15910.

Willig, K.I., Rizzoli, S.O., Westphal, V., Jahn, R., Hell, S.W., 2006. STED microscopy reveals that synaptotagmin remains clustered after synaptic vesicle exocytosis. Nature 440, 935–939.

Zhang, Q., Cao, Y.Q., Tsien, R.W., 2007a. Quantum dots provide an optical signal specific to full collapse fusion of synaptic vesicles. Proc Natl Acad Sci USA 104, 17843–17848.

Zhang, Q., Li, Y., Tsien, R.W., 2009. The dynamic control of kiss-and-run and vesicular reuse probed with single nanoparticles. Science 323, 1448–1453.

Zhang, X., Chen, Y., Wang, C., Huang, L.Y., 2007b. Neuronal somatic ATP release triggers neuron-satellite glial cell communication in dorsal root ganglia. Proc Natl Acad Sci USA 104, 9864–9869.

INDEX

Note: Page numbers followed by "f" and "t" refer to figures and tables respectively.

2D line imaging and deblurring 9–11

2PE-4pi microscopy, optical configuration for 126–127, 127f

3D object recognition 20

3D reconstructed cell resemblance 24–25

4pi microscopy 55
  alignment
    of bottom lens 59–60
    coverslip correction ring, optimizing 59
    of excitation by eye 58–59
    of second mirror 60
  deconvolution 61–62
  DNA visualization 66–77
    microbubbles with medicine 75–77
    Muntjac chromosomes 67
    single-photon excitation 68–70
    synaptonemal complex protein 3 (SYCP3) 70–75
  future of 77
  imaging 60–61
  microtubule and microtubule plus end imaging 65–66
  principle 56–58
  sample preparation 62–65
    fixation 63
    fluorescent dyes, selection of 64–65
  setup 56–58
  types 58
  single photon excitation 64–65
  two-photon excitation 64, 126–127
β-tubulin Alexa Fluor 488 staining, 4pi image of 58f

Abbe's diffraction equation 2–11, 5–6t

Abbe's spatial resolution formula 37–38

Aberrations 101–102, 109–110

Actin-binding phallotoxin phalloidin 88–89

Adaptive optics (AO) 50, 101–102

Adult-born neuron 201
  future directions 214–216
  imaging newborn neurons in 210
  in vivo imaging setup and acquisition 213
  in vivo live animal imaging of adult neurogenesis 210–211
  in vivo window preparation 211–213
  labeling techniques 202–203
  post-acquisition image processing and analysis 213–214
  retroviral labeling of 206f
  retrovirus-mediated labeling of 203–205, 204t
  retrovirus production and delivery 207–209
  single-cell genetic manipulation in 205–207
  viral-labeled cell toxicity and physiological changes 209–210

Adult neurogenesis 201–203, 205–207, 214–216
  in hippocampus 215–216
  in vivo live animal imaging of 210–211

Alexa Fluor 488 18–19, 26, 64

Alexa Fluor 555 64

Apposition 26, 28–30

Atto647N 49

Avalanche photo diodes (APDs) 57–58

Axial resolution 11–12

Binding-activated localization microscopy (BALM) 83–84, 88–89

Biotinylated dextran amine (BDA) 26
  neuroanatomical tracing with 32f

Blinking dyes 88

"Blur" imaging 4–8

Borrelia 163, 173

Bottom objective lens, alignment of 59–60

Brain microcircuits, optical dissection of 112–116
  illuminating neurons with 3D light patterns 115–116
  light-sensitive proteins, activation of 114–115
  photostimulation of neuronal cells with complex light patterns 114

Brain networks, optical investigation of 101

Brain slices, preparation of 228

BrdU DNA analogs 204t

Brightfield imaging 204t

Bromodeoxyuridine (BrdU) 202–203

BX51, from Olympus 42

Caged compounds 111–112

Calcium imaging 215–216

CARS microscopy, *see* Coherent anti-Stokes Raman scattering (CARS) microscopy

Catastrophe 66

Caulobacter crescentus 163
Cellular landscapes at molecular
    resolution 141–143
Cellular-level optical biopsy using
    full-field optical coherence
    microscopy 185
  FF-OCM technique 187–189
    detection sensitivity 189–191
    for high-resolution "optical
      biopsy" 193–195
    sample motion
      artifacts 192–193
    spatial resolution 191–192
Central nervous system 248
  light-sensitive molecular tools
    and 111–112
Channelrhodopsin-2 (ChR2)
    114–115
Charge-coupled device (CCD)
    146–147
Chromatin organization, SYCP3
    Axis as a marker for 70–75
Chromeo494 49
Coherence gates 190f
Coherent anti-Stokes Raman
    scattering (CARS)
    microscopy
  demyelination and
    remyelination enabled by
    231–236, 239
    electrical stimulation 234–235
    experimental autoimmune
      encephalomyelitis (EAE)
      235–236
    glutamate toxicity 234
    lysophosphatidylcholine-
      induced demyelination
      231–234
  for *ex vivo/in vivo* myelin
    imaging 227–231
    CARS microscope 227–228
    images of myelin 229,
      230–231f
    sample preparation 228–229
    spinal cord imaging
      229–231, 232f
  myelin imaging for 238–239
  principle and history of 224–
    226, 225f

technical characteristics of
    226–227
Colocalization 30–32
Complementation-activated
    localization microscopy
    (CALM) 83–84
Computer software to define
    contact 29
Confocal imaging 204t, 210–211,
    213–214
Confocal laser scanning
    microscope (CLSM) 2
  2D line and deblurring, imaging
    9–11
  3D reconstructed cell
    resemblance 24–25
  Abbe's diffraction equation 2–9
  actual experiment 26–29
    computer software to define
      contact 29
  advantages of 4
  axial resolution 11–12
  biological objects translated to
    pixels 19–20
  colocalization 30–32
  conventional 2–4
  cross talk, elimination of 18–19
  cross talk awareness 16–18
    excitation cross talk 18
  diameter, determination of 20–24
    automated objective
      threshold analysis 22–24
    best-fit object 3D recognition
      20–22
  "essential" component of 14–16
  need for 4–8
  Nyquist criterion 8–9
  orders of magnitude 13–14
  pinhole, depth of focus, and
    laser illumination 2–4
  pinhole in 14–16
  principal components and light
    path of 3f
  resolution and sampling 12–13
  resolution limits 13–14
  Shannon' sampling theorem 8–9
  signal separation 13–14
  synaptic contacts 29–30
  touch 25–26

Confocal microscopy 50, 124, 210
Contrast transfer function (CTF) 146
Convallaria majalis 125–126
Correlative and direct
    identification approaches
  labeling strategies for 161–163
Correlative electron microscope
    (CLEM) and FRAP 265–269
  for detailing synaptic vesicle
    dynamics 266–268
  inverse 268–269
Correlative LM and cryo-ET
    156–161
Coverslip correction ring,
    optimizing 59
Cross talk
  awareness 16–18
  elimination of 18–19
  excitation of 18
Cryo-electron tomography (ET)
    method 141
  correlative and direct
    identification approaches
    161–163
  correlative LM and cryo-ET
    156–161
  cryo preparation 143–145
  functional manipulations in
    163–169
    distinct shape, structures of
      163–165
    presynaptic cytomatrix
      165–169
  image processing, uncovering
    information by 169–175
    CTF correction 169
    denoising 169–170
    segmentation 170–171
    subtomogram averaging
      171–173
    template matching 173–175
  imaging in electron microscopy
    (EM) 145–147
  resolution, noise, and radiation
    damage 149–151
  tomography 148–149
  visual identification and
    cryo-ET of intact cells
    151–156

cytoskeleton 154–155
fibroblasts 155–156
neurons 153–154
Plasmodium 151–153
Cryo-LM imaging 157–160
Cryo-tomograms 169, 172
low SNR of 150
CTF correction 169
Cyan fluorescent protein (CFP) 63
Cytoskeletal filaments 163
Cytoskeleton 154–155

DAPI staining 64, 70
Deconvolution methods 9–11,
61–62
Demyelination, visualization of
231–236, 238, 239
electrical stimulation 234–235
experimental autoimmune
encephalomyelitis (EAE)
235–236
glutamate toxicity 234
lysophosphatidylcholine-
induced demyelination
231–234
Dendra2 255–257
photoswitching 258–259
and tracking of synaptic vesicle
dynamics 257–258
Dendronephthya sp. 255–257
Denoising algorithms 169–170
Diameter, high-probability
determination of 20–24
automated objective threshold
analysis 22–24
best-fit object 3D recognition
20–22
Diaminobenzidine (DAB) 260–262
Dictyostelium nuclei 154, 172
Diffractive optical elements
(DOEs) 103–104, 109
diffraction barrier 36
Diffusion tensor imaging (DTI)
222–224, 237
Direct detector devices (DDD)
146–147
Direct stochastic optical
reconstruction microscopy
(dSTORM) 83–84, 88

Discrete Fourier transforms
(DFTs) 104–107
DMI 6000, from Leica 42
DNA-binding dyes 88–89
DNA visualization 66–77
microbubbles with medicine
75–77
Muntjac chromosomes, 4pi
imaging of 67
single-photon excitation 68–70
synaptonemal complex protein
3 (SYCP3) 70–75
Dopamine (DA) neurons, redox
state measurement in
68–70
Dreiklang fluorescent protein 89
DSPC 75–76
DSPE-PEG2000 75–76

Electron energy loss spectroscopy
(EELS) 162
Electron microscopy (EM) 37,
141–142, 222–224, 249
cryo-electron microscopy
143–151
imaging in 145–147
Electron radiation sensitivity of
biological samples 150
Emiliani, Valentina 115–116
Epi-detected CARS (E-CARS)
224–228
Epifluorescence signal 29–30, 44,
68, 266
Epifluorescent imaging 204t
Escherichia coli 161–165
Ethynyl deoxyuridine (EdU) 203
Eukaryotic flagella 172
Exocytic–endocytic
coupling 265
Exo-endocytic recycling 249–251
Experimental autoimmune
encephalomyelitis (EAE)
model 235–236

Fibroblasts 155–156
Filopodia 154
FITC 64
Fluorescence and ultrastructure,
linking 259–263

FM-dye with DAB in
hippocampal cultures
262–263
photoconversion and
correlative microscopy
using FM1-43 260–262
Fluorescence microscopy 36, 82
Fluorescence photoactivated
localization microscopy
(FPALM) 83–84, 87
Fluorescence recovery after
photobleaching (FRAP) 249
with correlative electron
microscope (CLEM)
265–269
and phototoxicity 254–255
presynaptic vesicle dynamics,
high-resolution approaches
to 247
to study intersynaptic vesicle
mobility 253–254
to study intrasynaptic vesicle
mobility 251–253
at ultrastructural level 267f
using photoswitchable
fluorophores, variations on
255–259
Fluorescent probes 89
Fluorescent proteins 89–90
for 4pi imaging 64
labeling techniques 90
organic fluorophores 90
PALM and FPALM 87
single-molecule localization-
based superresolution
microscopy 87–89
Fluorochromated streptavidin 26
FM-dye labeling, in electron
micrographs 264f
FM-styryl-dyes, family of 260–262
FRAP-CLEM for detailing synaptic
vesicle dynamics 266–268
Full-field optical coherence
microscopy (FF-OCM)
186–189
advantages of 192
cellular-level optical biopsy
using 185
detection sensitivity 189–191

Full-field optical coherence
microscopy (FF-OCM)
(*Continued*)
for high-resolution "optical
biopsy" 193–195
sample motion
artifacts 192–193
spatial resolution 191–192
Full-field optical coherence
tomography (FF-OCT)
186–187
Functional imaging 112–113,
115–116

Gaussian deconvolution
algorithm 61
Generalized phase contrast
(GPC) 108
Glutamate toxicity 234
"Gratings and lenses" algorithm
104–107
Green fluorescent protein (GFP)
37, 63, 82, 202, 222–224
Ground-state depletion and single
molecule return (GDSIM)
83–84, 88

Halorhodopsin (HR) 114–115
"Haze" imaging 4–8
Hegerl-Hoppe dose fractionation
theorem 148–149
Hela cells, microtubules
visualization in 65–66
High-resolution "optical biopsy,"
FF-OCM technique for
193–195
Hippocampal neurons 257–258,
262
Hippocampal synapses,
structure–function
relationships of vesicle
pools in 263–265
Hoechst 64
Human breast tissues 194–195
biological features of 195f
FF-OCM images of 194f, 195f
Huygens-Fresnel diffraction
theory 103–104
Hypertonic sucrose 167

Immunohistochemistry (IHC) 202
proliferation markers 204t
Imspector 46, 61
Inertia-free SLM-focus control
115–116
"Input-output" algorithms
104–107
Interference microscopy 186–187
Inverse FRAP-CLEM 268–269, 269f
In vivo imaging of myelin 236–238
animal preparation for 228–229
nuclear MRI 237
positron emission tomography
(PET) 238
third-harmonic generation
(THG) microscopy 236
two-photon excited
fluorescence (TPEF)
microscopy 236–237
"Iso-STED" 52
Iterative Fourier transform
algorithms 104–107

Labeled fraction 86–87
Labeling approaches, properties
of 85t
Labeling density 87
Labeling techniques 90
Laser illumination 2–4
Laser-scanning microscopy 39
Leica 4pi microscope, *see* 4pi
microscopy
Leica Microsystems 56
Light microscopy (LM) techniques
36, 141–142, 157–160
Light-sensitive molecular tools, for
CNS investigation 111–112
Light-sensitive proteins, activation
of 114–115
Linkage error 86
Linnik microscope 187–188
Liquid crystal 103–104
Live-cell imaging 51, 95f
LLTech 194–195
Localization-based techniques
129–131
Localization microscopy
matters of concern 85–87
labeled fraction 86–87

labeling density 87
linkage error 86
precision 85
specificity 86
multicolor 90–93
application of, in neurons
92–93, 94–95f
approaches to 3D PALM 92
single particle tracking 91–92
principle of 83–84, 84f
single-molecule localization-
based superresolution
microscopy 87–89
using a pair of interacting
cyanine dyes 88
using blinking dyes 88
using fluorescent proteins 87
using high-intensity light 88
using targeted molecules
88–89
Lysophosphatidylcholine (LPC)
231–234
Lysophosphatidylcholine-induced
demyelination 231–234

Magnetization transfer imaging
(MTI) 222–224, 237
Malaria 151–152
mClavGR 89
mEOS$_2$ 89
Meiosis 70
Metal-binding proteins 161–162
Metallothionein 161–162
Microbubbles with medicine
75–77
Microendoscope technique
214–215
Microprism implantation into
hippocampus 215
Microtubule images 65–66
MIrisFP 89
Molecular resolution, cellular
landscapes at 141–143
MRI 222–224, 237
Multicolor localization
microscopy 90–93
3D PALM, approaches to 92
application of, in neurons
92–93, 94–95f

single particle tracking 91–92
Multifluorescence imaging 2–4
Multiphoton excitation (MPE)
  microscopy 121–123
Multiphoton fluorescence
  microscopy,
  superresolution techniques
    for 126–134
  localization techniques
    129–131
  multiple excitation spot optical
    microscopy 127–129
  STED microscopy 131–134
  two-photon 4pi microscopy
    126–127
Multiphoton imaging 121
  point spread function
    (PSF) for single- and
    123–126
Multiple sclerosis 222–224
Muntiacus muntjak 67
Muntjac chromosomes, 4pi
  imaging of 67, 68f
Murine leukemia retroviruses
  (MLVs) 205
myelin imaging 227–231
  CARS microscope 227–228
  images of 229, 230–231f
  sample preparation 228–229
  spinal cord imaging 229–231,
    232f
Myelin sheaths study
  by CARS microscopy, *see*
    Coherent anti-Stokes
    Raman scattering (CARS)
    microscopy
  electron microscopy 222–224
  in vivo imaging methods
    236–238
    nuclear MRI 237
    positron emission
      tomography (PET) 238
    third-harmonic generation
      (THG) microscopy 236
    two-photon excited
      fluorescence (TPEF)
      microscopy 236–237
  traditional myelin imaging
    methods 221–224, 223f

Nano resolution optical imaging
  through localization
  microscopy 81
  application of, in neurons
    92–93, 94–95f
  fluorescent probes 89
  fluorescent proteins 89–90
  multicolor localization
    microscopy 90–93
    approaches to 3D PALM 92
    single particle tracking 91–92
  outlook
    biological questions 96
    technical development 93–95
  superresolution microscopy
    techniques 82–87
    concept behind localization
      microscopy 83–84
    labeled fraction 86–87
    labeling density 87
    linkage error 86
    localization precision 85
    single-molecule localization-
      based 87–89
    specificity 86
Neuromuscular junction 251–253
Neuronal morphology 48
Neuronal nuclei (NeuN) 202–203
Neurons 153–154
Newborn neurons in adult brain
  201
  future directions 214–216
  imaging of 210
  in vivo imaging setup and
    acquisition 213
  in vivo live animal imaging
    210–211
  in vivo window preparation
    211–213
  labeling techniques 202–203
  post-acquisition image
    processing and analysis
    213–214
  retrovirus-mediated labeling of
    203–205, 204t
  retrovirus production and
    delivery 207–209
  single-cell genetic manipulation
    205–207

viral-labeled cell toxicity and
  physiological changes
  209–210
Nipkow disk 2–4
Nodes of Ranvier 221–222, 229
Nonlinear anisotropic diffusion
  169–170
Nonlinear structured illumination
  microscopy 38
Nuclear MRI, for diagnosis of
  myelin-related diseases 237
Nyquist criterion 8–9

Okadaic acid 167
One-photon confocal imaging 204t
Optical coherence microscopy
  (OCM) 186–187, *see also*
  Full-field optical coherence
  microscopy (FF-OCM)
Optical coherence tomography
  (OCT) 186
Orders of magnitude 13–14
Organic fluorophores 90, 91t

PAmCherry 89
PEG-40 stearate 75–76
Phase contrast 146
Phase contrast filter (PCF) 108
Photo-activated localization
  microscopy (PALM) 38,
  83–84, 87
  3D PALM, approaches to 92
  live-cell imaging with 95f
Photoactivation, mechanisms of 91t
Photobleaching 254
Photoblinking 90
Photostimulation of neuronal cells
  114
Photoswitchable fluorophore
  to monitor mobile elements fate
  256f
Pico-green 88–89
Pinhole 2–4
Pixels, biological objects
  translated to 19–20
Plasmodium 151–153
Point accumulation imaging
  of nanoscale topology
  (PAINT) 83–84

Point spread function (PSF) 39, 56–57
  for single- and multiphoton imaging 123–126
Positron emission tomography (PET) 222–224, 238
Postsynaptic density 29–30
Presynaptic cytomatrix 165–169
Presynaptic terminals 165–167, 249–251, 260, 266–267
Presynaptic vesicle dynamics, high-resolution approaches to
  FRAP and electron microscopy, using variants of 247
  with correlative electron microscope (CLEM) 265–269
  and phototoxicity 254–255
  presynaptic vesicle dynamics, high-resolution approaches to 247
  to study intersynaptic vesicle mobility 253–254
  to study intrasynaptic vesicle mobility 251–253
  using photoswitchable fluorophores, variations on 255–259
Propidium iodide (PI)
  for 4pi imaging 63, 64
PSCFP2 89
PSOrange 89

Quenching effect 41

Radial diffraction
  simulation of 10f
  three-dimensional representation of 9f
Reactive oxygen species (ROS) production 68–69
Readily releasable pool (RRP) 167
Red blood cells (erythrocytes) 152–153
Redox immunohistochemistry (RHC) 68–69
Redox state of dopamine neurons, measuring 68–70

Refractive index, optimizing 62
Remyelination, visualization of 231–236, 238
Resolution
  axial 11–12
  checking 47
  limits 13–14
  noise, and radiation damage 149–151
  optical 110–111
  and sampling 12–13
  spatial 50–51
  super techniques 126–134
  temporal 47–48
Retroviral labeling and imaging of adult-born neuron 201, 206f, 210
  future directions 214–216
  in vivo imaging setup and acquisition 213
  in vivo live animal imaging of adult neurogenesis 210–211
  in vivo window preparation 211–213
  labeling techniques 202–203
  post-acquisition image processing and analysis 213–214
  retrovirus-mediated labeling of 203–205, 204t
  retrovirus production and delivery 207–209
  single-cell genetic manipulation in 205–207
  viral-labeled cell toxicity and physiological changes 209–210
Retrovirus-mediated expression 204t
Reversible saturable optical fluorescence transitions (RESOLFT) 52
Rhodamine (Rh) amide, photoconversion of 130f
RIM proteins 168–169

Sample preparation
  for 4pi microscopy 62–65
  fixation 63

  fluorescent dyes, selection of 64–65
Saxton scheme 148
Scanning systems, coupling SLM with 109–110
Segmentation 170–171
Sequential scanning 64–65, 112–113
Shannon' sampling theorem 8–9
"Sharp" imaging 4–8
Signal separation 13–14
Signal-to-noise ratio (SNR) 47, 50, 61, 101–102, 142–143, 146, 150–151
  in 4pi images 61
  quantitative images 122–123
Silicon spatial light modulators (SLMs) 102–103
  central nervous system investigation 111–112
  optical setup 104–111
    computational aspects 104–107
    coupling with scanning systems 109–110
    generalized phase contrast 108
    optical resolution 110–111
    temporal focusing 107
  for optical dissection of brain microcircuits 112–116
    activation of light-sensitive proteins expressed over large areas 114–115
    complex light patterns 114
    illuminating neurons with 3D light patterns 115–116
  structuring light by phase modulation 103–104
Single-molecule localization-based superresolution microscopy 87–89
  using a pair of interacting cyanine dyes 88
  using blinking dyes 88
  using fluorescent proteins 87
  using high-intensity light 88
  using targeted molecules 88–89
Single particle tracking 91–92

Single-photon excitation 68–70
   transition probability for
      138–140
SNARE complex-dependent
      tethers 168–169
Specificity 86
Spinal cord
   ex vivo stripes 234
   imaging 229–231, 232f, 223f
   in vivo CARS imaging of
      229–231
   samples, preparation of 228
Spiroplasma melliferum 173
SPO11 70
sptPALM 91–92, 93
Stimulated emission depletion
      (STED) microscopy 35,
      82–83, 122–123, 131–134
   basic concept 39–42
   depth penetration and spatial
      resolution 50
   implementation of 42–47
      doughnut 42–44
      emission detection 44
      laser sources 42
      microscope base 42
      objective lenses 44
      resolution checking 47
      scanning schemes 45–46
      software 46
      stability considerations 46
      synchronization 44
   labeling strategies 48
   live-cell imaging 51
   multicolor imaging 49–50
   spatial resolution in z 50–51
   temporal resolution and
      imaging speed 47–48
Stochastic optical reconstruction
      microscopy (STORM) 38,
      83–84, 88
Structured illumination
      microscopy (SIM) 38,
      82–83, 101–102
Structured light 101–103
   by phase modulation, using
      SLMs 103–104
Subgranular zone (SGZ) 202
Subtomogram averaging 171–173

Subventricular zone (SVZ) 202
Superresolution microscopy
      techniques 82–87, 122–123
   concept behind localization
      microscopy 83–84
   matters of concern 85–87
      labeled fraction 86–87
      labeling density 87
      linkage error 86
      localization precision 85
      specificity 86
   for multiphoton fluorescence
      microscopy 126–134
      multiphoton localization
         techniques 129–131
      multiphoton STED
         microscopy 131–134
      multiple excitation spot
         optical microscopy 127–129
      two-photon 4pi microscopy
         126–127
   single-molecule
         localization-based 87–89
      using a pair of interacting
         cyanine dyes 88
      using blinking dyes 88
      using fluorescent proteins 87
      using high-intensity light 88
      using targeted molecules
         88–89
Surface plasmon-based
      nanoimaging 122–123
Synapse biology 49
Synaptonemal complex protein 3
      (SYCP3) 70–75
Synaptic contacts 29–30
Synaptic vesicle clustering
      167–169
Synaptobrevin 167–168
Synaptonemal complex protein 3
      (SYCP3) staining 70–75
Synaptosomes 165–167
Synaptotagmin 160–161

Template matching 173–175
Temporal focusing 107
Third-harmonic generation (THG)
      microscopy 236
Tomography 148–149

TPEF microscopy 227–228
   and transgenic animal models
      236–237
Transgenic animal models
      236–237
Transgenic mice 204t
TRITC 64
Tritiated thymidine nucleotide
      202–203
Tubulin 65–66
Two-photon 4pi microscopy 68,
      126–127
Two-photon excitation (2PE)
      fluorescence microscopy
      50, 112–114, 122–123, 125
Two-photon excited fluorescence
      (TPEF) microscopy 236–237
Two-photon live animal imaging
      204t
Two-photon scanning
      microscope 114

Ultraviolet-visual spectroscopy
      (UV-VIS) range 121–122
Uncovering information by image
      processing 169–175
   CTF correction 169
   denoising 169–170
   segmentation 170–171
   subtomogram averaging 171–173
   template matching 173–175
Universal point accumulation
      imaging of nanoscale
      topology (uPAINT) 83–84

Vesicle pools in hippocampal
      synapses, structure-function
      relationships of 263–265
Vesicular glutamate transporter 2
      (VGluT2) 30–31
Vesicular stomatitis virus (VSV-G)
      207–209
Vitrification 143–144
Visual identification and cryo-ET
      of intact cells 151–156
   cytoskeleton 154–155
   fibroblasts 155–156
   neurons 153–154
   Plasmodium 151–153

Visualization of microtubules, in
    mammalian cells 65–66
Visual spectroscopy-infrared
    (VIS-IR) light 121–122

Wavefront engineering using SLMs
    104–111

computational aspects 104–107
coupling with scanning systems
    109–110
generalized phase contrast 108
optical configuration for 105f
optical resolution 110–111
temporal focusing 107

Yellow fluorescent protein (YFP) 63
YOYO-1 88–89

Z-stepper 58–59
Zebrafish model 239
Zernike, Frederik 108
Zero pinhole diameter 2–4

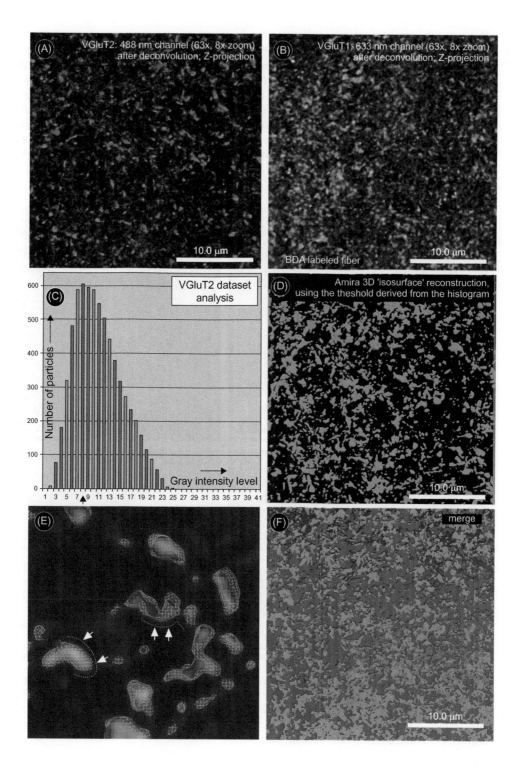

(A) VGluT2: 488 nm channel (63x, 8x zoom) after deconvolution; Z-projection

10.0 μm

(B) VGluT1: 633 nm channel (63x, 8x zoom) after deconvolution; Z-projection

BDA labeled fiber

10.0 μm

(C) VGluT2 dataset analysis

Number of particles

Gray intensity level

(D) Amira 3D 'isosurface' reconstruction, using the theshold derived from the histogram

10.0 μm

(E)

(F) merge

10.0 μm

◄**Figure 1.7** Automatic thresholding and counting of VGluT1 and VGluT2 immunofluorescent "particles" (clusters of synaptic vesicles). (A and B) *Z*-projections from a two-fluorochrome, two-channel CLSM experiment: VGluT2 (488 nm fluorochrome) and VGluT1 (633 nm fluorochrome). (C) Both datasets subjected to ImageJ threshold-particle analysis with the plug-in "3D object counter." Threshold 35 (eight step in a 5 gray intensity level stepping series) distinguishes most particles ($n = 604$). (D) This threshold is used to 3D reconstruct in Amira. (E) Combination of isosurface and orthoslice functions in Amira to show the fit of the isosurface wireframes with pixel aggregates (enlarged and brightness artificially increased). The dashed lines indicate tight fit as a human operator could have decided. (F) Merge of the VGluT2 and VGluT1 3D reconstructions (the VGluT1 dataset identically processed as the VGluT2 dataset).

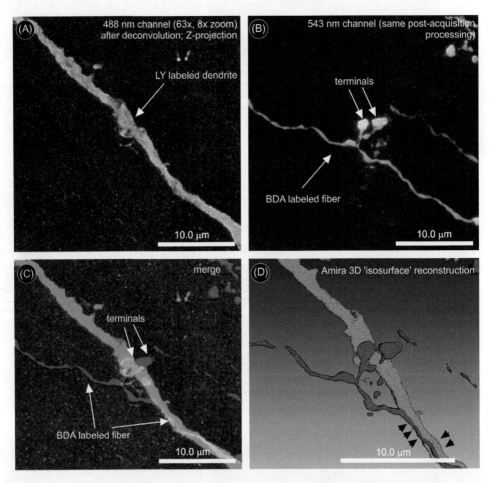

Figure 1.8 Contacts in the substantia nigra of the rat between labeled fibers and a dendrite of an intracellularly injected nigra neuron. Two-channel scanning with the 63× NA 1.3 immersion objective, 8× zoom. (A) *Z*-projection of the 488 nm dataset: a dendrite of a Lucifer yellow-injected cell. (B) *Z*-projection of thin labeled fibers that give rise to two big terminals (arrows). (C) Merge of both datasets. Note the intimate relationship of the BDA fiber and part of the dendrite. (D) 3D reconstruction isosurface in Amira.

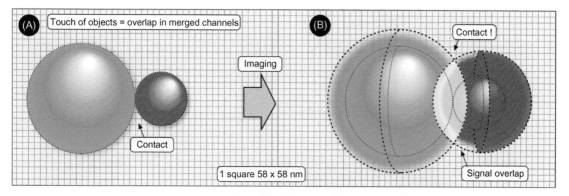

Figure 1.9 Simulation of two-channel imaging at high resolution of "biological" objects that are in physical contact. Lens=63× immersion NA 1.3; zoom=8; each square represented on a pixel. In the acquired images, each structure has a diffraction-induced "edge" of 3 pixels (green object, 488 nm fluorochrome) or 5 pixels (red object, 647 nm fluorochrome). The merged channels produce voxel overlap at the site where, in the specimen, the biological objects are in contact.

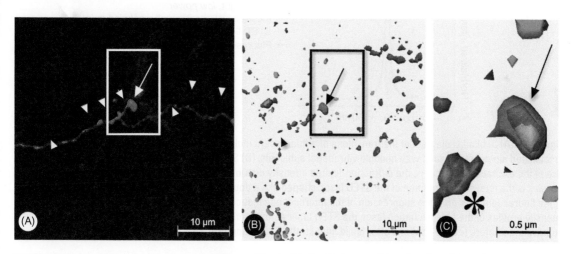

Figure 1.10 Neuroanatomical tracing with BDA combined with double-immunohistochemistry (detecting VGluT2 and calretinin). Three-channel scanning with the 63× NA 1.3 immersion objective, 8× zoom. (A) *Z*-projection of images acquired in the 543 nm channel combined with isosurface 3D reconstruction. The BDA labeled fiber (arrowheads) has a terminal bouton (arrow). (B) Amira 3D isosurface reconstructions of all there channels merged. VGluT2 shown in green, BDA in red, and calretinin in blue. The arrow indicates the BDA-labeled terminal (red). (C). Enlarged part of the 3D reconstruction of (B) where the BDA-3D reconstruction has been made transparent. Inside the terminal are small clusters of voxels associated with VGluT2 (green) and calretinin (blue). Note that the BDA and calretinin in the terminal on the left (asterisk) only partially overlap.

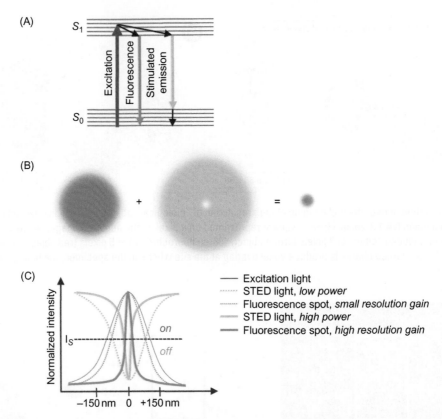

Figure 2.1 (A) Jablonksi diagram of the transitions of a fluorescent molecule, S_0 is the ground state, and S_1 is the first excited singlet state, each with multiple vibrational sublevels. (B) STED principle: the blue spot indicates the size of the excitation PSF, that is, the diffraction-limited intensity distribution of the blue excitation laser; the orange annulus is the intensity distribution of the STED laser, shaped like a doughnut, and the green spot denotes the size of the fluorescent spot after the suppression of the fluorescence induced by the STED doughnut. (C) Normalized intensity profiles of the blue excitation laser, the STED laser at low (dotted line) and high (solid line) intensity, and the fluorescence before (dotted line) and after (solid line) strong STED suppression.

(A)

(B)

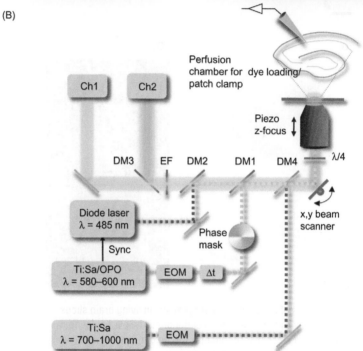

(C)

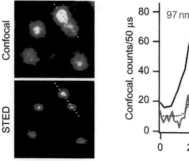

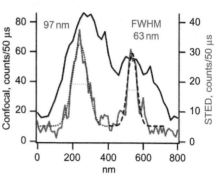

Figure 2.2 (A) Home-built STED microscope equipped for electrophysiological recordings and two-photon microscopy in living brain slices. (B) Schematic representation of the system in (A). (C) Imaging fluorescent beads to determine the spatial resolution of the STED microscope, direct comparison with confocal case. Line profiles across beads to determine their FWHM with the STED laser switched off (black trace) and on (red trace).Source: Reprinted from Tonnesen et al., 2011.

(A)

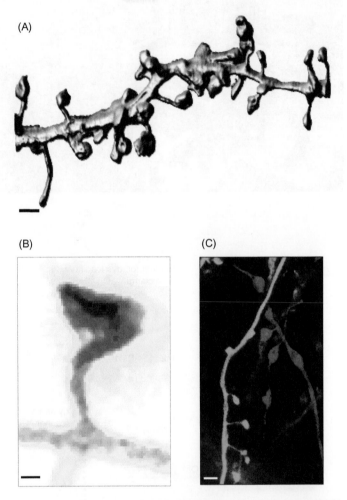

(B)   (C)

Figure 2.3 Examples of STED imaging of synapses in living brain slices. (A) Three-dimensional reconstruction of a stretch of dendrite of a pyramidal neuron filled with YFP based on a stack of STED images. Scale bar, 1 μm. Source: Reprinted from Nägerl et al., 2008. (B) Distribution of the cytoskeletal protein actin inside a dendritic spine imaged by STED using YFP-Lifeact as label for actin. Scale bar, 250 nm. Source: Reprinted from Urban et al., 2011. (C) Two-color STED microscopy of axons and dendrites using GFP and YFP as volume markers. Scale bar, 1 μm. Source: Reprinted from Tonnesen et al., 2011.

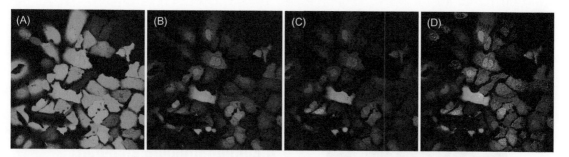

Figure 3.5 Fixation of free GFP-transfected cells with the addition of PI. (A) Before fixation, (B) directly after the addition of 4% paraformaldehyde. (PFA), (C) 2.5 hours after the addition of PFA, and (D) 3 hours after addition of paraformaldehyde.

Figure 3.11 Two different 3D reconstructions of oxidation (green) in endoplasmic reticulum-labeled (red) cells. In these images empty portions represent reduced areas.

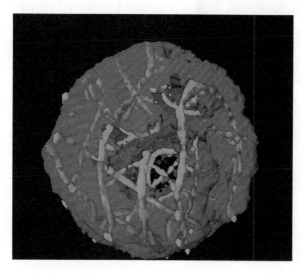

Figure 3.13 Surface-rendered SYCP03 Alexa Fluor 488 (green) and DNA PI (red) staining of a pachytene mouse spermatocyte.

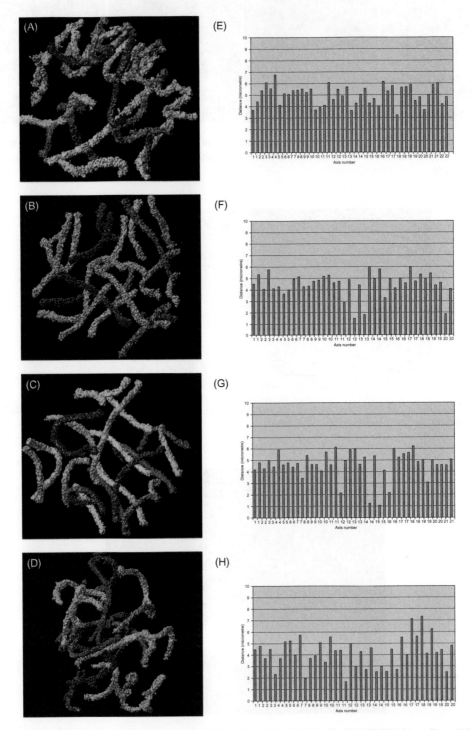

Figure 3.14 Surface rendering, segmentation (A, B, C, D), and analysis (E, F, G, H) of the SYCP3 Alexa Fluor 488 staining in mouse spermatocytes.

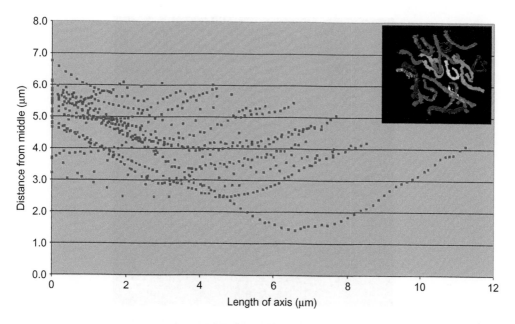

Figure 3.16 Quantification of the distance from different points of SYCP03 axis toward the center of the mouse spermatocyte. Almost all axes from this one cell are curved.

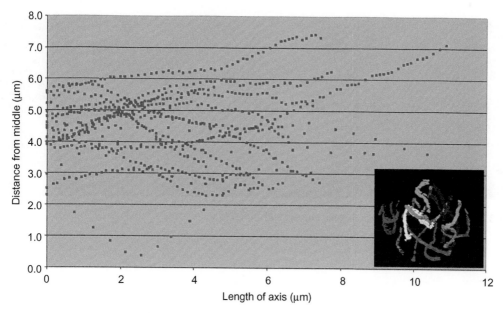

Figure 3.17 Quantification of the distance from different points of the SYCP3 axis toward the centre of the mouse spermatocyte. Almost all axes from this one cell are curved.

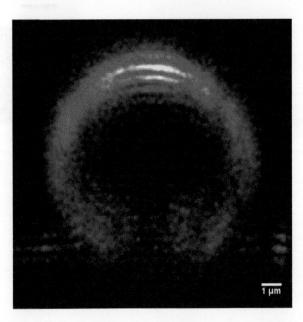

Figure 3.18 Typical 4pi side lobe showing up on *XZ* images of a lipid-labeled microbubble.

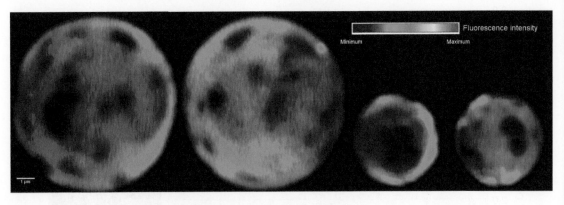

Figure 3.19 Lipid distribution of microbubble coating with different sizes.

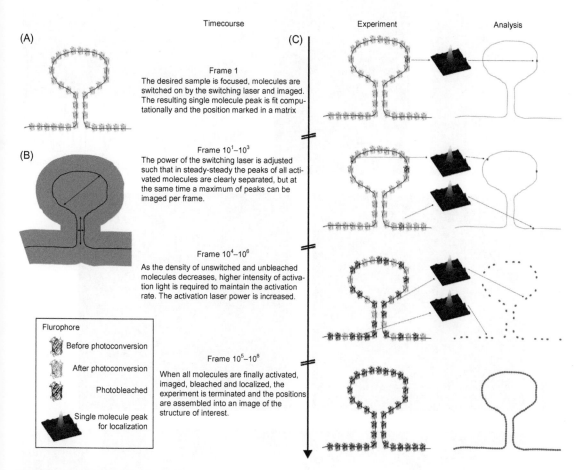

(A)

(B)

Timecourse

(C)

Frame 1
The desired sample is focused, molecules are switched on by the switching laser and imaged. The resulting single molecule peak is fit computationally and the position marked in a matrix

Frame 10^1–10^3
The power of the switching laser is adjusted such that in steady-steady the peaks of all activated molecules are clearly separated, but at the same time a maximum of peaks can be imaged per frame.

Frame 10^4–10^6
As the density of unswitched and unbleached molecules decreases, higher intensity of activation light is required to maintain the activation rate. The activation laser power is increased.

Frame 10^5–10^8
When all molecules are finally activated, imaged, bleached and localized, the experiment is terminated and the positions are assembled into an image of the structure of interest.

Experiment

Analysis

Flurophore

Before photoconversion

After photoconversion

Photobleached

Single molecule peak for localization

Figure 4.1 Principle of localization microscopy.

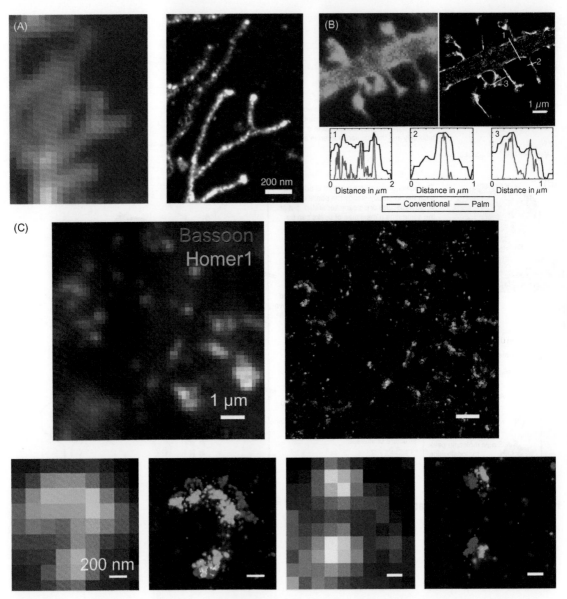

Figure 4.2 Applications of localization microscopy techniques in neurons. (A) Superresolution microscopy of dendritic filopodia. Rat hippocampal neurons were transfected with Myr/Palm mEOS2 and imaged in PALM (Ries and Ewers, unpublished). (B) Conventional (left) and localization microscopy (right) images of a rat hippocampal neuron expressing ABP-tdEosFP and fixed at day *in vitro* 25. Bottom, Profiles across the dendritic shaft (1), the spine neck (2), and the spine head (3), as indicated. *From* Izeddin et al., 2011. (C) Presynaptic protein Bassoon and postsynaptic protein Homer1 in the mouse MOB glomeruli were identified by immunohistochemistry using Cy3-A647- and A405-A647-conjugated antibodies, respectively. The conventional fluorescence image (top left) shows punctate patterns that are partially overlapping, whereas the STORM image (top right) of the same area clearly resolves distinct synaptic structures. Further zoom-in of the conventional images does not reveal detailed structure of the synapses, whereas the corresponding STORM images distinguish the presynaptic Bassoon and postsynaptic Homer1 clusters. *From Dani et al., 2010.*

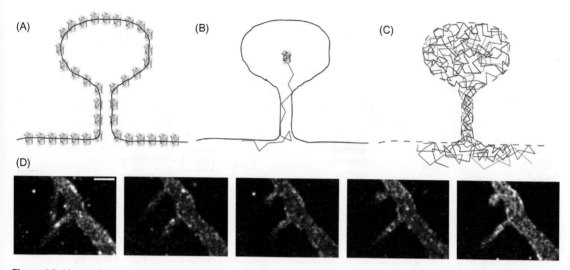

Figure 4.3 Live-cell imaging with PALM. (A) If a membrane-associated fusion construct with a photoswitchable fluorescent protein is expressed in neuronal cells, individual activated molecules can be localized several times as they move around in on the cell surface (B). Since the molecules can only move within the boundaries of the cell membrane, the outline of the neuron can be assembled from the localizations (C). (D) Experimental data from a neuron expressing Myr/Palm mEOS2 (Ries and Ewers, unpublished). For every image, 5000 consecutive frames have been accumulated. Scale bar, 300 nm.

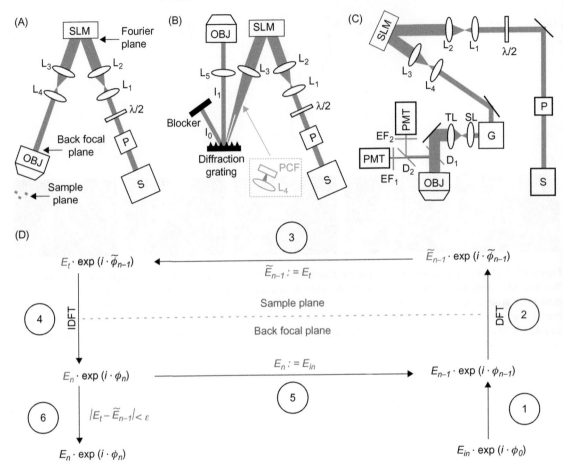

Figure 5.1 Optical configuration for wavefront engineering using SLMs. (A) A basic system consists of a laser source (S), Pockels cell (P), half-wave plate ($\lambda/2$), two telescopes (L_1, L_2 and L_3, L_4), spatial light modulator (SLM), and objective (OBJ). (B) To achieve temporal focusing, a diffraction grating is inserted between lenses L_3 and L_5 and a beam blocker is used to suppress the zero-order. By adding a phase distribution resembling that of a prism to the DOE, the first and zero diffracted orders can be spatially separated and further dispersed in different directions by the grating. The zero-order can then be eliminated by using an optical blocker, which results in a total power loss of approximately half of the initial power, while the first diffracted order is directed into the objective. To obtain generalized phase contrast, a PCF and a lens (L_4) are positioned along the optical path (blue inset, see also Papagiakoumou et al., 2010). (C) To perform simultaneous scanning imaging and inertia-free z-focusing, a more complex experimental setup is needed. G, galvanometric mirrors; SL, scan lens; TL, tube lens; D1, 660 nm long-pass dichroic mirror; D_2,575 nm long-pass dichroic mirror; EF$_1$ and EF$_2$, emission filters; PMTs, photomultiplier tubes. *Modified from Dal Maschio et al., 2011.* (D) Schematic representation of the procedural steps used to iteratively generate DOEs. E_{in}, input intensity; E_t, target intensity; Φ_0, initial estimate for the phase map.

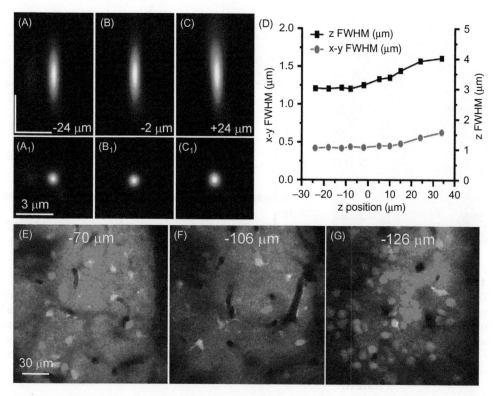

Figure 5.3 SLM-mediated z-focusing for inertia-free 3D imaging *in vivo*. (A–C) *x-z* profiles of illuminated fluorescent beads, 170 nm in diameter, obtained by moving the sample in the z-direction with a piezoelectric translator. Beads were dissolved in agarose and coverslipped. Profiles are obtained at different refocusing axial positions (+24 μm, −2 μm, and −24 μm) operated by the SLM control (see values in microns on the image). The corresponding *x-y* profiles are shown in A_1–C_1. Scale bars, 3 μm. (D) Plot of the *x-y* FWHM (gray circles) and z FWHM (black squares) as a function of the axial position of the laser beam. (E) Images of neocortical cells loaded with Oregon Green BAPTA and sulforhodamine in anesthetized mice at different z-positions (see values in microns on the images) obtained with the inertia-free SLM-based focus control. *Modified with permission from Dal Maschio et al., 2011.*

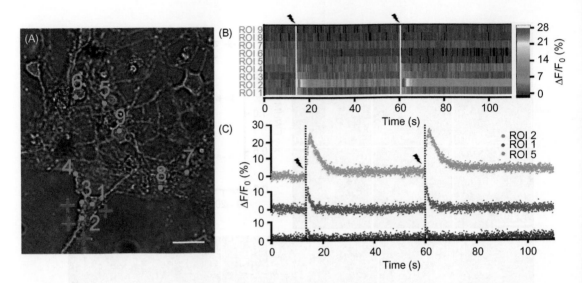

Figure 5.4 Fast holographic fluorescence imaging combined with galvo-steered uncaging. (A) Based on the image, ROIs corresponding to different cells in culture are identified (green dots numbered 1 to 9). DOEs enabling fluorescence imaging of calcium indicators over time in only those ROIs are then projected onto the SLM. Scale bar, 20 μm. (B) Values of $\Delta F/F_0$ for Fluo-4 fluorescence are shown as a function of time for the nine regions identified in A. The arrows indicate the time of delivery of the four photolysis stimuli (red crosses in A) to uncage the MNI-glutamate present in the bathing medium. (C) Time course of the fluorescence signal in ROIs 1, 2, and 5. ROI 1 and 2, but not 5, display a clear response to MNI-glutamate uncaging. *Modified with permission from Dal Maschio et al., 2010.*

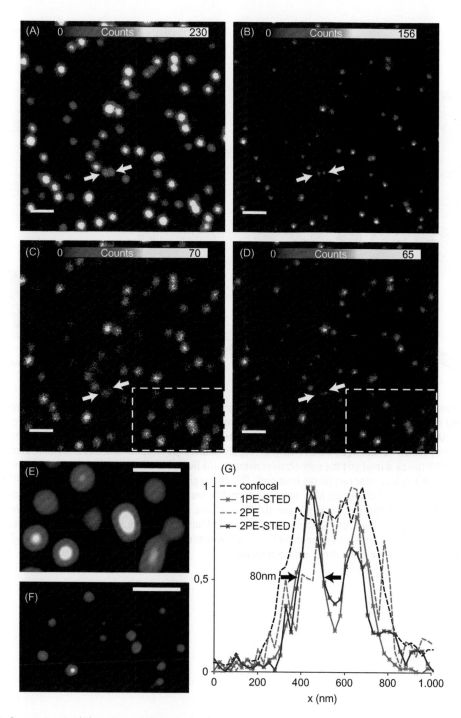

Figure 6.7 Comparison of (A) confocal, (B) STED, (C) 2PE, and (D) 2PE-STED images of 40 nm fluorescent microspheres (yellow-green fluorescent FluoSpheres carboxylate-modified microspheres; Invitrogen) mounted on a coverglass. (A–D) Raw pictures of confocal, STED, 2PE and 2PE-STED. (E) and (F) show linearly deconvolved enlargements of the squared areas marked in (C) and (D). Scale bars, 1 μm. Normalized line profiles taken between the arrows in (A–D) are plotted in (G).

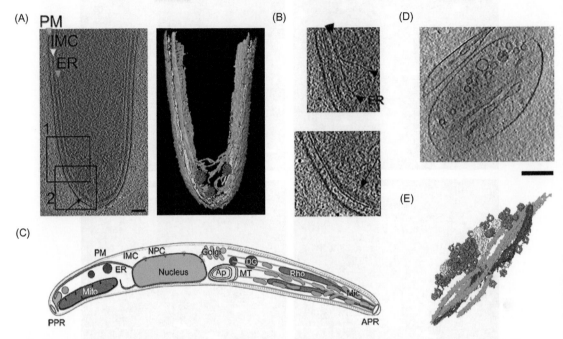

Figure 7.3 Tomography of whole eukaryotic cells. (A) Tomographic slice and a surface rendering of *Plasmodium* posterior end showing plasma membrane (PM; blue), inner membrane complex (IMC; yellow), endoplasmic reticulum (ER; green), cytoplasmic vesicles (red), and the proximal polar ring (black arrow). Boxes indicate regions enlarged in (B). Scale bar, 100 nm. (B) Enlarged boxes from (A). Top, ER membranes (black arrowheads). Bottom, continuity between the PPR (black arrow) and the subpellicular network. (C) Overall architecture of a *Plasmodium* sporozoite. Ap, apicoplast; APR, apical polar ring (at the front end); DG, dense granule; ER, endoplasmic reticulum; IMC, inner membrane complex; Mic, micronemes; Mito, mitochondrion; MT, microtubule; Rho, rhoptries; PPR, proximal polar ring (at the posterior end), PM, plasma membrane. (D) Tomographic slice showing a presynaptic terminal from a dissociated hippocampal culture (DIV 12). Scale bar, 200 nm. (E) Manual segmentation of the presynaptic terminal shown in (D). Green, microtubules; blue, smooth ER; orange, synaptic vesicles; yellow, larger vesicles. (*A–C reproduced from Kudryashev et al., 2010b, with permission.*)

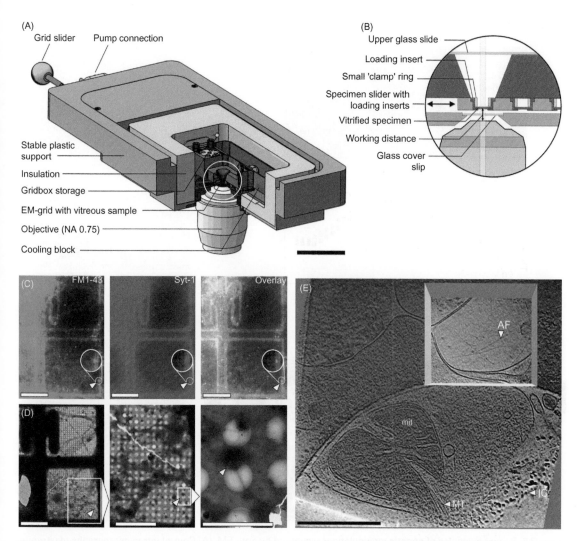

Figure 7.4 Correlative LM and cryo-ET. (A) Cut-away perspective view of the cryo-LM holder with a 63× long distance objective (NA 0.75), depicting housing and insulation requirements. Scale bar, 40 mm. (B). Magnified view of the central part (white circle, A), pointing at the imaging position and specimen slider with four loading positions for vitrified EM grids. (C) Cryo-fluorescent labeling of endocytic markers in a cultured hippocampal neuron (9 DIV). Left, FM 1–43; center, anti-synaptotagmin 1 antibody; right, overlay. Round insets show the spot of interest (white arrowhead) at higher magnification and with increased contrast. Scale bar, 50 μm. (D) Cryo-EM images, at increasing magnifications, at the position (indicated by the red cross) that was correlated to the fluorescence spot indicated in (C). Scale bars: left, 50 μm; center, 30 μm; right, 5 μm. (E) Slice from a cryo-electron tomogram of the neuronal varicosity indicated by an arrowhead in (D). Mitochondrion (mit), microtubule (MT). Indented inset: actin filaments (AF) in a different z-slice. Scale bar, 250 nm. (*A and B reproduced from* Rigort et al., 2010, *with permission. C–E reproduced from* Fernández-Busnadiego et al., 2011, *with permission.*)

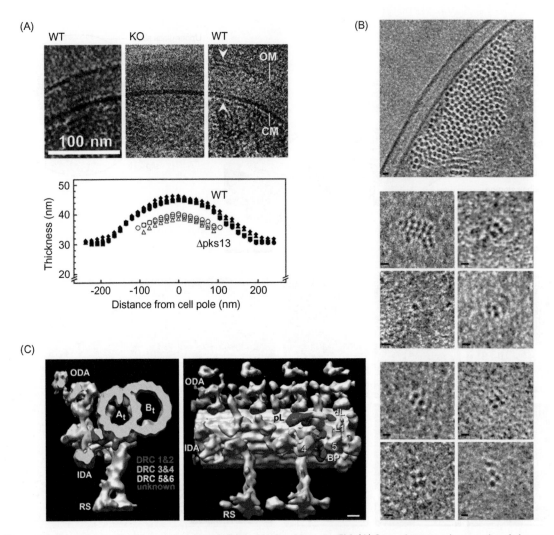

Figure 7.5 Recent examples of genetic manipulations visualized by cryo-EM. (A) Cryo-electron micrographs of vitreous sections from *Corynebacterium glutamicum*. Wild-type cells imaged at high (top left) and low (top right) defocus. The bilayer structure of the cytoplasmic membrane (CM) and of the outer membrane (OM) is resolved (arrowheads). Top middle, cryo-section of the mycolic acid-lacking mutant *C. glutamicum* Δpks13 at low defocus. Bottom, thickness of the cell walls determined from several cells as measured from the surface of the CM to the outer surface of the cell wall (filled symbols, wild type cells; open symbols, mutant cells). (*Reproduced from* Hoffmann et al., 2008, *with permission.*) (B) Cryo-electron micrographs of vitreous sections of cells overexpressing ParM (top) or expressing high (middle) or low (bottom) ParM copy plasmids showing end-on views of ParM filaments. (*Reproduced from* Salje et al., 2009, *with permission.*) (C) Architecture of the dynein regulatory complex (DRC) in cross-section (left) and as overview (right) visualized by subtomogram averaging. ODA; outer dynein arms; IDA, inner dynein arms; RS, radial spokes; At, A-tubule; Bt, B-tubule; BP, base plate protrusion; (dL), distal lobe; L1, linker protrusion 1; pL, proximal lobe. The locations of DRC subunits (see color legend, left) were obtained from mutants lacking specific DRC regions. Scale bars: A, 100 nm; B and C, 10 nm. (*Reproduced from* Heuser et al., 2009, *with permission.*)

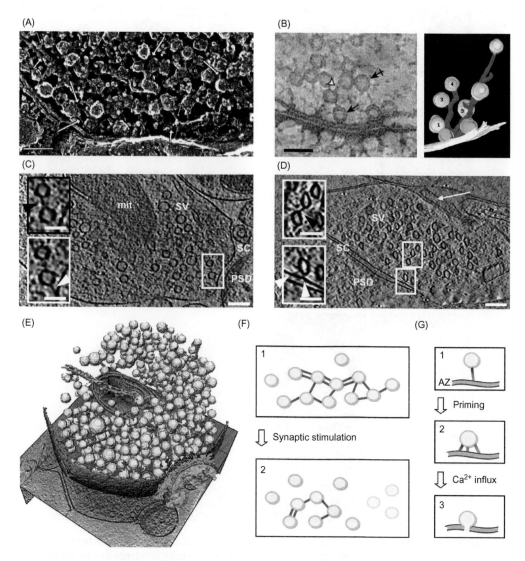

Figure 7.6 Presynaptic architecture by EM techniques. (A) Presynaptic terminal of a rat cerebellar mossy fiber upon quick-freezing, deep-etching, and platinum replication, showing abundant presynaptic filaments (arrows). (*Reproduced from* Hirokawa et al., 1989, *with permission.*) (B) Rat hippocampal presynaptic terminal upon high-pressure freezing, freeze substitution, and heavy-metal staining, imaged by ET. Filaments linking vesicles to each other and to the AZ can be seen in the tomographic slice (left) and 3D rendering (right). (*Reproduced from* Siksou et al., 2007, *with permission.*) (C–G). Presynaptic morphology visualized in tomograms of vitrified synapses visualized by cryo-ET. (*From* Fernández-Busnadiego et al., 2010). SV, synaptic vesicle; mit, mitochondrion; SC, synaptic cleft; PSD, postsynaptic density. (C) Synaptosome. Upper inset. connector linking two vesicles (black arrowhead). Lower inset, tether linking a vesicle to the AZ (white arrowhead; same vesicles as in the upper inset at another z-slice). (D) Organotypic slice. Synaptic vesicles are compressed along the cutting direction (white arrow). Upper inset, connector linking two vesicles (black arrowhead). Lower inset, tethers linking a vesicle to the AZ (white arrowheads). (E) Three-dimensional rendering of a synaptosome and corresponding tomographic orthoslices. Synaptic vesicles (yellow), connectors (red), tethers (blue), mitochondria (light blue), microtubule (dark green), plasma membrane (purple), cleft complexes (green), PSD (orange). Proposed models for connector (F) and tether (G) function in synaptic vesicle mobilization and fusion. The number and topology of connectors and tethers shown in (F) and (G) are based on average values. Shaded vesicles in (F) (bottom) represent the decrease in vesicle number upon prolonged stimulation. Scale bars: A–D: 100 nm in main panels, 50 nm in insets.

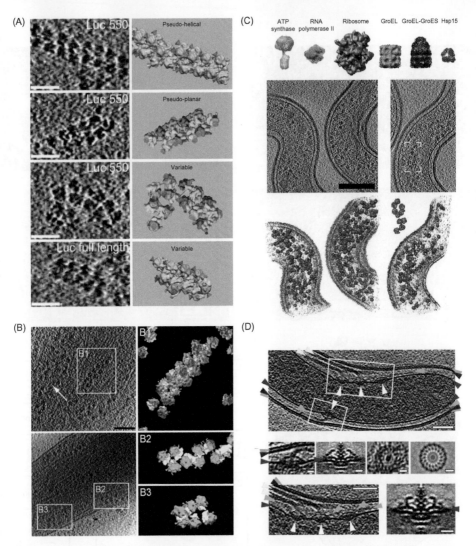

Figure 7.7 Recent cryo-ET examples illustrating image processing techniques. (A) Representative topologies of polyribosomes from *E. coli* lysates. Left, tomographic slices of polyribosomes; right, corresponding localization of ribosomes obtained by template matching (large ribosomal subunit, blue; small subunit, yellow; red cones point to the peptide exit tunnel; dark gray ribosomes were not unambiguously assigned to the polysomes). The three top cases represent polyribosomes stalled with Luc 550, and the bottom case represents a nonstalled full-length Luc. Scale bars, 50 nm. (*Reproduced from* Brandt et al., 2009, *with permission.*) (B) Same as in (A) except that ribosomes were imaged in intact human cells. (*Reproduced from* Brandt et al., 2010, *with permission.*) (C) Template matching in intact *Leptospira interrogans* cells: top, macromolecule template library; middle, representative tomographic subvolumes and a group of ribosomes resembling the pseudoplanar relative orientation can be seen in the bracketed region; bottom, corresponding localization of macromolecules. Scale bars: 200 nm. (*Reproduced from* Beck et al., 2009, *with permission.*) (D) Single and averaged *Borrelia garinii* flagellar motors. Top, cryo-electron tomogram slice showing four motors (white arrowheads). Colored arrowheads point to the surface layer (light blue), outer membrane sheath (dark blue), peptidoglycan (green), cytoplasmic membrane (magenta), and periplasmic flagellar filaments (yellow). Middle, side, and top views of single motors and the corresponding 16-fold rotationally averaged structures. Bottom, peptidoglycan binding to the motor. Enlargement of the three motors from the top left and average structure from the middle rotated by 11° around the vertical axis (right). Scale bars: 100 nm (top and bottom left) and 20 nm (middle and bottom right). (*Reproduced from* Kudryashev et al., 2010a. *with permission.*)

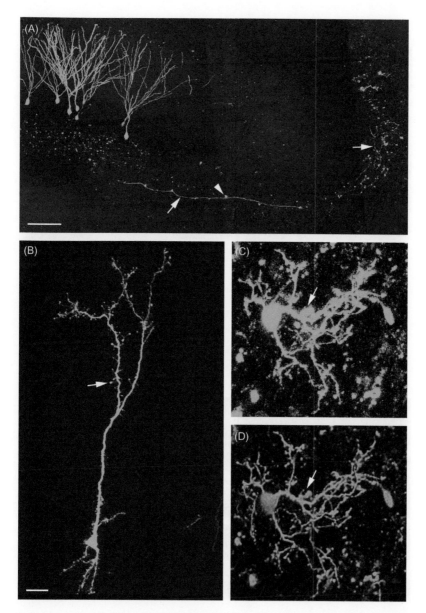

Figure 9.1 Retroviral labeling of adult-born neurons. (A) Adult-born dentate granule cells in the mouse hippocampus labeled by infection with GFP-expressing MLV retrovirus and sacrificed at 8 weeks after virus injection. Note that the complete dendritic structure of granule cells can be visualized along with their axons (arrows) and synaptic boutons (arrow head) projecting to the CA3 hippocampal subfield. (B) An adult-born granule cell in the mouse olfactory bulb labeled by infection with GFP-expressing MLV retrovirus and sacrificed 3 weeks after virus injection. The complete dendritic structure along with spines can be visualized. (C) An adult-born mouse olfactory bulb periglomerular neuron labeled by infection with GFP-expressing MLV retrovirus and two-photon *in vivo* imaged through an open skull cranial window at 5 weeks after viral injection. Also shown in D is the same neuron as in C with deconvolution. Note that with deconvolution, fine structures lost in the raw image (C, arrow) due to axial and lateral light scattering become defined (D, arrow) also with a significant decrease in overall noise. Scale bars: (A) 100 μm, (B–D) 15 μm.

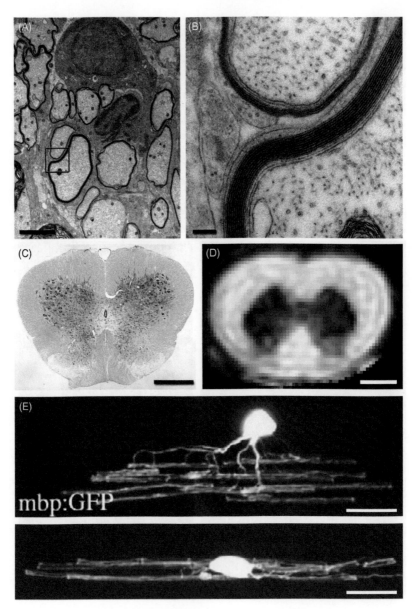

Figure 10.1 Traditional method for myelin observation. (A) EM data show an oligodendrocyte that exhibits a process and wraps on two axons ( ). Scale bar, 5 μm. (B) High-magnification imaging of the box in (A), showing the compact multilaminar membrane of myelin sheath. Scale bar, 100 nm. (C) Transverse section of spinal cord stained by Luxol Fast Blue. Myelin shows as light blue structure around the white matter. (D) DTI imaging showing myelin structure in the spinal cord of the rat. The strong myelin signal rises from the water diffusion and the anisotropy in the highly regulated myelin structure. Scale bars: (in C, D) 1 mm. (E) Two-photon fluorescence enhancement factor imaging of the Tg (mbp:EGFP) zebrafish reveals the *bis*-membrane morphology of the myelin sheath formed by the single oligodendrocyte. Scale bar, 10 μm. *E is adapted from Almeida et al., 2011.*

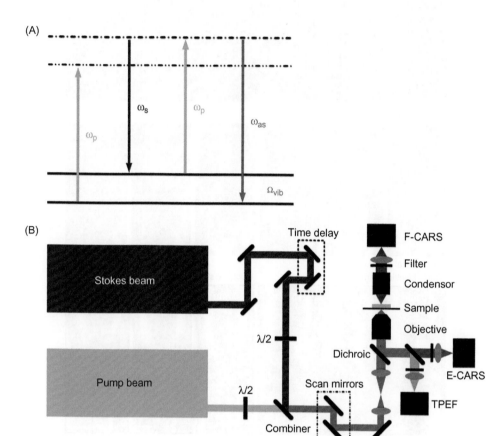

Figure 10.2 CARS process and CARS microscope. (A) Energy diagram of the CARS process. When the frequency difference between the pump and Stokes signals ($\omega_p-\omega_s$) matches the Raman-active molecular vibrational frequency, Ω_{vib}, the anti-Stokes signal is coherently generated at a frequency $\omega_{as}=2\omega_p-\omega_s$. (B) Schematic of a CARS microscope. The synchronized pump and Stokes picoseconds-pulse trains are collinearly combined and polarized. The combined beams are directed to a confocal microscope. A high-NA objective focuses the beams to the sample. The CARS signal is detected simultaneously in the forward and backward direction. The TPEF signal is spectrally separated in the same optical pathway from the E-CARS signal.

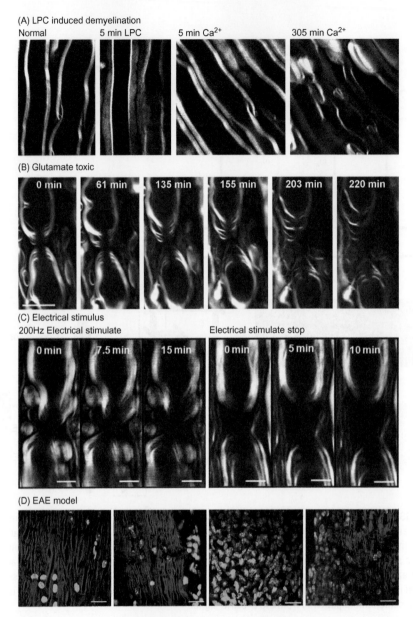

Figure 10.5 Demyelination visualized by CARS microscopy. (A) After injection of LPC, the compact myelin sheath loosens, starting from the outside of the myelin, and the demyelination region encroaches gradually into the inside. After being treated with LPC for 5 min, the intense CARS signal decreases significantly. The thickness of myelin increases because of the relaxation of the multilaminar compact myelin sheath. Scale bar, 10 µm. Adapted from Fu et al., 2007 with permission. (B) After the incubation of 1 mM glutamate, the paranodal myelin sheath retracts from the node of Ranvier, and the length of node significantly increases. The myelin loops split severely and the interval between paranodal myelin increases. Scale bar, 10 µm. (C) During electrical stimulation at 200 Hz, paranodal myelin retracts and the node of Ranvier lengthens. Retraction continues even after the electrical stimulation has been stopped. Scale bar, 5 µm. (D) In the naive whiter matter, myelin fibers are regularly distributed, with visible myelin structure and with only a few glial cells. At the onset stage of EAE, the myelin does not display any obvious abnormality, but the number of glial cells increases. In the acute stage of EAE, myelin loses its regular alignment and many glial cells aggregate at the lesion site. At the remission state, myelin restores and the parallel structures reappear. Scale bar, 20 µm. Adapted from Fu et al., 2011.

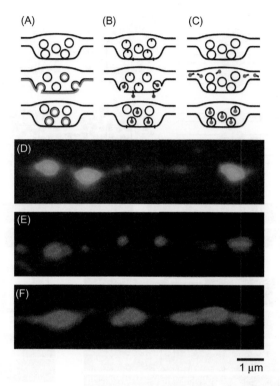

Figure 11.1 Example methods for labeling synaptic vesicle pools. (A) Schematic of FM-dye labeling method. Vesicles are labeled by the addition of FM-dye to extracellular solution followed by neuronal stimulation. Dye binds to the outer leaflet of plasma membrane and is inserted onto the lumenal membrane of the vesicle during endocytosis. Washing to remove extracellular FM-dye reveals selective labeling of recycled vesicles. (B) Schematic of synaptic vesicle labeling using antibodies directed to vesicular proteins. Native vesicular proteins with lumenal epitopes move to the extracellular membrane surface during exocytosis. Fluorescently tagged antibodies target proteins and are taken into vesicles during endocytosis. Washing to remove surface antibody reveals selective labeling of recycled vesicles. (C) Schematic of genetically encoded labeling of vesicles. cDNA coding for synaptic vesicle protein and a fluorescent protein tag is transfected into neurons. Protein is trafficked to target synapses and expressed in vesicles. (D) FM1-43 labeling with presynaptic terminals appearing as discrete fluorescent punctum. (E) Live antibody labeling with synaptotagmin I Oyster550. (F) Expression of Synaptophysin I-mCherry.

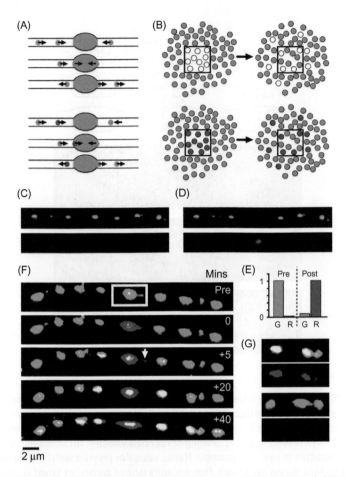

Figure 11.3 Using a photoswitchable fluorophore to monitor the fate of mobile elements. (A) Schematic showing how photoswitchable fluorophores provide additional information for tracking movement of fluorescent elements between compartments in time-lapse sequences. The fate of trafficking fluorescent particles (top) is unclear when they encounter a larger fluorescent structure. However, when photoswitched, the movement of a mobile element can be tracked, revealing a feature of the trafficking (a direction reverse) obscured using single-color imaging. (B). Schematic illustrating additional information gained by photoswitching method when used in a defined compartment. Top, conventional FRAP provides no information on fate of bleached elements. Using photoswitching, the fate of elements at photoswitched sites can be monitored alongside recovery of fluorescent elements into target region. (C, D) Experiment illustrating a critical requirement for a photoswitching experiment: that a discrete target structure can be photoswitched without influencing the fluorescence of neighbors. Images show a population of presynaptic terminals along an axon in hippocampal cultures expressing SytI-Dendra2 before (C) and after (D) photoswitching of the central synapse. Top shows green channel and bottom shows red channel. (E) Fluorescence intensity changes in a target synapse for green and red channels pre- and post-switching. (F) Example of photoswitching experiment to examine lateral vesicle mobility. A target synapse along an axon is selectively photoswitched and the spread of new red fluorescence to synaptic neighbors is monitored over time. Panels show the merge of green and red channels where overlap appears yellow. (G) Illustrates an important requirement for this experiment: that synapses in the same field of view but on different axons do not photoswitch. Top shows synapse in a target region 40 minutes after photoswitching (upper, red/green overlay; lower, red alone). Bottom show synapses in same field of view but on different axon (upper, overlay; lower, red alone). (*F and G are adapted from Staras et al., 2010, with permission.*)

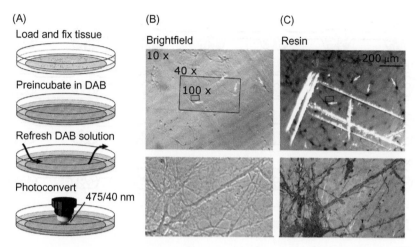

Figure 11.4 Steps in photoconversion and correlative EM for FM-dye-labeled hippocampal-cultured neurons. (A) Diagram illustrating basic steps in DAB photoconversion. Synapses are labeled with FM-dye and fixed. The sample is subsequently preincubated in DAB solution and then refreshed prior to photoconversion. The photoconversion process proceeds with illumination of the target region using the appropriate wavelength of light at the excitation peak for the fluorophore. (B) Illustrates correlative process. Top, at Brightfield level, multiple images of the target region at different wavelengths are taken to generate a montage. Bottom, detail of 100× field of view in top. (C) Top, equivalent area to (B), top after sample is embedded in resin. Bottom, equivalent to (B), bottom after thin sectioning and viewed in EM. Again, landmarks can be used to identify structures of interest. (*B and C are adapted from Darcy et al., 2006a,b.*)

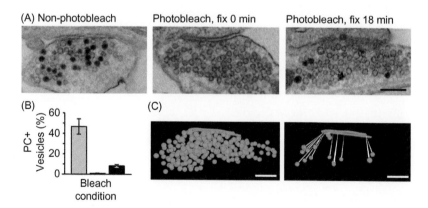

Figure 11.6 Readout of FRAP at the ultrastructural level — FRAP-CLEM. (A) Sample electron micrographs for synapses from samples labeled with FM-dye and treated in three different ways prior to fixing, photoconverting, serially sectioning, and viewing in EM. Left, not treated; middle, photobleached and immediately fixed; and right, photobleached and left for recovery period prior to fixing. Photoconverted vesicles have dark lumen. (B) Histogram summary of fraction of photoconverted vesicles in synapses for each condition from (A). (C) Three-dimensional reconstruction of synapse photobleached, fixed, and left for recovery period (right, A): the equivalent of the bleach-recovery signal shown in Figure 11.2E. Recycling vesicles are orange and nonrecycling vesicles appear blue. Vesicles are shown relative to the active zone (green). (*A–C are adapted from* Darcy *et al.*, 2006a,b.)

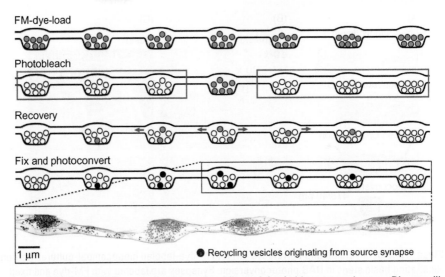

FM-dye-load

Photobleach

Recovery

Fix and photoconvert

1 μm

● Recycling vesicles originating from source synapse

Figure 11.7 "Inverse" FRAP-CLEM for pursuing fates of mobile vesicles in hippocampal axons. Diagram illustrates the principle of the experiment. FM-dye-labeled synapses flanking a target terminal (first diagram) are irreversibly photobleached (second diagram). Recovery of fluorescence (third diagram) at bleached sites occurs as vesicles traffic from target terminal. (fourth diagram) preparation is fixed, photoconverted, and prepared for EM. Photoconverted signal appears electron dense. Bottom, sample reconstruction from a real experiment showing a segment of axon indicated with rectangles. The "source" synapse contains many photoconverted vesicles. Neighboring terminals show progressively fewer photoconverted vesicles, illustrating that movement of vesicles originates from source synapse. (*Bottom diagram is adapted from Staras et al., 2010 , with permission.*)

Printed and bound by CPI Group (UK) Ltd, Croydon, CR0 4YY

08/05/2025

01865018-0001